Lecture Notes in Computer Science 4474

Commenced Publication in 1973
Founding and Former Series Editors:
Gerhard Goos, Juris Hartmanis, and Jan van Leeuwen

T0223139

Giuseppe Prencipe   Shmuel Zaks (Eds.)

Structural Information and Communication Complexity

14th International Colloquium, SIROCCO 2007
Castiglioncello, Italy, June 5-8, 2007
Proceedings

 Springer

Volume Editors

Giuseppe Prencipe
Università degli Studi di Pisa
Dipartimento di Informatica
Largo Bruno Pontecorvo 3, 56127 Pisa, Italy
E-mail: prencipe@di.unipi.it

Shmuel Zaks
Technion
Department of Computer Science
Haifa, Israel
E-mail: zaks@cs.technion.ac.il

Library of Congress Control Number: 2007927754

CR Subject Classification (1998): F.2, C.2, G.2, E.1

LNCS Sublibrary: SL 1 – Theoretical Computer Science and General Issues

ISSN   0302-9743
ISBN-10   3-540-72918-6 Springer Berlin Heidelberg New York
ISBN-13   978-3-540-72918-1 Springer Berlin Heidelberg New York

Springer is a part of Springer Science+Business Media

springer.com

© Springer-Verlag Berlin Heidelberg 2007
Printed in Germany

Typesetting: Camera-ready by author, data conversion by Scientific Publishing Services, Chennai, India
Printed on acid-free paper   SPIN: 12072859   06/3180   5 4 3 2 1 0

Preface

The Colloquium on Structural Information and Communication Complexity (SIROCCO) is an annual meeting focused on the relationship between algorithmic aspects of computing and communication. Over its 14 years of existence, SIROCCO has become an acknowledged forum bringing together specialists interested in the fundamental principles underlying the interplay between information, communication, and computing. SIROCCO covers topics such as distributed computing, high-speed networks, interconnection networks, mobile computing, optical computing, parallel computing, sensor networks, wireless networks, and autonomous robots. Its topics of interest include communication complexity, distributed algorithms and data structures, information dissemination, mobile agent computing, models of communication, network topologies, routing protocols, sense of direction, structural properties, and selfish computing.

SIROCCO 2007 was held in Castiglioncello (LI), Italy, June 5–8, 2007. The previous 13 SIROCCO colloquia took place in Ottawa (1994), Olympia (1995), Siena (1996), Ascona (1997), Amalfi (1998), Lacanau-Océan (1999), L'Aquila (2000), Val de Nuria (2001), Andros (2002), Umeå (2003), Smolenice Castle (2004), Mont Saint-Michel (2005), and Chester (2006).

The 66 contributions submitted to SIROCCO 2007 were subject to a thorough refereeing process, and 23 high-quality submissions were selected for publication. We would like to thank the authors of all the submitted papers. The excellent quality of the final program is also due to the dedicated and careful work of the Program Committee members. Our gratitude extends to them. We also thank the numerous sub-referees for their valuable help.

We had four invited speakers: Hans Bodlaender (Utrecht), Luisa Gargano (Salerno), S. Muthukrishnan (Google), and Alessandro Panconesi (Rome). We thank them for accepting our invitation to share their insights on new developments in their areas of interest.

We would like to express our gratitude to the conference Chair Pierre Fraigniaud (CNRS and Paris) for his enthusiasm and invaluable consultations.

Special thanks go to the local Organizing Team of the Dipartimento di Informatica, Università di Pisa, and in particular to Vincenzo Gervasi.

We acknowledge the use of the EasyChair system (for handling the submission of papers, managing the refereeing process, and generating these proceedings).

We thank the Università di Pisa, and its Dipartimento di Informatica, for their generous support. SIROCCO 2007 was co-located with FUN 2007 and with a meeting of the EU COST 293 action (GRAAL - Graphs and Algorithms in Communication Networks). The two sessions of the invited talks were held jointly with the GRAAL meeting. We thank the Management Committee of GRAAL, and especially their past and present Chairs Xavier Munoz and Arie Koster, for supporting the idea of this joint event and for their generous support.

June 2007

Giuseppe Prencipe
Shmuel Zaks

Conference Organization

Steering Committee

Paola Flocchini	University of Ottawa, Canada
Pierre Fraigniaud	CNRS, France (Chair)
Leszek Gasieniec	University of Liverpool, UK
Lefteris Kirousis	University of Patras, Greece)
Rastislav Královič	Comenius University, Slovakia
Evangelos Kranakis	Carleton University, Canada
Danny Krizanc	Wesleyan University, USA
Bernard Mans	Macquarie University, Australia
Andrzej Pelc	Université du Québec en Outaouais, Canada
David Peleg	Weizmann Institute, Israel
Giuseppe Prencipe	Università di Pisa, Italy
Michel Raynal	IRISA, France
Nicola Santoro	Carleton University, Canada
Paul Spirakis	University of Patras, Greece
Shmuel Zaks	Technion, Israel

Program Committee

Christoph Ambühl	University of Liverpool, UK
Ioannis Caragiannis	University of Patras, Greece
Lenka Carr-Motyckova	Palacky University, Czech Republic
Bogdan Chlebus	University of Colorado, USA
Thomas Erlebach	University of Leicester, UK
Rastislav Královič	Comenius University, Slovakia
Evangelos Kranakis	Carleton University, Canada
Jan van Leeuwen	Utrecht University, Netherlands
Euripides Markou	McMaster University, Canada
Toshimitsu Masuzawa	Osaka University, Japan
Yves Métivier	Université Bordeaux 1, France
Giuseppe Prencipe	Università di Pisa, Italy (Co-chair)
Guido Proietti	Università de L'Aquila, Italy
Andrzej Proskurowski	University of Oregon, USA
Geppino Pucci	Università di Padova, Italy
Tomasz Radzik	King's College London, UK
Sergio Rajsbaum	UNAM, Mexico
Michel Raynal	IRISA, France
Tami Tamir	The Interdisciplinary Center, Israel
Savio Tse	Bilkent University, Turkey
Peter Widmayer	ETHZ, Switzerland
Shmuel Zaks	Technion, Israel (Co-chair)

Local Organization

Vincenzo Gervasi, Dipartimento di Informatica, Università di Pisa

Sponsoring Institutions

Università di Pisa
EU COST 293 action (GRAAL)

External Reviewers

Ittai Abraham

Hagit Attiya

Alberto Bertoldo

Vittorio Bilò

Davide Bilò

Maria Blesa

Anat Bremler-Barr

Marco Bressan

Alfred Bruckstein

Keren Censor

Milind Dawande

Xavier Defago

Stefan Dobrev

Frederick Ducatelle

Toby Ehrenkrantz

Michael Elkin

Angelo Fanelli

Arthur Farley

Laura Feeney

Michele Flammini

Paola Flocchini

Luca Forlizzi

Leszek Gasieniec

Ornan (Ori) Gerstel

Vincenzo Gervasi

Sukumar Ghosh

Emmanuel Godard

Luciano Gualà

Erez Hadad

Panos Hilaris

Martin Hoefer

Jiung-yao Huang

David Ilcinkas

Michiko Inoue

Tomas Johansson

Hirotsugu Kakugawa

Panagiotis Kanellopoulos

Branislav Katreniak

Ralf Klasing

Lukasz Kowalik

Dariusz Kowalski

Richard Královič

Danny Krizanc

Qin Lv

Francesca Martelli

Russell Martin

Valia Mitsou

Gianpiero Monaco

Mohamed Mosbah

Luca Moscardelli

Alfredo Navarra

Francesco Nidito

Fukuhito Ooshita

Aris Pagourtzis

Evi Papaioannou

Dana Pardubska

Susanna Pelagatti

Andrzej Pelc

David Peleg

Enoch Peserico

Andrea Pietracaprina

Katerina Potika

Danny Raz

Nicola Santoro

Fabiano Sarracco

Christian Schindelhauer

Table of Contents

Session 5. Autonomous Systems: Location Problems

Session 6. Wireless Networks

Session 7. Communication Networks: Fault Tolerance

Session 8. Autonomous Systems: Fault Tolerance

Session 9. Communication Networks: Parallel Computing and Selfish Routing

Fast Distributed Algorithms Via Primal-Dual
(Extended Abstract)

Alessandro Panconesi

Informatica, Sapienza University
via Salaria 113, 00198 Roma, Italy
ale@di.uniroma1.it

> *When a trick works once, it is a trick.*
> *If it works twice, it is a technique.*
> *If it works three times, it is a method.*
>
> *Juris Hartmanis*

1 Introduction

We would like to discuss what seems to be a general methodology to develop fast distributed algorithms for optimization problems on graphs, based on the primal-dual schema. The kind of problems we have in mind are of the following type. We have a synchronous, message-passing network that is to compute a global function of its own topology. Examples of such functions are maximal independent sets, vertex and edge colorings, small dominating sets, vertex covers and so on. Crucially, nodes only know their neighbours and have very little or no global information. In what follows, the only global information allowed will be n, the number of nodes in the network (or an upper bound on it). In this setting the running time of a protocol is given by the number of communication rounds needed to compute the output. By the end of the algorithm each node or edge will have decided its final status: its own color, whether or not to be part of the dominating set etc. In many situations of interest the cost of communication is orders of magnitude larger than local computation cost, and the model provides a rough, but quite useful, quantitative framework to develop and analyze interesting algorithms.

The combinatorial objects that we are interested in computing are useful both on theoretical and practical grounds. Small dominating sets for instance are the method of choice to set up the routing infrastructure of ad-hoc networks (the so-called backbone). Edge colorings have been repeatedly used to parallelize data transfers in distributed architectures, and so on. Maximal independent sets on the other hand appear to be a basic building block of many distributed algorithms.

Note the basic challenge here: If a protocol runs for t many rounds, each node will be able to collect information only from nodes at distance t. If t is much smaller than the diameter of the network, what we are trying to do is to compute a global function of the entire network, based on local information alone.

Distributed algorithms for graph problems is a very wide and active area. In what follows we do not try to be complete or encyclopaedic. Rather, we focus on the issues

G. Prencipe and S. Zaks (Eds.): SIROCCO 2007, LNCS 4474, pp. 1–6, 2007.
© Springer-Verlag Berlin Heidelberg 2007

that are immediately relevant to our topic. This means that we will not do a proper job of acknowledging the large and relevant body of brilliant literature that exists, and we offer our apologies in advance for this lack of completeness. The papers we cite and the references therein provide a good entry point for the research areas we will be discussing.

2   Distributed Algorithms Via the Primal-Dual Schema

The primal-dual schema is a very powerful methodology to develop efficient algorithms for combinatorial optimization problems. In recent years it has been applied with good success to NP-hard problems for which it yields many sophisticated approximation algorithms with performance guarantee. The main thesis of this talk is that in general a primal-dual algorithm exhibits certain locality properties that make it amenable to a fast, distributed implementation. This point of view is cogently developed in the PhD dissertation [8] which is the basis of the discussion to follow. We shall outline the method by discussing the example of vertex cover. The algorithm we describe is the first example of this general methodology [6]. The algorithm was developed in the dark days when people were interested in PRAM algorithms, but the solution is in fact fully distributed.

As it is well-known, in the vertex cover problem we are given an undirected network and we are to compute a so-called cover, i.e. a set of vertices such that, for every edge at least one of the two endpoints lies in the cover. Among all covers, we are interested in computing one of the smallest possible size. When vertices have positive integer costs we seek a cover of the smallest possible aggregate cost. This problem is NP-hard even with unit costs. What we will do is the following: (a) develop a sequential 2-approximated algorithm for it, and then (b) show how to parallelize it efficiently by means of a distributed implementation. Note that there are two separate issues here: to make the process distributed *and* fast. To do so, one must be able to exploit the inherent parallelism.

We begin by formulating the problem as an integer program (IP):

$$\min \quad \sum_{v \in V} c(v) \cdot x_v \tag{IP}$$

$$\text{s.t} \quad x_v + x_u \geq 1 \qquad\qquad \forall e = (u, v) \in E \tag{1}$$

$$x_v \in \{0, 1\} \qquad\qquad \forall v \in V \tag{2}$$

The cover is defined to be the set of all vertices v such that the corresponding binary indicator variable $x_v = 1$. The set of constraints (1) ensure that it is indeed a cover, for at least one endpoint for every edge must be included in it.

We now let (LP) be the standard LP relaxation obtained from (IP) by replacing the constraints (2) by

$$x_v \geq 0 \qquad\qquad \forall v \in V \tag{3}$$

In the linear-programming dual of (LP) we associate a variable α_e with constraint (1) for every $e \in E$. The linear programming dual (D) of (LP) is then

$$\max \quad \sum_{e \in E} \alpha_e \tag{D}$$

$$\text{s.t} \quad \sum_{e=(u,v) \in E} \alpha_e \leq c(v) \qquad \forall v \in V \tag{4}$$

$$\alpha_e \geq 0 \qquad \forall e \in E \tag{5}$$

We will build a cover working with the dual variables. Note that we have a constraint of type (4) for every node v, denoted as $(4)_v$. Consider the following continuous process. We let all the variables α_e grow at uniform speed. Sooner or later a constraint of type (4) will be satisfied with equality. If $(4)_v$ is the constraint we say that it becomes *tight*. When $(4)_v$ becomes tight we add v to the cover that we are computing. We do this by setting $x_v = 1$ (initially all primal variables are set to 0). Then we freeze the values α_e of the edges incident to v. The α-values of frozen edges stop growing, so that the constraint considered remains tight. The process continues with the remaining edges, until all edges are frozen.

At this point we have a set of vertices C containing all vertices v such that $x_v = 1$. We want to show that (a) it is a cover and that (b) its size is at most twice the optimum cost. To see why it is a cover, suppose not. But then there is an edge $e = uv$ which is not covered, i.e. $x_u = x_v = 0$. This means that the constraints $(4)_u$ and $(4)_v$ corresponding to u and v are not tight and α_e can continue to grow, a contradiction.

The cost of C is upper-bounded by twice the cost of the dual solution:

$$\sum_{v \in V} c(v) x_v = \sum_{v \in C} c(v) \leq \sum_{v \in C} \sum_{e=(u,v) \in E} \alpha_e \leq 2 \sum_{e \in E} \alpha_e.$$

You can think of this chain of inequalities in the following way. At the end of the algorithm we have a value α_e for every edge in the network. By doubling each α_e we have enough "money" to pay for the cost of vertices we put in C. This is true locally: for every $v \in C$, since v is tight, the sum of the α_e's, where e is incident to v, is equal to its cost. Since we doubled every α_e, each edge e has enough cash to pay the cost of both vertices it is incident to. Thus, twice the sum of the α_e's covers the cost of the solution we computed.

That the solution computed is 2-approximate follows by weak duality. Weak duality states that any dual feasible solution is no more than any primal solution. That is, denoting with z^* the optimal value of the primal solution we have

$$\sum_{e \in E} \alpha_e \leq z^*.$$

The primal is a relaxation of the original integer program and thus $z^* \leq \mathsf{opt}$. The claim follows.

The continuous process above can be easily turned into a discrete one that runs in polynomial-time. The problem is how to make it both distributed and fast. It is apparent that the algorithm can be simulated by a message-passing mechanism in which the nodes exchange information in order to set the values of the α's. The technical details are slightly tricky, but the main idea is the following. The α's are initialized to a small quantum value q and then they are increased by means of synchronous jumps assuming values of the form $(1 + \epsilon)^k q$. At round k vertices can verify locally if they are tight and exchange information on which variables to freeze. The only complication is that it is unlikley that dual values of the form $(1 + \epsilon)^k q$ will add up exactly to $c(v)$. The problem is solved by relaxing the notion of tightness. A vertex v is weakly tight if $(1 - \epsilon)c(v) \leq \sum_{e=(u,v)\in E} \alpha_e \leq c(v)$. The solution consisting of weakly tight vertices will be $2 + \epsilon$ approximated, where ϵ can be as small as we want. Clearly, the smaller the ϵ the higher the running time.

This algorithm and analysis are some 15 years old. Recently, the method has been brought back to life and applied to problems that are much more complex. The first application concerns *vertex cover with capacities*. Here, every node v has a budget b_v and can cover at most b_v many edges. As before we are looking for the cover of smallest cost, where covers must comply with the budget requirements. Note that here we have not only to compute a cover, but also an edge assignment, i.e. we must specify which node covers which edge.

The problem comes in two different flavours. In the so-called soft version, we are allowed to put in the cover multiple copies of the same vertex, provided that we pay for each copy. The primal-dual methodology discussed here can be applied with success to this case. [2] gives an efficient distributed implementation of the sequential algorithm of [5], together with another application of the method to facility location. A much harder problem is the so-called hard version, where only one copy of each vertex is allowed to be considered in the solution. Indeed, this problem cannot be solved quickly in a distributed setting. Consider for instance the case of the ring, where each vertex has a budget of 1. Then the nodes must essentially synchronize either in clockwise or in counterclockwise order, a task that is easily seen to require linearly many rounds. Thus one considers the so-called semi-hard case where the budget constraints can be violated by a certain amount. An (α, β)-approximation algorithm for vertex cover with semi-hard capacities is an algorithm that computes a set of vertices S and an edge assignment such that (a) the cost of S is at most $\alpha \cdot$ opt and (b) budgets constraints are violated by at most a β factor (notice that opt refers to the original problem with hard capacities). These kind of algorithms are often referred to as bi-criteria approximation algorithms. In [3] a $(2+\epsilon, 4+\epsilon)$-approximation algorithm is given, whose running time is polylogarithmic in the size of the input. It might happen that the algorithm is not able to compute any solution. When this happens we have a certificate that the input instance does not have a feasible solution to the problem with hard capacities. The analysis of this algorithm is considerably more complex and nuanced than the original application to vertex cover, but one might object that the method is still being applied to a problem whose combinatorial structure is quite related. Consider then the following scheduling problem.

We have a bipartite graph with processors and resources on the two sides of the bipartition, respectively. Each processor has a list of jobs to be executed. Each job comes with its own length (processing time) and profit, and one or more time windows within which it is to be executed. In fact, both profit and length may vary according to the time the job is scheduled. A processor can schedule its jobs only on the resources that are adjacent to it in the graph. All resources are identical in the sense that a job can be scheduled on any of the neighbouring resources. The basic constraint is that a resource can process at most one job at any given time.

Processors that share a resource can communicate in one time step. The network is synchronous and message-passing: a process can communicate in one time step with all other processors with which it shares a resource.

The goal is to schedule a set of jobs in order to maximize the aggregate profit.

This very general problem is ageless and, apart from its combinatorial relevance, finds new applications as technology changes. For instance the resources could be internet hot spots for whose access several wireless devices are competing, they could be shared resources in a peer-to-peer network, or they could be the result of self-organization of a sensor network, with backbone nodes on one side of the bipartition.

It should not be hard to believe that this problem exhibits a combinatorial structure entirely different from the covering problems seen above. Indeed, its algorithmic solution requires an entire set of new ideas already in the sequential case. It is well-known that our scheduling problem is NP-hard. Therefore the best we can realistically hope for is to give approximated solutions. To put things in perspective recall that the best sequential approximation for this problem is within a factor of 2 [1]. In [7] a distributed algorithm is given such that, for any $\epsilon > 0$, it computes a $\left(\frac{1}{20+\epsilon} \right)$-approximation in a number of rounds that is polylogarithmic in the size of the input (when the weights are polynomial in the size of the network). Again the solution is derived starting from a sequential primal-dual algorithm. The basic idea is to parallelize the sequential primal-dual algorithm presented in [1]. It might be instructive to review briefly that algorithm. The algorithm is based on the interplay between the classical primal-dual mechanism to manipulate the dual variables, and a stack. Jobs are sorted by increasing end times and are pushed in this order onto the stack. For every job there is a dual constraint, which is saturated when the job is pushed onto the stack. The dual variables of the saturated constraint are frozen. Overlapping jobs have dual variables in common. Thus when a job enters the stack and its dual variables are raised this affects the constraints of overlapping jobs. If this causes a constraint to become saturated, the corresponding job is eliminated. The process continues with the remaining jobs (corresponding to unsatisfied constraints), until every constraint is satisfied. At that point every job is either in the stack or was eliminated. Since all constraints are satisfied the solution is dual feasible. The algorithm then computes a scheduling with the jobs in the stack as follows. It starts a reverse process by popping jobs out of the stack. When a job is popped it is scheduled unless it has a conflict with previosly scheduled jobs, in which case it is eliminated. This ensures feasibility. The analysis shows that the scheduling so constructed is also 2-approximated. If this brief outline makes little sense then we have conveyed our main point! This primal-dual algorithm is much more complex than the simple application we discussed for vertex cover. In fact, it is also much more complex than the primal-dual sequential algorithm that is used as the starting point of the solution for vertex cover with semi-hard capacities. This is an illustration of the variety of applications the primal-dual schema is capable of.

The distributed solution mimics all this by introducing a "parallel" stack into which jobs belonging to different processors are pushed in parallel. Jobs pushed at the same time on the stack belong to the same "layer". One of the main problems to solve with this approach is to ensure that the depth of the stack, which in general can be linear in the input size, be bounded by a polylogarithmic function in the size of the input, for the number of steps is at least the number of push and pop operations we perform.

Already from this very rough outline the versatility of the methodology should be apparent. The scheduling problem has a complex combinatorial structure that is very different from the vertex covering problems we discussed. Note also that it is a *maximization* (as opposed to minimization) problem, a fact that carries with it several complications even in the sequential case.

To summarize, the method applies with success to problems that are very different. This provides solid evidence of the validity of the general methodology, that of deriving efficient distributed algorithms starting from a sequential primal-dual solution. The primal-dual schema is a very powerful and versatile methodology for deriving algorithmic solutions, and thus the method might very well become the source for an entire class of new, sophisticated distributed algorithms for combinatorially hard problems.

References

[1] Bar-Noy, A., Bar-Yehuda, R., Freund, A., Naor, J., Schieber, B.: A unified approach to approximating resource allocation and scheduling. Journal of the ACM 48, 1069–1090 (2001)

[2] Chudak, F., Erlebach, T., Panconesi, A., Sozio, M.: Primal-Dual Distributed Algorithms for Covering and Facility Location Problems. Submitted

[3] Grandoni, F., Könemann, J., Panconesi, A., Sozio, M.: Primal-Dual based Distributed Algorithms for Vertex Cover with Semi-Hard Capacities. In: Proceedings of Twenty-Fourth Annual ACM SIGACT-SIGOPS Symposium on Principles of Distributed Computing (PODC 2005) (2005)

[4] Guha, S., Hassin, R., Khuller, S., Or, E.: Capacitated vertex covering with applications. In: Proceedings, ACM-SIAM Symposium on Discrete Algorithms, pp. 858–865 (2002)

[5] Guha, S., Hassin, R., Khuller, S., Or, E.: Capacitated vertex covering. Journal of Algorithms, pp. 257–270 (2003)

[6] Khuller, S., Vishkin, U., Young, N.: A primal-dual parallel approximation technique applied to weighted set and vertex covers. J. Algorithms 17(2), 280–289 (1994)

[7] Panconesi, A., Sozio, M.: Fast Distributed Scheduling via Primal-Dual. Submitted

[8] Sozio, M.: Efficient Distributed Algorithms via the Primal-Dual Schema. PhD dissertation, Sapienza University, Rome, Italy, October (2006)

Time Optimal Gathering in Sensor Networks

Luisa Gargano

Dip. di Informatica ed Applicazioni, Universitá di Salerno, 84084 Fisciano, Italy

Abstract. Efficient data gathering is an important issue in sensor networks. We will discuss the problem of time efficient data gathering, in which data sensed throughout the network must be collected at a sink node; the aim is to minimize the time needed to complete the process. The emphasis is on some algorithmic and graph theoretical problems arising in the area.

1   Sensor networks

Recent technological improvements have made a reality the deployment of small, inexpensive, low-power, wireless sensors. Sensor networks are dense wireless networks of sensor nodes that are used to collect and disseminate environmental data. By using wireless multihop communication, collected data are sent toward some selected data sinks in the network. In this way individual measurements converge to a global picture of an entire physical phenomenon.

Sensor networks constitute an important class of emerging networked systems providing diverse services to numerous important applications in industries, transportation, manufacturing, environmental oversight, safety and security.

The deployment of a large quantity of nodes in such systems has become feasible due to the availability of cheap wireless technology, and the emergence of microsensors based on MEMS technology [9]. These nodes are generally stationary after deployment and the connections between them are realized by wireless media as infrared devices or radios. The range of each radio will be much less than the size of the entire network, so that a multihop topology will result. Because of the non–mobile nature of the sensor nodes, as well as the finite energy resources (and therefore lifetime) of the nodes, there will be a distinct bootup phase in which the nodes self-organize to form the network. Information processing is allowed in a sensor network by merging sensing, signal processing, and communication functions.

One of the most important communication primitives that has to be provided by a sensor network is that of *data gathering*: In data gathering, the information collected at sensor nodes is sent to a selected data *sink* node, which is responsible of further processing for end-user queries, by a repeated use of wireless multihop communication.

Data gathering in sensor networks received much attention in the last few years, cfr. the surveys [1,5]. Most of the research has focused on the problem of reducing the energy consumption during the gathering process. However, an other important factor to consider in data gathering applications is the *delay* of

G. Prencipe and S. Zaks (Eds.): SIROCCO 2007, LNCS 4474, pp. 7–10, 2007.

the gathering process. Indeed, the data collected by a sensor node can frequently change, hence it is essential that data are received by the sink as soon as possible, without being delayed by collisions [12].

2   Time Efficient Data Gathering

Here we are concerned with efficiency limits of data gathering with respect to the time. Time performances of gathering algorithms are evaluated by considering the simple discrete mathematical model first adopted in [6].

2.1   The Model

The sensor network is a finite collection of identical nodes. Each sensor node u carries, after an observation period, a finite number r_u of unit data packets to be delivered to the base station (BS) s.

Nodes (including the BS) have common transmission range d_T and interference range d_I. The information transmitted by a sensor becomes available to nodes that are within its transmission range d_T if these nodes are in listening mode and they are not in the interference range d_I of a third (transmitting) sensor [2].

We assume that time is slotted and a one hop transmission consumes one time slot (TS). The network is further assumed to be synchronous. A node can only transmit/receive one data packet per TS.

Multiple transmissions may occur within the network in one TS under this interference model by virtue of spatial separation. A *collision* happens at a node u if two or more nodes having u in their interference range try to transmit at the same time. However, simultaneous transmissions among pair of nodes may occur whenever the interference model is respected.

Summarizing, the network can be represented by means of a graph where nodes represent the sensors and an edge exists between two nodes if the two sensors are in the range of each other; the collision–free data gathering problem can be then stated as follows [12].

> **Data Gathering.** Given a graph $G = (V, E)$ and a base station s, for each $v \in V - \{s\}$, schedule the multi-hop transmission of the data items sensed at v to s so that the whole process is collision–free, and the time when the last data is received by s is minimized.

Time–efficient gathering strategies have been studied under various assumptions. In the following, we briefly survey the main results on the subject.

2.2   Directional Antennas Systems

Each node is equipped with *directional antennas* allowing the transmission over a distance d_T. The use of directional antennas allows to select the neighbor to which the transmission is actually directed. In this model, a *collision* happens at a node u if two or more of its neighbors try to transmit to u during the same time slot.

In the hypothesis that each node has one packet of sensed data to deliver to the base station, optimal algorithms for any type of networks are known [8].

Under the general assumption that any node u has an arbitrary number $r_u \geq 0$ of packets to be delivered to the base station, provably optimal strategies for tree networks and an approximation algorithm with performance ratio 2 for general networks have been proposed in [6].

The possibility of having multiple channels between adjacent nodes has been considered in [12], where an approximation algorithm with performance ratio 2 is given for collision-free data gathering under the hypothesis that each node has one data packet to send to the base station.

Relationships between data gathering time and transmission range, packets size, and channel noise have also been analyzed [7].

2.3   Omni–Directional Antennas Systems

Nodes are equipped with *omni–directional antennas*: When a node transmits one data packet, all its neighbours (the nodes within distance d_T) can receive while nodes within its interference range (within distance d_I, $d_I \geq d_T$) cannot listen to other transmissions due to interference.

Under the assumption that no buffering is allowed during the data gathering process (this is a reasonable assumption due to the limited data storage of each sensor) optimal strategies are known for line and tree networks in case $d_I = d_T$ [6], [3].

In case buffering is allowed at intermediate nodes, the problem is known to be NP-hard and an approximation algorithm with approximation factor 4 is known [2]. Using the same model, an on-line algorithm which gives a 4-approximation is presented in [4].

Papers [10,11] consider the time needed to gather information in conjunction with the energy spent to complete the process. They present schemes that attempt to optimize the *energy* $\times$ *delay* cost function.

It should be noticed that the above studies are mostly concerned with centralized algorithms requiring cooperation between nodes, which is not necessarily compatible with the requirements of sensor networks. Therefore, when requirements are more stringent, these algorithms may no longer be practical. However, they still continue to provide a lower bound on the data collection time of any given collection schedule.

3   Open Problems

The study of time–efficient gathering poses various algorithmic and graph theoretical problems. Optimal or approximation algorithms have been proposed in the paper quoted above. However, several problems remain open in the area. In particular the time complexity of gathering is unknown under various hypothesis.

In case of unidirectional antennas, the complexity of the problem in the general setting, in which some nodes can also have no packets to deliver, is an open issue. We believe gathering in this general setting is an NP-complete problem.

In case of omni–directional antennas, when intermediate buffering is possible, the gathering problem is known to be NP-Hard. However, if no buffering is allowed at intermediate nodes, the complexity of the problem remains unknown.

References

1. Akyildiz, I.F., Su, W., Sankarasubramaniam, Y., Cayirci, E.: Wireless sensor networks:a survey. Computer Networks 38, 393–422 (2002)
2. Bermond, J.-C., Galtier, J., Klasing, R., Morales, N., Perennes, S.: Hardness and approximation of gathering in static radio networks. In: Proceedings FAWN06 (2006)
3. Bermond, J.-C., Gargano, L., Rescigno, A.: Time–Optimal Data Gathering in Omnidirectional Antennas Sensor Networks, manuscript (2007)
4. Bonifaci, V., Korteweg, P., Marchetti-Spaccamela, A., Stougie, L.: An Approximation Algorithm for the Wireless Gathering Problem. In: Proceedings of SWAT 2006, pp. 328–338 (2006)
5. Chong, C.-Y., Kumar, S.P.: Sensor networks: Evolution, opportunities, and challenges. In: Proceedings of the IEEE, vol. 91(8), pp. 1247–1256 (2003)
6. Florens, C., Franceschetti, M., McEliece, R.J.: Lower Bounds on Data Collection Time in Sensory Networks. IEEE Journal on Selected Areas in Communications 22 (6), 1110–1120 (2004)
7. Florens, C., Sharif, M., McEliece, R.J.: Delay issues in Linear Sensory Networks. ISIT 2004, vol. 82, Chicago (June27–July2) (2004)
8. Gargano, L., Rescigno, A.: Optimally Fast Data Gathering in Sensor Networks. In: Královič, R., Urzyczyn, P. (eds.) MFCS 2006. LNCS, vol. 4162, pp. 399–411. Springer, Heidelberg (2006)
9. Kahn, J.M., Katz, R.H., Pister, K.S.J.: Mobile Networking for Smart Dust. In: Proceedings of ACM MobiCom 99 (1999)
10. Lindsey, S., Raghavendra, C., Sivalingam, K.M.: Data gathering algorithms in sensor networks using energy metrics. IEEE Transactions on Parallel and Distributed Systems 13 (9), 924–935 (2002)
11. Yu, Y., Krishnamachari, B., Prasanna, V.: Energy-latency tradeoffs for data gathering in wireless sensor networks. In: Proceedings of IEEE INFOCOM 2004 (2004)
12. Zhu, X., Tang, B., Gupta, H.: Delay efficient data gathering in sensor networks. In: Jia, X., Wu, J., He, Y. (eds.) MSN 2005. LNCS, vol. 3794, pp. 380–389. Springer, Heidelberg (2005)

Treewidth: Structure and Algorithms

Hans L. Bodlaender

Institute of Information and Computing Sciences, Utrecht University, P.O. Box
80.089, 3508 TB Utrecht, the Netherlands
hansb@cs.uu.nl

Abstract. This paper surveys some aspects of the graph theoretic notion of treewidth. In particular, we look at the interaction between different characterizations of the notion, and algorithms and algorithmic applications.

1 Introduction

This paper aims at giving an overview of some aspects of the notion of *treewidth*. Treewidth is a graph parameter with several applications, many algorithmic, and some more graph theoretic.

Best known is the characterization of treewidth in terms of *tree decompositions*, introduced by Robertson and Seymour [71] in their fundamental graph theoretic work on graph minors. There are a number of other, equivalent characterizations of treewidth. In this short overview paper, we will look at several of such characterizations. Many different algorithms that use or compute treewidth use different characterizations. With each of the different characterizations, we sketch some algorithms that exploit the characterization.

First, in Section 2, we sketch some roots of the notion of treewidth. In Section 3, we give the definition of treewidth by means of tree decompositions, and discuss nice tree decompositions, recursive construction of graphs, and algorithms using tree decompositions, and algorithms for problems in Monadic Second Order Logic. In Section 4, we discuss the representation as subgraph of a chordal graph with small clique size and by elimination orderings. In Section 5, we look at the representation with help of search games, and in Section 6 to the notion of bramble. Graph minors are briefly discussed in Section 7.

2 Roots of Treewidth

2.1 Kirchhoff Laws and Series-Parallel Graphs

Before introducing treewidth, I would like to discuss some well known notions, where we can find roots of the ideas that now play an important role in the (algorithmic) theory of treewidth.

The first are the Kirchhoff laws. In a landmark paper in 1847, Kirchhoff gave rules for computing the resistance of electrical networks. Many learn two of

G. Prencipe and S. Zaks (Eds.): SIROCCO 2007, LNCS 4474, pp. 11–25, 2007.

these rules in high-school. The *series* rule tells that if we have two resistors with resistance R_1 and R_2 in series, then the resistance of the two together is $R_1 + R_2$. The *parallel* rule gives the formula $\frac{1}{R} = \frac{1}{R_1} + \frac{1}{R_2}$ for the resulting resistance R when the resistors are placed in parallel. See Figure 1.

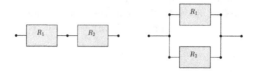

Fig. 1. Resistors in series and in parallel

A network where we can compute the resistance with only the series and parallel rule is a *series-parallel graph*. The following rules build all series-parallel graphs.

- A graph with two vertices, both terminals, s and t, and a single edge $\{s, t\}$ is a series-parallel graph.
- If G_1 with terminals s_1, t_1, and G_2 with terminals s_2, t_2 are series-parallel graphs, then the series composition of G_1 and G_2 is a series-parallel graph: take the disjoint union, and then identify t_1 and s_2. s_1 and t_2 are the terminals of the new graph.
- If G_1 with terminals s_1, t_1, and G_2 with terminals s_2, t_2 are series-parallel graphs, then the parallel composition of G_1 and G_2 is a series-parallel graph: take the disjoint union, and then identify s_1 and s_2, and identify t_1 and t_2; these two vertices are the terminals of the new graph.

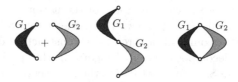

Fig. 2. Series and parallel composition

A commonly repeated mistake is that a graph is a series-parallel graph, if and only if it has treewidth at most two. However, the $K_{1,3}$ has treewidth one, but cannot be build with the series and parallel composition rules. Series-parallel graphs have treewidth at most two, and a graph is series-parallel, if and only if each of its biconnected components is series-parallel (see [20]).

Thus, the Kirchhoff laws for resistors in parallel and resistors in series allow to compute the resistance of a network, when it is series parallel.

Later, it was realized that many graph problems, that are NP-hard on arbitrary graphs, become polynomial, and often linear size solvable on series-parallel graphs, see e.g., [30,37,49,50,56,67,70,78,79,82].

2.2   Trees

For trees, it also has been observed long ago that many problems that are NP-hard on arbitrary graphs have efficient (e.g., linear) time algorithms when they are restricted to trees, see e.g. [29,33].

Graphs of bounded treewidth form a natural generalization to trees and series-parallel graphs. Trees can be formed by gluing trees together at one vertex, and series-parallel graphs are formed by gluing series-parallel graphs together at two *terminal* vertices.

We can generalize this, by gluing graphs with some bounded number k of terminals together. This gives e.g., k-terminal recursive graphs [84,83].

2.3   Gauss Elimination and Chordal Graphs

Consider Gauss elimination on a sparse symmetric matrix M. To a symmetric n by n matrix M, we associate the undirected graph G_M in the following natural way. We take n vertices, $v_1, \ldots, v_n$. Vertex v_i represents row and column i of M. We take an edge $\{v_i, v_j\}$, if and only if $M_{ij} \neq 0$. Now consider what happens in one step of Gauss elimination. We eliminate one row and its corresponding column, say i, and each pair of values M_{jk} and M_{kj} can become non-zero only if $M_{ij} \neq 0$ and $M_{ik} \neq 0$. Translated to G_M, this operation takes some vertex v_i, turns the neighborhood of v_i into a clique ($M_{ij} \neq 0$ and $M_{ik} \neq 0$ implies $\{v_i, v_j\} \in E$ and $\{v_i, v_k\} \in E$), and then removes v_i.

The sequence of pairs of rows and columns that are eliminated during the Gauss elimination process can be expressed as a permutation of the vertices, usually called *elimination ordering* in this context. Applying an elimination ordering π on a graph $G = (V, E)$ is the process where we, in the order where vertices v appear in π, turn the neighborhood of v into a clique, and remove v.

An important special case is when each vertex v is *simplicial* when it is eliminated: a vertex is simplicial, when its neighbors for a clique. I.e., we do not create new non-zero entries during Gauss elimination. An elimination ordering with each vertex simplicial at its elimination time is called a *perfect elimination ordering*.

Graphs that have a perfect elimination ordering are well studied for over thirty years, and are known as *chordal graphs*, or *triangulated graphs*. Chordal graphs are a special form of perfect graphs. See e.g., [45] for an algorithmic and graph theoretic survey of chordal graphs.

Chordal graphs have different equivalent characterizations, see [43,73].

- A graph is chordal, if it does not contain a chordless cycle of length at least four.
- A graph is chordal, if it is the intersection graph of subtrees of a tree.
- A graph is chordal, if it has a perfect elimination scheme.

The representation as intersection graph of subtrees of a tree can be viewed as a special *tree decomposition*. From the perfect elimination scheme representation, we can see that we can define the class of chordal graphs as follows: a clique is chordal, and if we have a chordal graph $G = (V, E)$ with W a clique in G, then we obtain a chordal graph y adding one new vertex v to G that is incident exactly to the vertices in W. Arnborg and Proskurowski [6], see also [1,2], define k-trees as follows: a clique with k vertices is a k-tree, and if we have a k-tree G with W a clique with k vertices in G, then the graph obtained by adding a new vertex incident to W is again a k-tree. We directly see that the k-trees form a subclass of the chordal graphs.

A graph is a *partial k-tree*, if it is a subgraph of a k-tree. It is well known that a graph is a partial k-tree, if and only if it has treewidth at most k. We discuss chordal graphs and their connection to treewidth further in Section 4

3   Tree Decompositions

Around 1980, several groups of researchers found independently notions that are equivalent, or strongly related to treewidth. The currently most used of these is the notion of *treewidth*, defined in terms of *tree decompositions*, introduced by Robertson and Seymour [71].

A *tree decomposition* of a graph $G = (V, E)$ is pair $(\{X_i \mid i \in I\}, T = (I, F))$, with $T = (I, F)$ a tree, and with a set of vertices $X_i \subseteq V$ associated to each $i \in I$, called a *bag*, such that

- $\bigcup_{i \in I} X_i = V$,
- for all $\{v, w\} \in E$, there exists an $i \in I$ with $\{v, w\} \subseteq X_i$, and
- for all $v \in V$, the set $\{i \in I \mid v \in X_i\}$ induces a subtree of T.

The *width* of a tree decomposition $(\{X_i \mid i \in I\}, T = (I, F))$ equals $\max_{i \in I} |X_i| - 1$, and the treewidth of a graph equals the minimum width of a tree decomposition of the graph.

The third condition of the definition can be replaced by the following, equivalent condition.

- for all $i_1, i_2, i_3 \in I$, if i_2 is on the path from i_1 to i_3 in T, then $X_{i_1} \cap X_{i_3} \subseteq X_{i_2}$.

Tree decompositions of bounded width have many applications. Perhaps the most striking one is that they allow linear or polynomial time algorithms for problems that are NP-hard on arbitrary graphs. Such algorithms use in general *dynamic programming* as main technique. They exploit the fact that 'most' bags X_i in the tree decomposition are separators. In particular, some bag $i_r \in I$ is taken as root of T. Now, for each bag, a table is computed. The bags in T are processed in postorder, and to compute a table for a node i, only local information (e.g., edges between vertices in X_i) and the tables for the bags of the children of i are used. If $v \in X_j$ for some descendant j of i, but $v \notin X_i$, then all neighbors of v are in bags that are descendants of i. This allows us to abstract away much of 'what happens below' i, and thus keep the amount of information needed for and stored in a table relatively small.

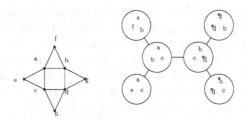

Fig. 3. A graph and a tree decomposition

For more details, see e.g., [6,61,80], or see [59, Chapter 10.4].

A notion related to treewidth is *pathwidth*, defined in terms of *path decompositions*, A tree decomposition $(\{X_i \mid i \in I\}, T = (I, F))$ is a *path decomposition*, if T is a path. The *pathwidth* of a graph G is the minimum width over all its path decompositions.

3.1 Nice Tree Decompositions

The notion of *nice tree decomposition* is a useful tool for the design of dynamic programming algorithms on graphs of bounded treewidth. A tree decomposition $(\{X_i \mid i \in I\}, T = (I, F))$ is *nice*, if we have some node $i_r \in I$ as root, and then each bag $i \in I$ is of one of the following four types:

- **Leaf**: i is a leaf of T and $|X_i| = 1$.
- **Join**: i has two children j_1 and j_2 with $X_i = X_{j_1} = X_{j_2}$.
- **Introduce**: i has one child j such that there is a v with $X_i = X_j \cup \{v\}$.
- **Forget**: i has one child j such that there is a v with $X_j = X_i \cup \{v\}$.

Theorem 1. *A graph $G = (V, E)$ has treewidth at most k, if and only if it has a nice tree decomposition of width at most k.*

The use of nice tree decompositions is mainly algorithmic. When we design (dynamic programming) algorithm to solve a certain problem on graphs with bounded treewidth, it suffices to give methods to compute the tables for each of the four types of nodes, and a method to obtain the desired result from the table for the root nodes. Typically, the design of such algorithms is trivial for Leaf nodes, almost trivial for Forget nodes, but can be quite complex for Join and Introduce nodes. See for example [18].

3.2 Recursive Families of Graphs

Nice tree decompositions can also be seen as algebraic expressions that generate a graph. A *terminal graph* is a triple (V, E, X) with (V, E) an undirected graph and $X \subseteq V$ an ordered set of *terminals*, i.e., vertices with a special role. Each of the four types of nodes of a nice tree decomposition can be translated to an operation on terminal graphs, as follows. Let k be given. LEAF gives as a constant a terminal graph with one vertex that is also terminal and no edges. JOIN is the binary operation, that when applied to two terminal graphs with the same number of terminals, yields the graph obtained by first taking the disjoint union, and then identifying for each i, the ith terminal of both graphs. For a subset of indices $S \subseteq \{1, \ldots, k\}$, INTRODUCE$_S$ is the unary operation, that when given a terminal graph (V, E, X) with at most k terminals, adds one new terminal vertex and makes it adjacent to the ith terminal in X, for each $i \in S$. The new terminal is assumed to be the last in the ordering. For an $i \in \{1, \ldots, k+1\}$, FORGET$_i$ is the operation, that given a terminal graph (V, E, X), yields the terminal graph $(V, E, X - \{v_i\})$, with v_i the ith terminal in X, i.e., the ith terminal is now no longer considered to be a terminal.

To each bag $i \in I$, in a nice tree decomposition, we can associate the terminal graph (V_i, E_i, X_i), with V_i the union of all bags X_j with $j = i$ or j a descendant of i, and E_i the set of edges in G between vertices in V_i. One can note that the operations exactly form the graphs G_i. By ending with a number of Forget operations, we thus can obtain:

Theorem 2. *A graph $G = (V, E)$ has treewidth at most k, if and only if $(V, E, \emptyset)$ can be formed by* LEAF, JOIN, INTRODUCE$_S$, *and* FORGET$_i$ *operations, with $S \subseteq \{1, \ldots, k\}$, and $i \in \{1, \ldots, k+1\}$.*

The construction shown above is a special case of a more general framework where graphs are formed by algebraic expressions on terminal graphs. The recursive families of graphs were introduced by by Borie [22], see [24,25,23]. Similar frameworks were made by Wimer [83,84], and by Courcelle (see e.g., [31]).

3.3 Monadic Second Order Logic

A very powerful tool to establish that problems are linear time solvable on graphs of bounded treewidth is *Monadic Second Order Logic*, or MSOL. MSOL is a language in which we can express properties of graphs. We can use logic operations ($\vee$, $\wedge$, $\neg$, $\Leftrightarrow$, etc.), quantification over vertices, edges, sets of vertices, and sets of edges, ($\exists W \subseteq V$, $\forall w \in V$, $\exists e \in E$, $\forall F \subseteq E$, etc.), membership tests ($v \in W$, $e \in F$), and incidence tests (v endpoint of e, $\{v, w\} \in E$).

Courcelle [31] has shown that for each problem that can be expressed in MSOL, and each fixed k, there is a linear time algorithm that, given a graph G of treewidth at most k, decides the problem on G. Moreover, for each MSOL property P and fixed k, it is decidable if P holds on all (or some, or no) graphs with treewidth at most k.

The result of Courcelle has been extended a number of times. One extension is *Counting Monadic Second Order Logic*, where we add statements of the form $|W| \bmod k = r$ and $|F| \bmod k = r$, to the language, and obtain the same results as for MSOL. In fact, Counting MSOL captures exactly the properties that can be solved on graphs of bounded treewidth by a table-based dynamic programming algorithm with $O(1)$ bits per table. (Such algorithms can be seen as *finite state tree automata.*) This was shown by Lapoire [62].

One can also extend MSOL to optimization or counting problems. For instance, consider the problem of determining the minimum (or maximum) size of a set $W \subseteq V$, such that $P(W)$ holds, with P in MSOL with one free vertex set variable. Many well known graph problems can be written in this format. It has been shown by Borie et al. [25], and by Arnborg et al. [4] that such problems can be solved in linear time on graphs of bounded treewidth. See for a further generalization [32].

4 Chordal Graphs and Elimination Schemes

4.1 Minimal Triangulations

Consider a tree decomposition $(\{X_i \mid i \in I\}, T = (I, F))$ of G. Suppose we turn each bag X_i into a clique, i.e., we add an edge between each pair of non-adjacent vertices that has a common bag. Let H be the resulting graph. H is chordal, as H is the intersection graph of subtrees of a tree (see [43]), as for each $v, w \in V$, $v \neq w$: $\{v, w\} \in E$, if and only if the subtrees of T with vertex sets $\{i \in I \mid v \in X_i\}$ and $\{i \in I \mid w \in X_i\}$ intersect. The properties of chordal graphs and tree decompositions imply that for each clique W, there is a bag $i \in I$ with $W \subseteq X_i$. These are the main ingredients of the following characterization of treewidth.

Theorem 3 (Folklore). *G has treewidth at most k, if and only if G is the subgraph of a chordal graph with maximum clique size at most $k + 1$.*

Thus, many algorithms to compute the treewidth of a graph do not work with tree decompositions, but instead use the somewhat easier to handle concept of chordal graphs. A chordal supergraph H of G is called a *triangulation* of G. A *minimal triangulation* is a triangulation that does not contain a triangulation with fewer edges as subgraph. We also have:

Theorem 4 (Folklore). *G has treewidth at most k, if and only if G has a minimal triangulation with maximum clique size at most $k + 1$.*

Important is the result of Bouchitté and Todinca [26,27] that shows that the treewidth of a graph can be computed in time, polynomial in the number of *minimal separators* of the graph. Several well known classes of graphs have a polynomial number of minimal separators (e.g., permutation graphs, circular arc graphs, weakly chordal graphs), and thus the algorithm of Bouchitté and Todinca [26,27] gives a polynomial time solution for treewidth for these classes. Central in this work is the notion of *potential maximal clique*, abbreviated pmc. A set $W \subseteq V$ is a pmc, if there is a minimal triangulation H with W is a maximal clique in H. Bouchitté and Todinca show that we can list all pmc's in time, polynomial in the number of minimal separators, and that given these, we can see if they can be 'composed' together to a chordal graph with the desired maximum clique size. See also [10,60,66].

Fomin et al. [41] and Villanger [81] have build upon the Bouchitté and Todinca algorithm to obtain efficient *exact* algorithms for treewidth, using $O(1.8899^n)$ time for graphs with n vertices.

4.2   Elimination Orderings

The representation of chordal graphs by perfect elimination schemes also leads to a different characterization of treewidth, in terms of permutations of the vertices. Recall the elimination process discussed in Section 2.3: eliminating a vertex v means that we turn the neighborhood of v into a clique and then remove v.

Suppose we have an elimination ordering π of the vertices of $G = (V, E)$, and suppose we eliminate the vertices in order of π. Let the *width* of π be the maximum over all vertices v of the number of (not yet eliminated) neighbors of v when v is eliminated.

Theorem 5. *The treewidth of G is at most k, if and only if there is a vertex ordering π of G with width at most k.*

The representation of treewidth by vertex orderings has been exploited in several cases. Gogate and Dechter [44] and Bachoore and Bodlaender [7] made branch and bound algorithms to compute treewidth, that build vertex orderings from left to right. A tabu search algorithm for treewidth, with the set of vertex orderings as search space, was made by Clautiaux et al. [28].

A useful property is the following. Suppose we eliminate a subset of the vertices $W \subseteq V$. Then the graph that results does not depend on the order in which we have eliminated the vertices $W \subseteq V$. This property was exploited by Clautiaux et al. [28] to improve upon the neighborhood structure of their tabu search; we want to avoid moving to different vertex orderings that represent the same triangulation of G. It was also exploited by Bodlaender et al. [16] for obtaining a practical exact algorithm for treewidth that uses dynamic programming in the same fashion as the classic Held-Karp algorithm for TSP [51], and an exact algorithm for treewidth using polynomial memory.

Pathwidth also has a representation with help of vertex orderings, see [57]. Many algorithms to compute the pathwidth use instead the notion of vertex separation number, see e.g., the algorithms to compute the pathwidth of trees [38,77].

5   Search Games

Treewidth and several related notions can also be characterized in terms of search games. In search games, there are a *robber* and *cops* that are on some vertices (or edges) of the graph. The robber tries to escape capture by the cops, and the cops try to catch the robber, using certain rules of capture and movement. The parameter of interest is the minimum number of cops, such that they have a strategy in which they always capture the robber.

The search game that models treewidth has the following rules. We only give here an informal description, see Seymour and Thomas [75] for a formal description. The robber stands on a vertex, and can at each point run to another vertex of the graph, with infinite speed. He runs along any path in the graph that does not contain a vertex with a cop on it. Each of the cops is either on a vertex, or in a helicopter. A cop in a helicopter can land on a vertex. The robber can see when the cop lands, and before its landing run to a new vertex. The cops capture the robber when a cop lands on the vertex with the robber.

Theorem 6 (Seymour and Thomas [75]). *A graph has treewidth at most k, if and only if $k + 1$ cops can capture the robber.*

A special form a search strategies are the *monotone* strategies. The idea of a monotone search strategy is that the robber is not given positions where he cannot move to back at later points. Seymour and Thomas also have shown that there is always a monotone search strategy with the minimum number of cops.

A related search game with an *inert* robber, (i.e., the robber can only move just before the searcher visits the vertex that he occupies), was studied by Dendris et al. [35]. This gives again a notion that is equivalent to treewidth, also when we require the search strategy to be monotone.

Pathwidth has a similar type of representation as a search game. In this game, the robber is on edges, not seen by the cops. The robber is caught when there are cops at both endpoints of its edge. It can be also explained as follows. In a search strategy, a move can be either placing a cop on a vertex, or deleting a cop from a vertex. Initially, each edge is contaminated. When there are cops on both endpoints of an edge, the edge becomes cleared. However, cleared edges become recontaminated when they have a path without vertices with cops to a contaminated edge. The goal is to clear all edges. It can be shown that there is always a search strategy that is *progressive*, or *monotone*, i.e., in which no edge ever becomes recontaminated. The successive sets that contain a cop form in fact a path decomposition. In fact, the minimum number of cops for this search game equals the pathwidth plus one. See Kirousis and Papadimitriou [58]. See also [38], [13, Chapter 10]. An extensive overview can be found in [11].

Several results for treewidth and pathwidth can often be as well, and sometimes easier be explained and understood when we look to the search game characterizations. One other notable example for a notion related to treewidth where a game characterization is of great importance is the *ratcatcher* algorithm by Seymour and Thomas [76]. This algorithm computes the *branchwidth* of planar graphs in polynomial time. Branchwidth is strongly related to treewidth: the treewidth and branchwidth of a graph differ by a factor of at most 1.5.

In [76], Seymour and Thomas show, with a complicated proof, that the problem of computing branchwidth on planar graphs can be transformed to the problem of determining the sound level the ratcatcher needs to catch a rat in the a game, where, following certain rules, the ratcatcher and rat move, and the ratcatcher can blow a whistle which prevents the rat visiting the parts of the graph where the whistle can be heard. This game allows an algorithm (e.g., using a form of dynamic programming), to determine the sound level for which there is a winning strategy. See also [52,53,46].

An interesting application of search games for distributed systems was given by Franklin et al. [42].

6   Brambles

Seymour and Thomas [75] have given a characterization of treewidth with help
of a combinatorial structure called *brambles*. Consider a graph $G = (V, E)$. Two
sets of vertices $W_1, W_2 \subseteq V$ are said to *touch*, if they intersect or a vertex in W_1
is adjacent to a vertex in W_2. A *bramble* of G is a collection of mutually touching
connected subsets of G. The *order* of a bramble is the minimum size of a set W
that intersects each set in the bramble.

Theorem 7 (Seymour and Thomas [75]). *A graph G has treewidth at least
k, if and only if it has a bramble of order at least $k + 1$.*

See also Reed [68], and see Bellenbaum and Diestel [9] for a shorter proof of
Theorem 7.

Most of the characterizations exploit structures that give an upper bound on
the treewidth: the treewidth is the minimum width of a tree decomposition, the
minimum number of searchers, or the minimum width of an elimination ordering.
A bramble however gives a lower bound on the treewidth: the treewidth is one
smaller than the maximum order of a bramble. It should be noted however
that computing the order of a given bramble is already NP-hard, and there is
no polynomial bound on the size of a bramble that is needed to obtain the
treewidth.

Still, brambles may be a useful tool to obtain lower bounds for treewidth.
In [17], two heuristics for obtaining brambles were proposed and evaluated. It
appears that for planar graphs, and graphs that are close to being planar, good
treewidth lower bounds can be obtained in this way.

7   Graph Minors

A graph G is a *minor* of a graph H, if G can be obtained from H by a series
of zero or more vertex deletions, edge deletions, or edge contractions. Robertson
and Seymour have shown that each class of graphs $\mathcal{G}$ that is closed under taking
of minors has a finite set of graphs $M_\mathcal{G}$, such that a graph G belongs to $\mathcal{G}$, if and
only if no graph in $M_\mathcal{G}$ is a minor of G.

This gives an $O(n^3)$ time algorithm to test membership in a minor closed
class of graphs (see e.g., [71,72,36,39,40]). If in addition $\mathcal{G}$ does not contain all
planar graphs, then the graphs in $\mathcal{G}$ have bounded treewidth, and the minorship
test can be done in linear time. As treewidth does not increase when we take
a minor, this gives, a *non-constructive* algorithm to test if a given graph has
treewidth at most k, for fixed k.

An example where these results were used is the following. In [21], k-label
Interval Routing Schemes are studied. In this application, we want to route
messages through a graph. The vertices in the graph must be labeled in such a
way, that each outgoing edge is labeled with a collection of at most k intervals;
with messages forwarded on an edge when the label of the destination is in one of
these intervals. (See [21] for precise details.) For fixed k, the class of graphs with

such a labeling scheme is closed under taking of minors, and does not contain all planar graphs. Thus, we know a linear time algorithm to test for a given graph whether it has a k-label interval routing scheme exists.

8 Conclusions

In addition to the characterizations of treewidth, there are also some that we only briefly discuss here.

Habel and Kreowski [48,47] have introduced hyperedge replacement grammars. See also [8]. These graph grammars rewrite a hypergraph, following certain rules. If we generalize the notion of tree decomposition to hypergraphs, by requiring that for each hyperedge e, there must be a bag containing all vertices in e, we have that each hyperedge replacement grammar generates hypergraphs with bounded treewidth, and there is a hyperedge replacement grammar, generating exactly the graphs of treewidth at most k, for each k. This was shown by Lautemann [63,64].

For each fixed k, there is a finite set of *reduction rules*, that, when given a graph, reduce it to an empty graph, if and only if G has treewidth at most k, see [3]. This leads, for each fixed k, to a linear time algorithm for recognizing the graphs of treewidth at most k; however, it may use more than linear memory. For $k \leq 4$, this gives practical algorithms to test if the graph has treewidth at most k, see [5,19,65,74].

From the overview above, we see that treewidth has a large number of different, equivalent, characterizations. Different characterizations are useful for different algorithmic applications. The interaction between graph theory and algorithm design is an interesting, and attractive feature of the research on treewidth. For other overviews on treewidth or related notions, see e.g., [1,12,13,14,15,34,54,55,68,69].

Acknowledgment

I am very grateful to many colleagues for discussions, help, and cooperations.

References

1. Arnborg, S.: Efficient algorithms for combinatorial problems on graphs with bounded decomposability – A survey. BIT 25, 2–23 (1985)
2. Arnborg, S., Corneil, D.G., Proskurowski, A.: Complexity of finding embeddings in a k-tree. SIAM J. Alg. Disc. Meth. 8, 277–284 (1987)
3. Arnborg, S., Courcelle, B., Proskurowski, A., Seese, D.: An algebraic theory of graph reduction. J. ACM 40, 1134–1164 (1993)
4. Arnborg, S., Lagergren, J., Seese, D.: Easy problems for tree-decomposable graphs. J. Algorithms 12, 308–340 (1991)
5. Arnborg, S., Proskurowski, A.: Characterization and recognition of partial 3-trees. SIAM J. Alg. Disc. Meth. 7, 305–314 (1986)

6. Arnborg, S., Proskurowski, A.: Linear time algorithms for NP-hard problems restricted to partial k-trees. Disc. Appl. Math. 23, 11–24 (1989)
7. Bachoore, E.H., Bodlaender, H.L.: A branch and bound algorithm for exact, upper, and lower bounds on treewidth. In: Cheng, S.-W., Poon, C.K. (eds.) AAIM 2006. LNCS, vol. 4041, pp. 255–266. Springer, Heidelberg (2006)
8. Bauderon, M., Courcelle, B.: Graph expressions and graph rewritings. Mathematical Systems Theory 20, 83–127 (1987)
9. Bellenbaum, P., Diestel, R.: Two short proofs concerning tree-decompositions. Combinatorics, Probability, and Computing 11, 541–547 (2002)
10. Berry, A., Bordat, J.-P., Cogis, O.: Generating all the minimal separators of a graph. Int. J. Found. Computer Science 11, 397–404 (2000)
11. Bienstock, D.: Graph searching, path-width, tree-width and related problems (a survey). DIMACS Ser. in Discrete Mathematics and Theoretical Computer Science 5, 33–49 (1991)
12. Bodlaender, H.L.: A tourist guide through treewidth. Acta Cybernetica 11, 1–23 (1993)
13. Bodlaender, H.L.: A partial k-arboretum of graphs with bounded treewidth. Theor. Comp. Sc. 209, 1–45 (1998)
14. Bodlaender, H.L.: Discovering treewidth. In: Vojtáš, P., Bieliková, M., Charron-Bost, B., Sýkora, O. (eds.) SOFSEM 2005: Theory and Practice of Computer Science. 31st Conference on Current Trends in Theory and Practice of Computer Science Liptovský, Ján, Slovakia, January 22-28, 2005. LNCS, vol. 3381, pp. 1–16. Springer, Heidelberg (2005)
15. Bodlaender, H.L.: Treewidth: Characterizations, applications, and computations. In: Fomin, F.V. (ed.) Graph-Theoretic Concepts in Computer Science. 32nd International Workshop, WG 2006, Bergen, Norway, June 22-24, 2006. LNCS, vol. 4271, pp. 1–14. Springer, Heidelberg (2006)
16. Bodlaender, H.L., Fomin, F.V., Koster, A.M.C.A., Kratsch, D., Thilikos, D.M.: On exact algorithms for treewidth. In: Azar, Y., Erlebach, T. (eds.) Algorithms – ESA 2006. 14th Annual European Symposium, Zurich, Switzerland, September 11-13, 2006. LNCS, vol. 4168, pp. 672–683. Springer, Heidelberg (2006)
17. Bodlaender, H.L., Grigoriev, A., Koster, A.M.C.A.: Treewidth lower bounds with brambles. In: Brodal, G.S., Leonardi, S. (eds.) Algorithms – ESA 2005. 13th Annual European Symposium, Palma de Mallorca, Spain, October 3-6, 2005. LNCS, vol. 3669, pp. 391–402. Springer, Heidelberg (2005)
18. Bodlaender, H.L., Kloks, T.: Efficient and constructive algorithms for the pathwidth and treewidth of graphs. J. Algorithms 21, 358–402 (1996)
19. Bodlaender, H.L., Koster, A.M.C.A., Eijkhof, F.v.d.: Pre-processing rules for triangulation of probabilistic networks. Computational Intelligence 21(3), 286–305 (2005)
20. Bodlaender, H.L., van Antwerpen-de Fluiter, B.: Parallel algorithms for series parallel graphs and graphs with treewidth two. Algorithmica 29, 543–559 (2001)
21. Bodlaender, H.L., van Leeuwen, J., Tan, R., Thilikos, D.: On interval routing schemes and treewidth. Information and Computation 139(1), 92–109 (1997)
22. Borie, R.B.: Recursively Constructed Graph Families. PhD thesis, School of Information and Computer Science, Georgia Institute of Technology (1988)
23. Borie, R.B.: Generation of polynomial-time algorithms for some optimization problems on tree-decomposable graphs. Algorithmica 14, 123–137 (1995)
24. Borie, R.B., Parker, R.G., Tovey, C.A.: Deterministic decomposition of recursive graph classes. SIAM J. Disc. Math. 4, 481–501 (1991)

25. Borie, R.B., Parker, R.G., Tovey, C.A.: Automatic generation of linear-time algorithms from predicate calculus descriptions of problems on recursively constructed graph families. Algorithmica 7, 555–581 (1992)
26. Bouchitté, V., Todinca, I.: Treewidth and minimum fill-in: Grouping the minimal separators. SIAM J. Comput. 31, 212–232 (2001)
27. Bouchitté, V., Todinca, I.: Listing all potential maximal cliques of a graph. Theor. Comp. Sc., 276:17–32, 2002.
28. Clautiaux, F., Moukrim, A., Négre, S., Carlier, J.: Heuristic and meta-heuristic methods for computing graph treewidth. RAIRO Operations Research 38, 13–26 (2004)
29. Cockayne, E.J., Goodman, S.E., Hedetniemi, S.T.: A linear algorithm for the domination number of a tree. Information Processing Letters 4, 41–44 (1975)
30. Colbourn, C.J., Stewart, L.K.: Dominating cycles in series-parallel graphs. Ars. Combinatorica 19A, 107–112 (1985)
31. Courcelle, B.: The monadic second-order logic of graphs I: Recognizable sets of finite graphs. Information and Computation 85, 12–75 (1990)
32. Courcelle, B., Mosbah, M.: Monadic second-order evaluations on tree-decomposable graphs. Theor. Comp. Sc. 109, 49–82 (1993)
33. Daykin, D.E., Ng, C.P.: Algorithms for generalized stability numbers of tree graphs. J. Austral. Math. Soc. 6, 89–100 (1966)
34. Demaine, E.D., Hajiaghayi, M.: The bidimensionality theory and its algorithmic applications. Unpublished manuscript (2006)
35. Dendris, N.D., Kirousis, L.M., Thilikos, D.M.: Fugitive-search games on graphs and related parameters. Theor. Comp. Sc. 172, 233–254 (1997)
36. Downey, R.G., Fellows, M.R.: Parameterized Complexity. Springer, Heidelberg (1998)
37. Duffin, R.J.: Topology of series-parallel graphs. J. Math. Anal. Appl. 10, 303–318 (1965)
38. Ellis, J.A., Sudborough, I.H., Turner, J.: The vertex separation and search number of a graph. Information and Computation 113, 50–79 (1994)
39. Fellows, M.R., Langston, M.A.: Nonconstructive advances in polynomial-time complexity. Information Processing Letters 26, 157–162 (1987)
40. Fellows, M.R., Langston, M.A.: Nonconstructive tools for proving polynomial-time decidability. J. ACM 35, 727–739 (1988)
41. Fomin, F.V., Kratsch, D., Todinca, I.: Exact (exponential) algorithms for treewidth and minimum fill-in. In: Díaz, J., Karhumäki, J., Lepistö, A., Sannella, D. (eds.) Automata, Languages and Programming. 31st International Colloquium, ICALP 2004, Turku, Finland, July 12-16, 2004. LNCS, vol. 3142, pp. 568–580. Springer, Heidelberg (2004)
42. Franklin, M., Galil, Z., Yung, M.: Eavesdropping games: A graph-theoretic approach to privacy in distributed systems. J. ACM 47, 225–243 (2000)
43. Gavril, F.: The intersection graphs of subtrees in trees are exactly the chordal graphs. J. Comb. Theory Series B. 16, 47–56 (1974)
44. Gogate, V., Dechter, R.: A complete anytime algorithm for treewidth. In: Proceedings of the 20th Annual Conference on Uncertainty in Artificial Intelligence UAI-04, pp. 201–208, Arlington, Virginia, USA, AUAI Press (2004)
45. Golumbic, M.C.: Algorithmic Graph Theory and Perfect Graphs. Academic Press, New York (1980)

46. Gu, Q.-P., Tamaki, H.: Optimal branch-decomposition of planar graphs in $O(n^3)$ time. In: Caires, L., Italiano, G.F., Monteiro, L., Palamidessi, C., Yung, M. (eds.) Automata, Languages and Programming. 32nd International Colloquium, ICALP 2005, Lisbon, Portugal, July 11-15, 2005. LNCS, vol. 3580, pp. 373–384. Springer, Heidelberg (2005)

47. Habel, A., Kreowski, H.J.: Characteristics of graph languages generated by edge replacement. Theor. Comp. Sc. 51, 81–115 (1987)

48. Habel, A., Kreowski, H.J.: May we introduce to you: hyperedge replacement. In: Ehrig, H., Nagl, M., Rosenfeld, A., Rozenberg, G. (eds.) Proc. Graph-Grammars and Their Application to Computer Science. LNCS, vol. 291, pp. 15–26. Springer, Heidelberg (1987)

49. Hare, E., Hedetniemi, S., Laskar, R., Peters, K., Wimer, T.: Linear-time computability of combinatorial problems on generalized-series-parallel graphs. In: Johnson, D.S., Nishizeki, T., Nozaki, A., Wilf, H.S. (eds.) Proc. of the Japan-US Joint Seminar on Discrete Algorithms and Complexity, Orlando, Florida, Academic Press, Inc. San Diego (1987)

50. Hassin, R., Tamir, A.: Efficient algorithms for optimization and selection on series-parallel graphs. SIAM J. Alg. Disc. Meth. 7, 379–389 (1986)

51. Held, M., Karp, R.: A dynamic programming approach to sequencing problems. J. SIAM 10, 196–210 (1962)

52. Hicks, I.V.: Planar branch decompositions I: The ratcatcher. INFORMS J. on Computing 17, 402–412 (2005)

53. Hicks, I.V.: Planar branch decompositions II: The cycle method. INFORMS J. on Computing 17, 413–421 (2005)

54. Hicks, I.V., Koster, A.M.C.A., Kolotoğlu, E.: Branch and tree decomposition techniques for discrete optimization. In: Smith, J.C. (ed.) TutORials 2005, INFORMS Tutorials in Operations Research Series, ch.1, pp. 1–29. INFORMS Annual Meeting (2005)

55. Hliněný, P., Oum, S., Seese, D., Gottlob, G.: Width parameters beyond tree-width and their applications. Paper to appear in this special issue (2006)

56. Kikuno, T., Yoshida, N., Kakuda, Y.: A linear algorithm for the domination number of a series-parallel graph. Disc. Appl. Math. 5, 299–311 (1983)

57. Kinnersley, N.G.: The vertex separation number of a graph equals its path width. Information Processing Letters 42, 345–350 (1992)

58. Kirousis, L.M., Papadimitriou, C.H.: Interval graphs and searching. Disc. Math. 55, 181–184 (1985)

59. Kleinberg, J., Tardos, É.: Algorithm Design. Addison-Wesley, Boston (2005)

60. Kloks, T., Kratsch, D.: Listing all minimal separators of a graph. SIAM J. Comput. 27(3), 605–613 (1998)

61. Koster, A.M.C.A., van Hoesel, S.P.M., Kolen, A.W.J.: Solving partial constraint satisfaction problems with tree decomposition. Networks 40(3), 170–180 (2002)

62. Lapoire, D.: Recognizability equals definability, for every set of graphs of bounded tree-width. In: Meinel, C., Morvan, M. (eds.) STACS 98. LNCS, vol. 1373, pp. 618–628. Springer, Heidelberg (1998)

63. Lautemann, C.: Efficient algorithms on context-free graph languages. In: Lepistö, T., Salomaa, A. (eds.) ICALP'88. Proceedings of the 15th International Colloquium on Automata, Languages and Programming. LNCS, vol. 317, pp. 362–378. Springer, Heidelberg (1988)

64. Lautemann, C.: The complexity of graph languages generated by hyperedge replacement. Acta Informatica 27, 399-421 (1990)

65. Matoušek, J., Thomas, R.: Algorithms for finding tree-decompositions of graphs. J. Algorithms 12, 1–22 (1991)
66. Parra, A., Scheffler, P.: Characterizations and algorithmic applications of chordal graph embeddings. Disc. Appl. Math. 79, 171–188 (1997)
67. Pfaff, J., Laskar, R., Hedetniemi, S.T.: Linear algorithms for independent domination and total domination in series-parallel graphs. Congressus Numerantium 45, 71–82 (1984)
68. Reed, B.A.: Tree width and tangles, a new measure of connectivity and some applications. LMS Lecture Note Series, vol. 241, pp. 87–162. Cambridge University Press, Cambridge, UK (1997)
69. Reed, B.A.: Algorithmic aspects of tree width. In: Reed, B.A. (ed.) CMS Books Math. / Ouvrages Math. SMC. 11, pp. 85–107. Springer, Heidelberg (2003)
70. Richey, M.B.: Combinatorial optimization on series-parallel graphs: algorithms and complexity. PhD thesis, School of Industrial and Systems Engineering, Georgia Institute of Technology (1985)
71. Robertson, N., Seymour, P.D.: Graph minors. II. Algorithmic aspects of tree-width. J. Algorithms 7, 309–322 (1986)
72. Robertson, N., Seymour, P.D.: Graph minors. XIII. The disjoint paths problem. J. Comb. Theory Series B. 63, 65–110 (1995)
73. Rose, D.J., Tarjan, R.E., Lueker, G.S.: Algorithmic aspects of vertex elimination on graphs. SIAM J. Comput. 5, 266–283 (1976)
74. Sanders, D.P.: On linear recognition of tree-width at most four. SIAM J. Disc. Math. 9(1), 101–117 (1996)
75. Seymour, P.D., Thomas, R.: Graph searching and a minimax theorem for tree-width. J. Comb. Theory Series B. 58, 239–257 (1993)
76. Seymour, P.D., Thomas, R.: Call routing and the ratcatcher. Combinatorica 14(2), 217–241 (1994)
77. Skodinis, K.: Construction of linear tree-layouts which are optimal with respect to vertex separation in linear time. J. Algorithms 47, 40–59 (2003)
78. Syslo, M.M.: Series-parallel graphs and depth-first search trees. IEEE Trans. on Circuits and Systems CAS-31(12), 1029–1033 (1984)
79. Takamizawa, K., Nishizeki, T., Saito, N.: Linear-time computability of combinatorial problems on series-parallel graphs. J. ACM 29, 623–641 (1982)
80. Telle, J.A., Proskurowski, A.: Algorithms for vertex partitioning problems on partial k-trees. SIAM J. Disc. Math. 10, 529–550 (1997)
81. Villanger, Y.: Improved exponential-time algorithms for treewidth and minimum fill-in. In: Correa, J.R., Hevia, A., Kiwi, M. (eds.) LATIN 2006: Theoretical Informatics. Proceedings of the 7th Latin American Theoretical Informatics Symposium (LATIN 2006), Valdivia, Chile, March 20-24, 2006. LNCS, vol. 3887, pp. 800–811. Springer, Heidelberg (2006)
82. Wimer, T.V.: Linear algorithms for the dominating cycle problems in series-parallel graphs, 2-trees and Halin graphs. Congressus Numerantium, vol. 56 (1987)
83. Wimer, T.V.: Linear Algorithms on k-Terminal Graphs. PhD thesis, Dept. of Computer Science, Clemson University (1987)
84. Wimer, T.V., Hedetniemi, S.T., Laskar, R.: A methodology for constructing linear graph algorithms. Congressus Numerantium 50, 43–60 (1985)

Fast Periodic Graph Exploration with Constant Memory*

Leszek Gąsieniec[1], Ralf Klasing[2], Russell Martin[1], Alfredo Navarra[2,3],
and Xiaohui Zhang[1]

[1] Department of Computer Science, University of Liverpool, Ashton Street,
Liverpool L69 3BX, UK
{leszek, martin, cloud}@csc.liv.ac.uk
[2] LaBRI - CNRS - Université de Bordeaux 1, 351 cours de la Liberation,
33405 Talence, France
{Ralf.Klasing, Alfredo.Navarra}@labri.fr
[3] Dipartimento di Matematica e Informatica, Universitá degli Studi di Perugia,
Via Vanvitelli 1, 06123 Perugia, Italy
navarra@dipmat.unipg.it

Abstract. We consider the problem of periodic exploration of all nodes
in undirected graphs by using a finite state automaton called later a *robot*.
The robot, using a constant number of states (memory bits), must be able
to explore any unknown anonymous graph. The nodes in the graph are
neither labelled nor colored. However, while visiting a node v the robot
can distinguish between edges incident to it. The edges are ordered and
labelled by consecutive integers $1, \ldots, d(v)$ called *port numbers*, where
$d(v)$ is the degree of v. Periodic graph exploration requires that the
automaton has to visit every node infinitely many times in a periodic
manner. Note that the problem is unsolvable if the local port numbers
are set arbitrarily, see [8]. In this context, we are looking for the minimum
function $\pi(n)$, such that, there exists an efficient deterministic algorithm
for setting the local port numbers allowing the robot to explore all graphs
of size n along a traversal route with the period $\pi(n)$. Dobrev *et al.*
proved in [13] that for oblivious robots $\pi(n) \leq 10n$. Recently Ilcinkas
proposed another port labelling algorithm for robots equipped with two
extra memory bits, see [20], where the exploration period $\pi(n) \leq 4n - 2$.
In the same paper, it is conjectured that the bound $4n - O(1)$ is tight
even if the use of larger memory is allowed. In this paper, we disprove
this conjecture presenting an efficient deterministic algorithm arranging
the port numbers, such that, the robot equipped with a constant number
of bits is able to complete the traversal period in $\pi(n) \leq 3.75n - 2$ steps
hence decreasing the existing upper bound. This reduces the gap with
the lower bound of $\pi(n) \geq 2n - 2$ holding for any robot.

---

* This research was partially funded by the project "ALPAGE" of the ANR "Masse de
données: Modélisation, Simulation, Applications", the project "CEPAGE" of INRIA,
the European projects COST Action 293, "Graphs and Algorithms in Communication Networks" (GRAAL), COST Action 295, "Dynamic Communication Networks"
(DYNAMO), the Nuffield Foundation grant NAL/32566, "The structure and efficient utilization of the Internet and other distributed systems", and by a visiting
fellowship from LaBRI/ENSEIRB.

G. Prencipe and S. Zaks (Eds.): SIROCCO 2007, LNCS 4474, pp. 26–40, 2007.
© Springer-Verlag Berlin Heidelberg 2007

1   Introduction

We consider the task of graph exploration by a mobile entity equipped with small (constant number of bits) memory. The mobile entity may be, e.g., an autonomous piece of software navigating through an underlying graph of connections of a computer network. The mobile entity is expected to visit all nodes in the graph in a periodic manner. For the sake of simplicity, we call the mobile entity a *robot* and model it as a finite state automaton. The task of periodically visiting all nodes of a network is particularly useful in network maintenance, where the status of every node has to be checked regularly.

We consider here undirected graphs that are anonymous, i.e., the nodes in the graph are neither labelled nor colored. However, while visiting a node the robot can distinguish between edges incident to it. At each node v the incident edges are ordered and labelled by consecutive integers $1, \ldots, d(v)$ called *port numbers*, where $d(v)$ is the degree of v. We will refer to port ordering as a *local orientation*.

Following formalism from [20], we model robots as *Mealy automata*. A Mealy automaton uses only input actions, i.e., output depends on input and the current state. In our case, the Mealy automaton has a transition function f and a finite number of states governing the actions of the robot. More precisely, if the automaton enters a node v of degree $d(v)$ through port i in state s, it switches to state s' and exits the node through port i'. This corresponds to $f(s, i, d(v)) = (s', i')$.

Periodic graph exploration requires that the automaton has to visit every node infinitely many times in a periodic manner. In this paper, we are interested in minimising the length of the exploration period. In other words, we want to minimise the maximum number of edge traversals performed by the robot between two consecutive visits of a generic node, in the same state and entering the node by the same port. Budach [8] proved that no finite automaton can explore all graphs if the local orientation is given by an adversary. In this context, we want to determine the minimum function $\pi(n)$, such that, there exists an efficient deterministic algorithm for setting the local port numbers allowing the robot to explore all graphs of size n along a traversal route with the period $\pi(n)$. Dobrev *et al.* proved in [13] that for oblivious robots $\pi(n) \leq 10n$. Very recently Ilcinkas proposed another port labelling algorithm for robots equipped with two extra memory bits, see [20], with exploration period $\pi(n) \leq 4n - 2$. The traversal route constructed by Ilcinkas' algorithms is based on edges of an arbitrary spanning tree encoded neatly by the port numbers. The automaton proposed by Ilcinkas is not oblivious but it has only three states. Moreover, it performs periodic exploration independently of its starting position and the initial state. In addition, if required, the robot is able to stop at the root of the spanning tree after finishing each period of the traversal route, and wait there, e.g., for the next wake-up message. In the same paper, Ilcinkas conjectured that the bound $4n - O(1)$ is tight even if the use of larger memory is allowed.

1.1   Our Results

We present an efficient deterministic algorithm arranging port numbers in the graph, such that, the robot equipped with a constant number of bits is able to accomplish each period of the traversal route in $\pi(n) \leq 3.75n - 2$ steps. This invalidates Ilcinkas' conjecture and shows that the problem of determining the minimum function $\pi(n)$ remains wide open. In addition, our result reduces the gap with the lower bound of $\pi(n) \geq 2n - 2$ holding for any robot.[1]

The improvement is a consequence of the construction of the traversal route on the basis of a very specific (rooted) spanning tree, rather than an arbitrary tree used by Ilcinkas. The main idea resides in powering the robot in such a way that it can recognise, or more precisely it can expect to meet, specific subtrees in the chosen spanning tree, hence saving in the number of so-called penalties. We say the robot pays a *penalty* at the node v if, starting from v, it traverses an edge not belonging to the spanning tree.

We introduce the concept of *Extended Leaves, paired Extended Leaves* and *paired Leaves*. An extended leaf of the spanning tree is a path of length 1 in which one endpoint is a leaf of the tree and the other is an internal node of the tree which has no other children than the leaf. A paired extended leaf is an extended leaf connected at a node whose children contain at least two nodes that are the endpoints of two extended leaves. A paired leaf is a leaf rooted at a node whose children contain at least two leaves.

Intuitively, our arrangement of the port numbers at each vertex means once the robot has met an extended leaf it expects to visit a paired one, and once it meets a leaf it expects to visit a paired one. In doing so, it saves penalties at each second, third, etc., paired extended leaf as well as at each second, third, etc., paired leaf since it knows what the topology should be, and hence does not have to explore further edges beyond such a leaf or extended leaf.

Our robot requires few (constant number) states allowing it to take advantage of the specific topology of the spanning tree.

1.2   Related Work

Graph exploration by robots has recently attracted growing attention. The unknown environment in which the robots operate is often modelled as a graph, assuming that the robots may only move along its edges. The graph setting is available in two different forms.

In [1,4,5,11,15], the robot explores strongly connected directed graphs and it can move only in one pre-specified direction along each edge. In [2,6,9,14,17,18,21], the explored graph is undirected and the agent can traverse edges in both directions. Also, two alternative efficiency measures are adopted in most papers devoted to graph exploration, namely, the *time* of completing the

---

[1] Note that this lower bound is obtained, e.g., when the graph to be explored is a tree. And indeed, in a full periodic exploration of a tree every edge of the tree must be traversed at least twice.

task [1,2,4,5,6,11,14], or the number of *memory bits* (states in the automaton) available to the agent [9,12,16,17,18,19].

Graph exploration scenarios considered in the literature differ in an important way: it is either assumed that nodes of the graph have unique labels which the agent can recognise, or it is assumed that nodes are anonymous. Exploration of directed graphs assuming the existence of labels was investigated in [1,11,15]. In this case, no restrictions on the agent moves were imposed, other than by directions of edges, and fast exploration algorithms were sought. Exploration of undirected labelled graphs was considered in [2,3,6,14,21]. Since in this case a simple exploration based on depth-first search can be completed in time $2e$, where e is the number of edges, investigations concentrated either on further reducing the time for an unrestricted agent, or on studying efficient exploration when moves of the agent are restricted in some way. The first approach was adopted in [21], where an exploration algorithm working in time $e + O(n)$, with n being the number of nodes, was proposed. Restricted agents were investigated in [2,3,6,14]. It was assumed that the agent is a robot with either a restricted fuel tank [2,6], forcing it to periodically return to the base for refuelling, or that it is a tethered robot, i.e., attached to the base by a rope or cable of restricted length [14]. It was proved in [14] that exploration can be done in time $O(e)$ under both scenarios.

Exploration of anonymous graphs by robots with limited memory presents different types of challenges. In this case, it is impossible to explore arbitrary graphs if no marking of nodes is allowed [8]. Hence, the scenario adopted in [4,5] was to allow *pebbles* which the agent can drop on nodes to recognise already visited ones, and then remove them and drop in other places. The authors concentrated attention on the minimum number of pebbles allowing efficient exploration of arbitrary directed n-node graphs. (In the case of undirected graphs, one pebble suffices for efficient exploration.) In [5], the authors compared the exploration power of one agent with pebbles to that of two cooperating agents without pebbles. In [4], it was shown that one pebble is enough, if the agent knows an upper bound on the size of the graph, and $\Theta(\log \log n)$ pebbles are necessary and sufficient otherwise.

In [9,12,16,17,18], the adopted measure of efficiency was the memory size of the agent exploring anonymous graphs. In [16,18], the agent was allowed to mark nodes by pebbles, or even by writing messages on whiteboards with which nodes are equipped. In [9], the authors studied special schemes of labelling nodes, which facilitate exploration with small memory. Another aspect of distributed graph exploration by robots with bounded memory was studied in [12,19], where the topology of graphs is restricted to trees. In [12] Diks *et al.* proposed a robot requiring $O(\log^2 n)$ memory bits to explore any tree with at most n nodes. They also provided the lower bound $\Omega(\log n)$ if the robot is expected to return to its original position in the tree. Very recently the gap between the upper bound and the lower bound was closed in [19] by Gąsieniec *et al.* who showed that $O(\log n)$ bits of memory suffice in tree exploration. However it is known, see [17], that in arbitrary graphs the number of memory bits required by any robot is $\Omega(D \log d)$, where D is the diameter and d is the maximum degree in the graph. In comparison, in the fully centralised model where the graph topology is represented by a random access matrix, Reingold [22] proved recently that $SL = L$, i.e., any decision problem which can be solved by a deterministic Turing machine using logarithmic memory (space) is log-space reducible to the USTCON (st-connectivity in undirected graphs) problem. This, e.g., proves the existence of a robot equipped with $O(\log n)$ bits being able to explore any n-node graph in the centralised model.

In this paper, we are interested in robots characterised by very low memory utilisation. In fact, the robots are allowed to use only a constant number of memory bits. This restriction permits modelling robots as finite state automata. Budach [8] proved that no finite automaton can explore all graphs. Rollik [23] showed later that even a finite team of finite automata cannot explore all planar cubic graphs. This result is improved in [10], where Cook and Rackoff introduce a powerful tool, called the *JAG*, for Jumping Automaton for Graphs. A JAG is a finite team of finite automata that permanently cooperate and that can use *teleportation* to move from their current location to the location of any other automaton. However, even JAGs cannot explore all graphs [10].

1.3   Outline of the Paper

Section 2 presents the spanning tree construction that will constitute the main route of the robot during its exploration. The same section also shows how to assign port numbers at each vertex. Section 3 includes specification of the robot by means of a finite state machine. Section 4 states the main result of the paper, that is, the analytical proof of the new upper bound of $3.75n - 2$ for the length of traversal period $\pi(n)$ required by robots equipped with constant memory. Further comments on the main themes of this paper can be found in Section 5.

2   The Spanning Tree Construction and the Port Numbering

In this section, we describe the spanning tree construction and how the port labelling is performed. This will define the route allowing the robot to periodically visit all the nodes of G. During the tree construction we make use of coloring strategies. These will be useful to analyze the length, π, of the closed walk P adopted by our robot to visit G. However, the robot is not aware of such a coloring.

We construct the spanning tree starting from a special subtree, from now on called the *backbone* and denoted by $B = (V_B, E_B)$. Procedures *Color()* and *Backbone_ Construction()* realize this structure (see Figure 1). Procedure *Backbone_ Construction()* takes graph G as an input and generates a tree, *the backbone*. Later, using procedure *Color()*, it colors all the nodes of G. The backbone is formed of *Red* and *Yellow* nodes, with the property that if a path connecting two *Red* nodes contains only yellow nodes, it contains exactly two of them. The

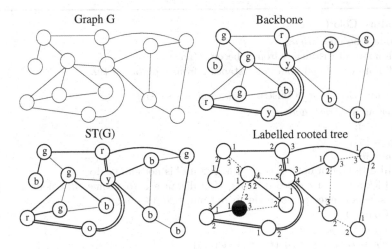

Fig. 1. The construction of $ST(G)$ and its labelling. The path characterised by double lines and joining the red nodes constitutes the backbone structure. The black node is the chosen root of $ST(G)$.

remaining nodes are colored *Green* or *Blue* if their distance from a *Red* node is 1 or 2, respectively. This implies another important property of the backbone, namely, every node outside the backbone is at distance at most two from it.

On the basis of the backbone, procedure *Tree_ Construction()* builds the spanning tree of G, from now on called $ST(G) = (V_T, E_T)$ (see Figure 1). The procedure also recolors some *Yellow* nodes (not having *Blues* as neighbors) into *Orange*. Again, such a recoloring will be used only for the purpose of analysing the length of P. Having constructed the spanning tree $ST(G)$, we then need to set the port numbers.

Let v be a node of a rooted tree T. We assume that the children $c_1, \ldots, c_k$ of v are listed in a nonincreasing order according to the sizes of $T(c_1), \ldots, T(c_k)$, where $T(c_i)$ is the subtree of T rooted in c_i. The main idea is that we set the port numbers so that the robot will explore the large subtrees at v first, followed by *Extended Leaves* (at v), and then (regular) *Leaves* (at v), allowing it to avoid paying penalties for most of the extended leaves and (regular) leaves. In what follows we use $d_G(v)$ (respectively, $d_T(v)$) to denote the degree of a node v in the graph G (respectively, the tree T). Procedures *Set_Port()* and *Labelling()* first label ports on edges of the input tree and then provide a consistent labelling to the remaining edges in G (see Figure 1).

3   The Automaton

In this section we provide a formal definition of the automaton that governs the robot's behavior. The automaton has eleven states, namely, I (Initial), RS (Root Search), RB (Root Backtrack), FF (Forward), N (Normal), T (Test), B (Backtrack), L (Leaf), LB (Leaf Backtrack), E (Extended Leaf), EB (Extended Leaf

procedure Color(node: v)

1: Node v becomes *Red*;
2: All nodes at distance 1 from v become *Green*;
3: All not yet colored nodes at distance 2 from v become *Blue*;

procedure Backbone_Construction(graph: G) → Tree

1: Pick an arbitrary node $v \in V$;
2: Color(v);
3: $V_B = \{v\}$;
4: $E_B = \emptyset$;
5: **while** the set of not yet colored nodes in G is not empty **do**
6:    Pick a not yet colored node $v \in V$ at distance 1 from some *Blue* node w_1 which is itself connected via node w_2 to some *Red* node v';
7:    Color(v);
8:    $V_B = V_B \bigcup \{w_1, w_2, v\}$;
9:    $E_B = E_B \bigcup \{(v', w_2), (w_2, w_1), (w_1, v)\}$;
10:   Nodes w_1 and w_2 become *Yellow*;
11: **end while**

procedure Tree_Construction(graph: G, backbone of G: B) → Tree

1: $V_T = V_B$;
2: $E_T = E_B$;
3: **for each** node $v \in V \setminus V_T$ at distance 1 from a red node $v' \in V_B$ **do**
4:    $V_T = V_T \bigcup \{v\}$;
5:    $E_T = E_T \bigcup \{(v, v')\}$;
6: **end for**
7: **for each** node $v \in V \setminus V_T$ at distance 1 from some node $v' \in V_T$ **do**
8:    $V_T = V_T \bigcup \{v\}$;
9:    $E_T = E_T \bigcup \{(v, v')\}$;
10: **end for**
11: Each *Yellow* node v not connected to any *Blue* node v' by an edge $(v, v') \in E_T$ becomes *Orange*;

procedure Set_Port(node: v)   //   under assumption $|T(c_1)| \geq |T(c_2)| \geq .. \geq |T(c_{d_T(v)-1})|$

1: **for each** edge $\{v, c_i\} \in E_T$ with $i = 1, ..., d_T(v) - 1$ **do**
2:    Set the port incident to v as $i + 1$;
3:    Set the port incident to c_i as 1;
4: **end for**

procedure Labelling(graph: G, spanning tree of G: T)

1: Pick an arbitrary leaf in T and set it as root r;
2: Set the port on the edge $\{r, v\}$ incident to the root r to 1 on both ends;
3: **for each** $v \in V_T \setminus \{r\}$ **do**
4:    Set_Ports(v);
5: **end for**
6: Set the remaining ports arbitrarily but consistently with degrees of the nodes;

Backtrack). Moreover, the automaton is powered by an extra two-bits counter c. We also denote by d the degree of the currently visited node and by i the entering port number.

Regardless of its starting vertex, the robot begins in state I, exiting port number 1 to visit the nodes of the input graph G. Its first goal is to find the right direction of the visit, i.e., from the root down to the leaves. In the worst case this process requires a constant number of steps. For example, it can happen, that the robot starts exploration going through a special edge $e_{1,1}$, the only edge in G with ports labeled 1 at both its endpoints. In this case the robot has to figure out which endpoint is the root. This task is realised by checking which endpoint is a leaf.[2] Once the robot finds out which endpoint leads to the rest of the tree, this initialisation phase is complete and the *proper exploration* starts. Let v be the node to which the root is connected. The robot visits the tree following the order of the ports from 2 to $d_{ST}(v)$. It first goes further in a depth-first-search manner (state T) as long as the entering port is 1. If different, it switches its state to L since a leaf has been discovered (recall that order of ports leading to subtrees of an internal node reflects the sizes of the subtrees). At this stage, the robot does not know whether such a leaf is a regular leaf or an extended leaf. It retreats through the same edge remembering (by setting the counter c to 1) that a leaf has been discovered, and it looks for another paired leaf while being in state L.

If now the robot encounters another entering port 1, it updates its leaves counter c to 2 and continues the search for more leaves. Note that, in this way the robot does not pay a penalty (it does not go beyond the leaf) at the second leaf. Moreover, the counter of leaves is not updated anymore, i.e., it is used to count at most two paired leaves. As soon as the robot arrives at some node via a port different from 1 or the whole degree of the node to which the paired leaves are connected has been explored, it retreats further in the tree, it sets c to 0 and switches its state to N.

Alternatively, if after visiting the first leaf the encountered port number was not 1, the robot goes backwards in the tree, sets the leaves counter to 0, and switches its state to E, since an extended leaf has been discovered. Now the robot searches for extended leaves, i.e., for two consecutive entering ports set to 1.

If after the first port 1 (that is counted by c), it finds a second one, a new extended leaf has been discovered and no penalties are paid at the two visited nodes. However if after the first port 1, the second one is different, the robot goes backwards switching its state to L since a regular leaf was found. Note that, in this case, c is already set to 1. And finally, if the first entering port is not 1, the robot retreats further in the tree while being in the state N.

The way the robot goes backwards depends on its current state, i.e., it must take into account whether it is currently looking for the root, leaves, extended leaves, whether it has just to leave the current subtree, or finally whether it has to go back since the traversed edge does not belong to the spanning tree (i.e.,

---

[2] In the case of the simple graph composed by just one edge, the first met node is considered as the root.

the robot paid a penalty). This is the reason why the robot must be equipped with five different backward states (RB, LB, EB, N, and B respectively).

Formally, the transition function f is defined as follows. If the robot enters a node v of degree $d(v)$ through port i in state s with current value of the two-bit counter c, it switches to state s' and exits the node through port i' with the counter set to c'. This corresponds to $f(s, i, d(v), c) = (s', i', c')$. Please note that function f has now four arguments in contrary to the more standard definition of Mealy automata presented in Section 1. This change is necessary to incorporate the introduction of the counter c.

Table 1 shows the transition function f. The first four transitions are used in the initialisation step of the search, while the remaining ones are devoted to the proper (periodic) graph exploration.

Table 1. The transition function f

$$f(I, i, d, c) = \begin{cases} (FF, 1, 0) & \text{if } i = d = 1 & (1) \\ (RS, 2, 0) & \text{if } i = 1 \ \& \ d \neq 1 & (2) \\ (T, i, 0) & \text{if } i \neq 1 & (3) \end{cases}$$

$$f(FF, i, d, c) = \begin{cases} (RB, 1, 0) & \text{if } d = 1 & (4) \\ (T, 2, 0) & \text{if } d \neq 1 & (5) \end{cases}$$

$$f(RS, i, d, c) = \begin{cases} (LB, 1, 1) & \text{if } i = d = 1 & (6) \\ (T, 2, 0) & \text{if } i = 1 \ \& \ d \neq 1 & (7) \\ (RB, i, 0) & \text{if } i \neq 1 & (8) \end{cases}$$

$$f(RB, i, d, c) = (FF, 1, 0) \qquad\qquad\qquad\qquad\qquad (9)$$

$$f(T, i, d, c) = \begin{cases} (LB, 1, 1) & \text{if } i = d = 1 & (10) \\ (T, 2, 0) & \text{if } i = 1 \ \& \ d \neq 1 & (11) \\ (B, i, 1) & \text{if } i \neq 1 & (12) \end{cases}$$

$$f(B, i, d, c) = \begin{cases} (N, 1, 0) & \text{if } c = 0 \ \| \ i \neq 2 & (13) \\ (LB, 1, 1) & \text{if } c \neq 0 \ \& \ i = 2 & (14) \end{cases}$$

$$f(N, i, d, c) = \begin{cases} (FF, 1, 0) & \text{if } i = 1 & (15) \\ (N, 1, 0) & \text{if } i = d \neq 1 & (16) \\ (T, i+1, 0) & \text{if } i \neq 1 \ \& \ i \neq d & (17) \end{cases}$$

$$f(L, i, d, c) = \begin{cases} (LB, 1, 2) & \text{if } i = 1 & (18) \\ (EB, i, 1) & \text{if } i \neq 1 \ \& \ c = 1 & (19) \\ (B, i, 0) & \text{if } i \neq 1 \ \& \ c \neq 1 & (20) \end{cases}$$

$$f(LB, i, d, c) = \begin{cases} (EB, 1, 0) & \text{if } i = d = 2 & (21) \\ (N, 1, 0) & \text{if } i = d \neq 2 & (22) \\ (L, i+1, c) & \text{if } i \neq d & (23) \end{cases}$$

$$f(E, i, d, c) = \begin{cases} (LB, 1, 1) & \text{if } i = d = 1 \ \& \ c = 0 & (24) \\ (E, 2, 1) & \text{if } i = 1 \ \& \ d \neq 1 \ \& \ c = 0 & (25) \\ (EB, 1, 2) & \text{if } i = 1 \ \& \ c \neq 0 & (26) \\ (B, i, c) & \text{if } i \neq 1 & (27) \end{cases}$$

$$f(EB, i, d, c) = \begin{cases} (FF, 1, 0) & \text{if } i = 1 & (28) \\ (EB, 1, 0) & \text{if } i \neq 1 \ \& \ c = 2 & (29) \\ (N, 1, 0) & \text{if } i = d \neq 1 \ \& \ c = 0 & (30) \\ (E, i+1, 0) & \text{if } i \neq 1 \ \& \ i \neq d \ \& \ c = 0 & (31) \\ (EB, 1, 0) & \text{if } i = 3 \ \& \ c = 1 & (32) \\ (N, 1, 0) & \text{if } i \neq 1 \ \& \ i \neq 3 \ \& \ c = 1 & (33) \end{cases}$$

Theorem 1 (Correctness). *Let G be a graph of size n, and let $ST(G)$ be a spanning tree of G constructed and labelled as previously described in Section 2. Starting at any node of G, the robot begins in the initial state I and follows exit port number 1. Then*

(a) After at most 8 steps, the robot enters a closed walk P and then periodically explores G forever.

(b) Moreover, suppose that a node v is connected to k extended leaves in $ST(G)$. Then the robot avoids paying possible penalties at the second, third, etc. extended leaf (so it avoids $2(k-1)$ penalties from these extended leaves). Similarly, if v is connected to j paired leaves, it only pays a single penalty for the whole set of these paired leaves.

Proof. First we show how the robot recognises the direction of the exploration from the root towards the leaves. At the beginning the robot is in the initial state I and follows port number 1. As shown in the first equation of the transition function f, three different cases can occur:

- If the entering port is 1 and the entering node has degree 1 then such a node is considered as the root since it is a leaf connected to the edge $e_{1,1}$. The only case in which it is unclear whether the considered node is the root is when G is composed of a single special edge $e_{1,1}$. In this case indeed, both nodes can be considered indifferently as the root since the resulting exploration is symmetric. The task is performed by switching the robot state first from I to FF (rule 1) and then repetitively from FF to RB and vice versa (rules 4 and 9).
- If the entering port is 1 and the entering node has degree larger than 1, then the robot has to verify whether such a node is a leaf or not (rule 2), and this requires two steps. If a leaf is found the robot retreats and starts proper exploration. If not, then it continues in the current direction assuming that the proper exploration was done from the beginning.
- If the entering port is different from 1 then the direction is determined since, by construction of $ST(G)$, the node connected via the exiting port on the edge is clearly a child of the node connected via the entering port. Hence, the robot has to reverse its direction (rule 3). In fact, in the worst case we may have to perform six steps more before the proper exploration starts. It can happen, e.g., that the robot starts from a leaf or an extended leaf of $ST(G)$ for which in P no penalties are paid. However, since the robot has just started and does not know whether the current leaf or extended leaf is a paired one or not, it pays some extra penalties that at the successive traversals will not be paid. Namely, it needs at most six (respectively, two) steps to find out whether it is exploring an extended leaf (respectively, a leaf).

Once this initialisation phase has been completed, the proper (periodic) graph exploration starts. Apart from cases where G coincides with $e_{1,1}$ and when the "tricks" of skipping penalties at paired leaves and paired extended leaves are used, the exploration follows the idea presented in [20]. Thus, in order to conclude the proof we have to show that while handling the above mentioned special cases the robot does not skip any nodes in the spanning tree and that it switches between standard and special cases safely.

While descending in the tree, the robot is in state T until a leaf is met. At the leaf, unless its degree is 1, a penalty is paid and the robot retreats in state B setting the leaf counter c to 1 (rule 12). Further it goes back in state LB

(rule 14) and, if the degree of the parent of the just visited leaf is larger than 2, it goes to the next child expecting the entering port number to be 1 (rule 23). If such a port is encountered the next leaf is found. The robot does not go beyond the leaf; it retreats, looking for more leaves (again by applying rules 14 and 23). This saves us one penalty on each leaf like this. Since in $ST(G)$ the order of ports reflects the sizes of subtrees, the case when after visiting a leaf the robot is expected to visit a larger subtree is not feasible. Hence the exploration is correctly performed.

For the extended leaves a similar argument can be used. Once a leaf of an extended leaf is met, the robot goes back (rule 14), it realises that it is an extended leaf if either the parent has degree 2 or if there are no paired leaves connected (rule 21 and 19 respectively). Further, it goes back again in the EB state and $c = 0$ since it is on a path of length 2 (rule 29 or 32). Now, while being in state E (by means of rule 31) it looks for paired extended leaves represented by two consecutive entering ports labelled by 1 (i.e., another path of length 2). If this happens (rules 31 and 25), the robot does not pay for the penalties on the two nodes of the found paired extended leaf. Again, by the construction of $ST(G)$ there cannot be the situation in which the second subtree is larger than an extended leaf, hence the exploration continues in the correct way. Alternatively, instead of another extended leaf the robot could find a smaller subtree, i.e., a leaf (a subtree of size 1), or a penalty edge (an empty subtree). In the first case, the robot switches its state to LB (rule 24) to retreat one edge and then starts searching for regular leaves. In the second case the robot retreats in state N going back to standard exploration (rules 30 or 33).

Finally, note that each time the edge $e_{1,1}$ is traversed, the robot recognises that situation and returns to the state FF (rules 15 and 28). This can be useful if a finite number of explorations is required. In fact, another counter can be added in order to stop the exploration of the graph after the required number of traversals. □

Concerning the length π of P, in the next section we show, in general, how many penalties at most can be paid by our robot.

4 Analysis

In this section, we show our upper bound on the length of the periodic exploration of the graph G. Our analysis uses the coloring of the nodes of G introduced in Section 2. Let RED be the set of Red nodes. We remind the reader that the robot is said to pay a *penalty* at node v if it traverses an edge, starting from v, that does not belong to $ST(G)$. The main goal is to give a bound on the number of penalties in order to prove our result.

Theorem 2. *The period $\pi(n)$ needed by the robot described in Section 3 to visit a graph G of n nodes, according to the spanning tree $ST(G)$ with the assigned port numbering, is less than $3.75n - 2$.*

Proof. Consider $v \in RED$. By construction, by removing the edges of the backbone connected to v, the remaining subtree T' rooted at v has depth at most 2. We decompose the children of v into 3 different types (see Figure 2):

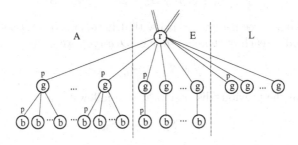

Fig. 2. The subdivision of the children of a red node. The other possible nodes connected to it can only be part of the backbone structure. The p associated to a node indicates where a penalty might be paid.

- **Type L:** Leaves
- **Type E:** Endpoints of extended leaves
- **Type A:** Neither leaves nor endpoints of extended leaves

Let $s_L(v)$ (resp. $s_E(v)$) be 0 if the number of children of v of Type L (resp. E) is 0, 1 otherwise. Let $s_A(v)$ be the number of children of v of Type A. For all the children of type L, the robot might incur at most 1 penalty, since all those children (if any) are paired leaves and the robot pays only at the first one (see Section 3). For all the children of type E, the robot might incur at most 2 penalties, since all those children (if any) are the endpoints of paired extended leaves and the robot pays only at the first one. For children of type A, by construction, each subtree rooted at a node of type A consists of two or more leaves. This implies that for each subtree rooted at a node of type A the robot might incur at most 2 penalties, i.e., one for the node of type A and the other for the first paired leaf.

Concerning the backbone structure, let S_Y (respectively, S_O, S_R) be the number of $Yellow$ (resp., $Orange$, Red) nodes in the tree. Consider a $Yellow$ node w, and its $Blue$ children. Note that there can be at most 2 penalties in this subtree. In fact, the $Blue$ children of w are paired leaves in this subtree, hence the robot does not pay penalties from the second paired leaf. Consider an $Orange$ node z. Note that z has no $Blue$ children in the tree, hence there can be at most 1 penalty in this subtree (made up of z itself). Concerning Red nodes, no penalties are paid on them, since by construction they are used with their full degree in $ST(G)$.

Let

$$S_L = \sum_{v \in RED} s_L(v), \quad S_E = \sum_{v \in RED} s_E(v), \quad S_A = \sum_{v \in RED} s_A(v),$$

and p be the number of nodes with penalties in the tree. As a consequence of the observations above, we have

$$p \leq 2S_A + 2S_E + S_L + 2S_Y + S_O,$$
$$n \geq 3S_A + 2S_E + S_L + 2S_Y + S_O + S_R.$$

As the backbone tree has the property that between two *Red* nodes there are exactly two nodes (each of them *Yellow* or *Orange*), we have

$$S_Y + S_O = 2S_R - 2.$$

Moreover, from the previous discussion, it also follows that

$$S_L \leq S_R, \quad S_E \leq S_R,$$

because for each *Red* node, at most one child of type L (resp. E) is counted in S_E (resp. S_L). Hence, we can bound the number of penalties with respect to the total number of nodes as follows:

$$
\begin{aligned}
\frac{p}{n} &\leq \frac{2S_A + 2S_E + S_L + 2S_Y + S_O}{3S_A + 2S_E + S_L + 2S_Y + S_O + S_R} \\
&\leq \frac{2S_A + 2S_E + S_L + 2(S_Y + S_O)}{3S_A + 2S_E + S_L + 2(S_Y + S_O) + S_R} \\
&< \frac{2S_A + 2S_E + S_L + 4S_R}{3S_A + 2S_E + S_L + 4S_R + S_R} \quad [S_Y + S_O = 2S_R - 2] \\
&\leq \frac{2 \cdot 0 + 2S_R + S_R + 4S_R}{3 \cdot 0 + 2S_R + S_R + 4S_R + S_R} \quad [S_A \geq 0, S_E, S_L \leq S_R] \\
&\leq \frac{7}{8}.
\end{aligned}
$$

By construction of the automaton, whenever a penalty is paid, i.e., whenever a non-tree edge is traversed, the robot incurs another penalty because it has to traverse again the same edge backward. As the automaton traverses every tree edge twice (once in each direction) it follows that

$$\pi(n) = 2(n-1) + 2p < 2n - 2 + 2 \cdot \frac{7}{8}n = 3.75n - 2. \qquad \square$$

5 Conclusion

In this paper, we studied the problem of periodic graph exploration by means of a simple robot that uses constant memory. We disproved the conjecture given in [20] claiming that it was not possible to obtain a period less than $4n - O(1)$. The new proved upper bound is in fact $3.75n - 2$, hence the gap with the lower bound of $2n - 2$ is reduced. The improvement is obtained by means of a careful construction of the route (i.e., the spanning tree) that the robot has to follow during the graph exploration, and by designing a smart automaton (still using constant memory) able to recognise some specific subtrees.

The main open problem left is whether it is possible to further close the gap of the bounds. For the tree-based approach, a more powerful robot able to recognise more structured topologies (of a constant number of nodes) than just our paired leaves and extended leaves seems not to be helpful in decreasing the number of penalties. However, a modified tree construction and/or port numbering may still lead to further improvements. Another interesting question is if there is a better strategy than tree-based exploration.

Acknowledgements

Comments and suggestions of anonymous referees are gratefully acknowledged.

References

1. Albers, S., Henzinger, M.R.: Exploring unknown environments. SIAM Journal on Computing 29, 1164–1188 (2000)
2. Awerbuch, B., Betke, M., Rivest, R., Singh, M.: Piecemeal Graph Exploration by a Mobile Robot. Information and Computation 152(2), 155–172 (1999)
3. Awerbuch, B., Kobourov, S.G.: Polylogarithmic-Overhead Piecemeal Graph Exploration. In: Proc. 11th Annual Conference on Computational Learning Theory (COLT 1998), pp. 280–286 (1998)
4. Bender, M.A., Fernandez, A., Ron, D., Sahai, A., Vadhan, S.: The Power of a Pebble: Exploring and Mapping Directed Graphs. Information and Computation 176(1), 1–21 (2002)
5. Bender, M.A., Slonim, D.: The power of team exploration: Two robots can learn unlabeled directed graphs. In: Proc. 35th Ann. Symp. on Foundations of Computer Science (FOCS 1994), pp. 75–85 (1994)
6. Betke, M., Rivest, R., Singh, M.: Piecemeal learning of an unknown environment. Machine Learning 18, 231–254 (1995)
7. Bhatia, R., Khuller, S., Pless, R., Sussmann, Y.J.: The Full Degree Spanning Tree Problem. In: Proc. 10th Annual ACM-SIAM Symposium on Discrete Algorithms (SODA 1999), pp. 864–865 (1999)
8. Budach, L.: Automata and labyrinths. Math. Nachrichten 86, 195–282 (1978)
9. Cohen, R., Fraigniaud, P., Ilcinkas, D., Korman, A., Peleg, D.: Label-guided graph exploration by a finite automaton. In: Caires, L., Italiano, G.F., Monteiro, L., Palamidessi, C., Yung, M. (eds.) Automata, Languages and Programming. Proc. 32nd International Colloquium, ICALP 2005, Lisbon, Portugal, July 11-15, 2005. LNCS, vol. 3580, pp. 335–346. Springer, Heidelberg (2005)
10. Cook, S., Rackoff, C.: Space lower bounds for maze threadability on restricted machines. SIAM J. on Computing 9(3), 636–652 (1980)
11. Deng, X., Papadimitriou, C.H.: Exploring an unknown graph. Journal of Graph Theory 32, 265–297 (1999)
12. Diks, K., Fraigniaud, P., Kranakis, E., Pelc, A.: Tree exploration with little memory. Journal of Algorithms 51, 38–63 (2004)
13. Dobrev, S., Jansson, J., Sadakane, K., Sung, W.-K.: Finding Short Right-Hand-on-the-Wall Walks in Graphs. In: Pelc, A., Raynal, M. (eds.) Structural Information and Communication Complexity. Proc. 12 International Colloquium, SIROCCO 2005, Mont Saint-Michel, France, May 24-26, 2005. LNCS, vol. 3499, pp. 127–139. Springer, Heidelberg (2005)

14. Duncan, C.A., Kobourov, S.G., Kumar, V.S.A.: Optimal constrained graph exploration. In: Proc. 12th Ann. ACM-SIAM Symp. on Discrete Algorithms (SODA 2001), pp. 807–814 (2001)
15. Fleischer, R., Trippen, G.: Exploring an Unknown Graph Efficiently. In: Brodal, G.S., Leonardi, S. (eds.) Algorithms – ESA 2005. Proc. 13th Annual European Symposium, Palma de Mallorca, Spain, October 3-6, 2005. LNCS, vol. 3669, pp. 11–22. Springer, Heidelberg (2005)
16. Fraigniaud, P., Ilcinkas, D.: Digraphs exploration with little memory. In: Diekert, V., Habib, M. (eds.) STACS 2004. Proc. 21st Annual Symposium on Theoretical Aspects of Computer Science, Montpellier, France, March 25-27, 2004. LNCS, vol. 2996, pp. 246–257. Springer, Heidelberg (2004)
17. Fraigniaud, P., Ilcinkas, D., Peer, G., Pelc, A., Peleg, D.: Graph exploration by a finite automaton. Theoretical Computer Science 345, 331–344 (2005)
18. Fraigniaud, P., Ilcinkas, D., Rajsbaum, S., Tixeuil, S.: The Reduced Automata Technique for Graph Exploration Space Lower Bounds, Essays in Memory of Shimon Even. In: Goldreich, O., Rosenberg, A.L., Selman, A.L. (eds.) Theoretical Computer Science. LNCS, vol. 3895, pp. 1–26. Springer, Heidelberg (2006)
19. Gąsieniec, L., Pelc, A., Radzik, T., Zhang, X.: Tree exploration with logarithmic memory. In: Proc. 18th Annual ACM-SIAM Symposium on Discrete Algorithms (SODA 2007) (2007)
20. Ilcinkas, D.: Setting port numbers for fast graph exploration. In: Flocchini, P., Gąsieniec, L. (eds.) Structural Information and Communication Complexity. Proc. 13th International Colloquium, SIROCCO 2006, Chester, UK, July 2-5, 2006. LNCS, vol. 4056, pp. 59–69. Springer, Heidelberg (2006)
21. Panaite, P., Pelc, A.: Exploring unknown undirected graphs. Journal of Algorithms 33, 281–295 (1999)
22. Reingold, O.: Undirected ST-Connectivity in Log-Space. In: Proc. 37th ACM Symposium on Theory of Computing, (STOC 2005), pp. 376–385 (2005)
23. Rollik, H.: Automaten in planaren Graphen. Acta Informatica 13, 287–298 (1980)

Why Robots Need Maps*

Miroslaw Dynia[1], Jakub Łopuszański[2], and Christian Schindelhauer[3]

[1] DFG Graduate College "Automatic Configuration in Open Systems",
University of Paderborn, Germany
mdynia@uni-paderborn.de
[2] Institute of Computer Science, University of Wroclaw, Poland
jakub.lopuszanski@ii.uni.wroc.pl
[3] Computer Networks and Telematics, University of Freiburg, Germany
schindel@informatik.uni-freiburg.de

Abstract. A large group of autonomous, mobile entities e.g. robots initially placed at some arbitrary node of the graph has to jointly visit all nodes (not necessarily all edges) and finally return to the initial position. The graph is not known in advance (an online setting) and robots have to traverse an edge in order to discover new parts (edges) of the graph. The team can locally exchange information, using wireless communication devices.

We compare a cost of the online and optimal offline algorithm which knows the graph beforehand (competitive ratio). If the cost is the total time of an exploration, we prove the lower bound of $\Omega(\log k / \log \log k)$ for competitive ratio of any deterministic algorithm (using global communication). This significantly improves the best known constant lower bound. For the cost being the maximal number of edges traversed by a robot (the energy) we present an improved $(4 - 2/k)$-competitive online algorithm for trees.

1 Introduction and Our Results

We are interested in the issue of coordination of a team of k autonomous robots. We would like the team to be driven by a distributed algorithm stored in the local memory and executed using the local computational power of a robot. The team's goal is to jointly visit all nodes of an unknown, but labeled graph G. To let the team cooperate we must allow it to exchange information about new findings and agree upon the strategy of the exploration. We can allow full communication scheme (global communication), yet it is more realistic to allow only local communication. Robots are equipped with wireless radio devices with a bounded communication radius which allows only the neighboring robots to communicate.

* This research is partially supported by the DFG-Sonderforschungsbereich SPP 1183: "Organic Computing. Smart Teams: Local, Distributed Strategies for Self-Organizing Robotic Exploration Teams" and by MNiSW grant number N206 001 31/0436, 2006-2008.

G. Prencipe and S. Zaks (Eds.): SIROCCO 2007, LNCS 4474, pp. 41–50, 2007.

The team is initially placed in a node of the unknown graph G modeling the network or e.g. an unknown terrain, where nodes correspond to the interesting locations, and edges model the accessibility between locations. Additionally, we assume that all edges and nodes of G are labeled and thus can be locally distinguished by a robot. A goal for the robots is to jointly visit all nodes of the graph and finally return to the initial node. We consider two cost models. In the first model we assume the cost measure to be the total time of the exploration where in the second model it is the maximal energy (number of edges) used by a robot.

It is clear that knowing the graph beforehand (*offline setting*), the team would agree on the best strategy before the exploration and then fully explore the graph without using communication at all. Intuitively, the cost of such an exploration should be smaller than the cost needed in the *online setting*, where the graph is not known in advance.

The *competitive ratio* is the ratio between the cost of the online and the optimal offline algorithm. Competitive ratio of 1 would mean that robots can efficiently explore even though the "map of the graph" is not known. In fact, for the time model we show that this ratio is significantly larger even assuming the global communication (see Sect. 3). We show the lower bound of $\Omega\left(\log k/\log\log k\right)$ for the competitive ratio of any deterministic graph exploration algorithm using k robots. This is a significant improvement comparing to the constant factor bounds known so far. For the energy model (Sect. 4) we show the $(4 - 2/k)$-competitive algorithm which explores trees and uses strictly local communication (in fact robots communicate only in the root of a tree).

2   Prior and Related Work

There are many results (e.g. [1,2,3,4,5,6]) concerning exploration of a graph using small number of robots. Authors of [5] present strategies for a robot which has to traverse all edges minimizing the number of edge traversals. They bound competitive ratio (an overhead related to the lack of topology's knowledge) of their algorithms for several classes of graphs. Authors of [2] show a strategy for two robots to explore (in polynomial time) all nodes of an unlabeled, strongly connected, directed graph.

The real impact of robot's cooperation can be observed by studying the algorithms which use larger number of robots ($k > 2$). Robots can collectively perform many tasks (e.g. black-hole search [7,8] or rendezvous [9]) but here we focus on the problem of an exploration. Dealing with a group of robots, there are many coordination problems e.g. gathering or pattern formation [10] which a team might exercise during the exploration of an unknown terrain (or unlabeled graph like in [2,11]). Those problems might arise from the sensor inaccuracy, odometry error related to the movement or from some computational problems. However, in many publications this is overcome assuming that either the exploration concerns the network or the terrain is represented by the labeled graph in which these problems do not occur.

The problem of collective tree exploration is addressed in [12]. Authors present the lower bound of $2 - 1/k$ for competitive ratio of an arbitrary exploration algorithm using k robots. Additionally, they prove that if no communication is allowed, it is not possible to explore efficiently (competitive ratio $\Omega(k)$). Their online algorithm for tree exploration uses global communication and achieves a competitiveness of $O(k/\log k)$. Additionally, in [13] we present online algorithms for exploration of so called sparse trees and e.g. for trees D in height, which can be embedded in 2-dimensional grids, the algorithm achieves competitive ratio of $O(\sqrt{D})$. A problem with the cost model related to the maximal energy used by a robot is addressed in [14]. Authors present the lower bound of $3/2$ and distributed 8-competitive algorithm for trees which uses only local communication.

In [15] the group of simple robots (could be represented by the primitive final automata) fills the integer grid subject to minimize the makespan. Initially, the robots are standing outside the grid (by so called doors) and can consecutively enter the grid, with the additional assumption that only one robot can occupy one node. They develop optimal solution for the single-door case and $O(\log(k + 1))$-competitive algorithm for multi-door case.

3   The Time Model

Consider an arbitrary graph $G = (E, V)$ with a distinguished node $s \in V$. The team of k autonomous mobile robots starts in s, visits all nodes of G and finally returns to s. It takes one time step for the robot to traverse an edge and since there are many synchronized robots, there might be many edges which are traversed in parallel by the team during one time step.

In this chapter we focus on the total time of such an exploration and furthermore we compare it to the time needed by the optimal offline algorithm. We show that for each deterministic algorithm there exists a tree-like graph which cannot be efficiently explored. Although there is a local communication granted for robots we show that having even global communication does not help if we do not know the map of the graph in advance. This means that in the worst case no online algorithm can explore efficiently, if compared to the time needed by the optimal offline algorithm.

First, in the Sect. 3.1 we introduce the Jellyfish tree which is used in Sect. 3.2 to prove the lower bound of $\Omega\,(\log k/\log\log k)$ for competitive ratio of an arbitrary deterministic exploration algorithm.

3.1   The Jellyfish Tree

Assume $t > k$ and take some permutation σ of set 1 through k. *Jellyfish tree* $J(k, t, \sigma)$ consists of k subtrees (*tentacles*) numbered from 1 trough k, connected by the root (see Fig. 1).

The tentacle consists of a *poison* of a certain size which is attached to the single path of the length t. Each level of poison (but the first one) consists of t nodes, all connected to the *main node* of the previous level. The main node is the one which was visited by any robot as the last one on this level. Therefore, all nodes on the level l have to be visited before any other node on the level $l+1$ is visited.

Consider a team of $k' \leq k$ robots positioned in the main node v on some level of a poison. It can traverse in parallel at most $k' \cdot t$ edges in t time steps. Since there are t edges to be explored on each level, there are at most k' additional main nodes discovered by the team during t time steps.

Lemma 1. *Assume that a team of $k' \leq k$ robots explores some tentacle for t time steps. Denote respectively by d and d' the maximal distance to the main node in this tentacle before and after this exploration, then $d' \leq d + k'$.*

The poison contained in a tentacle has some certain total number of levels, i.e. a size. The size of the poison contained in the i-th tentacle is defined by

$$s_{\sigma(i)} := \left\lceil \frac{k}{\log k} \cdot \frac{1}{i} \right\rceil ,$$

where σ is the permutation which allows to rearrange the order of poisons. In Sect. 3.2 the adversary defines the permutation $\sigma_{\mathcal{A}}$ for the algorithm $\mathcal{A}$ in such a way, that the sizes of poisons are in the inconvenient configuration for the algorithm.

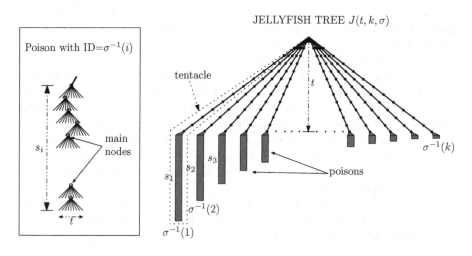

Fig. 1. Definition of a poison and the Jellyfish tree $J(t, k, \sigma)$

3.2   The Lower Bound

Suppose that we are given an arbitrary algorithm $\mathcal{A}$ for k robots to explore an arbitrary graph from the node s. Consider a large team of robots (it must be at least $\log k \geq 5$) and take an arbitrarily $t > k$. We analyze the performance of $\mathcal{A}$

on the Jellyfish tree $J(t, k, \sigma_{\mathcal{A}})$ rooted at s and where $\sigma_{\mathcal{A}}$ dictates an adversarial order of poisons. Recall here, that the main nodes are visited by $\mathcal{A}$ as the last nodes on certain level of a poison.

Algorithm $\mathcal{A}$ starts the exploration in the node s of the Jellyfish $J(t, k, \sigma_{\mathcal{A}})$. During the first t steps it can traverse the simple paths connecting poisons to the root but it will not enter any poison. After t steps there are $k_{\text{lazy}} \leq k$ robots positioned in s and $f_j^{(1)}$ robots in the j-th tentacle. Now consider the time interval $I_1 = [(t + 1), \ldots, 2t]$ which consists of t time steps. There are at most $f_j^{(1)}$ robots which can explore the poison contained in the j-th tentacle in I_1. All other robots initially are just too far from this poison to be able to get there within t time steps. Therefore, the value $f_j^{(1)}$ upper-bounds the potential exploration power the algorithm has within the interval.

Order the sequence $f_j^{(1)}$ in the increasing order of element values. Adversary reorders the poison's sizes in the reverse order, such that the tentacles with the large number of robots contain the smallest poisons. Despite $\mathcal{A}$ having the large exploration power in some tentacle, it turns out that the actual poisons size is small, and there is not much to explore. In this way the adversary makes the algorithm waste its resources, which results in an interesting lower bound for the algorithm efficiency.

For $y^{(r)} := t + \lceil (2 \log k)^r \rceil$, there is no node at distance greater than $y^{(1)}$ which is visited by a robot before the end of I_1. Assume that v' is a node at distance $t + 11 > y^{(1)}$ and that it is visited during the first time interval I_1. The node v' lies within the poison of the j-th tentacle and during the time interval there are $f_j^{(1)} \geq 11$ robots exploring it (from Lemma 1 we know that 10 robots does not suffice). There are at least $k/2$ tentacles which contain a poison smaller than 11 and each of them is also explored by at least 11 robots (adversarial order of poison's size). This gives $11 \cdot k/2 > k$ robots exploring the smallest tentacles during this interval.

Now assume that at the end of the I_{r-1}, all nodes in the tree at distance $y^{(r-1)}$ or smaller are visited by a robot. At the beginning of I_r the $f_j^{(r)}$ is a number of robots positioned in the j-th tentacle. The adversary sorts the still unexplored poisons in the reversed order of their sizes to the order of $f_j^{(r)}$. Each such a poison is of size greater than $y^{(r-1)}$ since for all smaller poisons the permutation $\sigma_{\mathcal{A}}$ is already fixed by the reorderings made in the previous intervals.

Suppose that there is a node v' which is visited during the interval I_r and which lies at the distance of $y^{(r)} + 1$ from the root. The node v' is contained in the tentacle for which $\mathcal{A}$ has had many robots exploring it. Certainly, by Lemma 1, $\mathcal{A}$ needs much "exploration power" to rapidly explore many levels in t time steps. To explore additional $y^{(r)} - y^{(r-1)}$ levels of the j-th tentacle it needs $f_j^{(r-1)} \geq y^{(r)} - y^{(r-1)}$ robots, and therefore we have

$$f_j^{(r-1)} \geq (\log k)^{r-1} \cdot (2 \log k - 2) .$$

For $i = \lceil k/\log k \cdot 1/(2\log k)^{r+1} \rceil$ we have $s_i \leq y^{(r)}$ and for $i' = \lfloor k/\log k \cdot 1/(2\log k)^r \rfloor$ we have $s_{i'} \geq y^{(r-1)}$ and so there are at least

$$i' - i \geq \frac{k}{\log k \cdot (2\log k)^{r-1}} \cdot (1 - 1/\log k)$$

tentacles which end between levels $y^{(r-1)}$ and $y^{(r)}$. All those tentacles have at least $f_j^{(r-1)}$ robots exploring it (because the adversary has sorted the tentacles in a reversed order). This means that there are at least

$$f_j^{(r)} \cdot (i' - i) \geq k \cdot (2 - 4/5) > k$$

robots exploring the tree in the r-th round. This contradicts the fact that $\mathcal{A}$ uses only k robots to explore the Jellyfish tree.

Lemma 2. *At the end of the I_r time interval of the $\mathcal{A}$ algorithm exploring $J(t, k, \sigma_\mathcal{A})$ no node at distance greater than $y^{(r)} := t + \lceil (2\log k)^r \rceil$ from s is visited.*

If the largest tentacle (containing a poison of size $k/\log k$) is completely explored during I_R interval, then $y^{(R)} \geq t + k/\log k$, and so $R = \Omega(\log k/\log\log k)$. By the definition, each interval (but probably the last one) takes t time steps, and therefore we can find the lower bound on the total exploration time.

Lemma 3. *Algorithm $\mathcal{A}$ explores $J(t, k, \sigma_\mathcal{A})$ (with $\sigma_\mathcal{A}$ being an adversarial permutation of poisons) in*

$$\Omega\left(t \cdot \frac{\log k}{\log\log k}\right)$$

time steps.

On the other hand, if the topology is known beforehand, the Jellyfish tree can be efficiently explored.

Lemma 4. *Assuming that the tree is known beforehand (i.e. $t > k$ and σ are known), there exists the algorithm with k robots which explores $J(t, k, \sigma, M)$ in $O(t)$ steps.*

Proof. The graph $J(t, k, \sigma_\mathcal{A})$ is $h := t + k/\log k$ in height and has at most

$$2tk + (tk/\log k) \cdot \sum_{i=1}^{k} 1/i = O(t \cdot k)$$

nodes. Therefore we have $h = O(t)$ and $n/k = O(t)$ and using e.g. the approximation algorithm from [14] we can recompute the offline routes for each of k robots so that the total exploration time is $O(t)$.   □

Combining the results of Lemma 3 and Lemma 4 we can state the following lower bound on the competitive ratio of an arbitrary online exploration algorithm.

Theorem 1. *For every online collective graph exploration algorithm $\mathcal{A}$ using k robots, there exists a tree for which the total time of the exploration is at least*

$$\Omega\left(\frac{\log k}{\log\log k}\right)$$

times greater than the time needed by the optimal offline algorithm.

4 Algorithm for the Energy Model

In this section we consider a tree $T = (E, V)$ rooted at $v_0 \in V$ consisting of n uniform labeled edges and D in height, measured by the number of edges on the longest path from v_0 to a leaf. All k robots with an unique ID drawn from the set $1, 2, \ldots, k$ are initially placed in v_0. Robots can communicate when they are in the same node and a goal of such a team is to jointly explore the unknown tree.

Assume that whenever a robot traverses an edge, it incurs a cost of one energy unit. We are interested in costs of the exploration defined as the maximal energy used by a robot. Once again we compare the cost of the online algorithm to the cost of the optimal offline algorithm to obtain a competitive ratio. In [14] the lower bound of $3/2$ as well as the 8-competitive online algorithm exploring tree using a team of k robots were shown. Here we improve this result by introducing the $4 - 2/k$-competitive algorithm. This confirms that the energy model is strictly weaker than the time model for which the first non-constant lower bound is presented in the previous section.

In the energy model robots do not care about an overall time of the exploration. In some situations halting and waiting for a new information may be more desirable for a robot than further exploration. This is exactly what happens in our algorithm. We have a group of k robots $(R_1, R_2, \ldots R_k)$ and during a round, there is only one robot which is active. All other robots are waiting in the root v_0 of the tree. The active robot first goes to the node where the robot active in the previous round has given up its exploration. Then it continues this exploration for a certain number of steps, and finally returns to the root v_0. The algorithm is described in details on the Fig. 1. The "while" loop corresponds to one round r, and there is only one robot R_{id} which moves during this round. Variable h_r denotes the height of the subtree visited by a robot in all rounds from 1 through r and the variable e_i holds the energy used so far by the i-th robot.

In the r-th round, the active robot first travels to the node v_{r-1} at which the previous robot stopped exploring (it takes at most h_{r-1} energy units). Then it continues to traverse consecutive DFS edges ("repeat" loop) until it collects $2h$ of it, where h is the height of the subtree visited by any robot (also the actual active one) so far. During this loop the robot traverses $2h_r$ edges and then finally returns to v_0 (which certainly takes at most h_r energy units). This gives us an upper bound of $h_{r-1} + 3h_r$ on the energy used by a robot during the round r. In the first round ($r = 1$) the active robot is the first working robot ever, so

---

Algorithm 1. PushDfs

$h \leftarrow 1$
$r \leftarrow 1$
$e_i \leftarrow 0$ for all $1 \leq i \leq k$

while (T is not yet explored) **do**
   $id \leftarrow argmin\{e_i : 1 \leq i \leq k\}$
   R_{id} travels to v_{r-1}
   $e \leftarrow 0$
   repeat
      R_{id} follows a DFS step
      $e \leftarrow e + 1$
      $h \leftarrow$ height of the visited subtree
   until $(e \geq 2h)$
   $h_r \leftarrow h$
   $v_r \leftarrow$ actual position of R_{id}
   R_{id} returns to v_0 //this takes at most h_r edges
   $e_{id} \leftarrow e_{id} + |path(v_0, v_{r-1})| + 2h_r + |path(v_r, v_0)|$
   $r \leftarrow r + 1$
end while

---

it uses only $0 + 3h_1$ energy units (we lay $h_0 = 0$). The last round q is perhaps shorter and it takes $h_{q-1} + w$ because a robot gets to the root already during the "repeat" loop and thus does not have to pay any extra costs for the return.

Let e_i be the energy of the robot R_i after completely exploring the tree and $E = \sum_{i=1}^{k} e_i$ be the total energy used by all robots. We have

$$E \leq 3h_1 + \sum_{r=2}^{q-1}(h_{r-1} + 3h_r) + h_{q-1} + w = 4 \cdot \sum_{r=1}^{q-1} h_r + w$$

as the upper bound for this energy. On the other hand we know that the robot active in the round $r \leq q - 1$ has done exactly $2h_r$ steps of the DFS tour (the last one exactly w steps), which for the whole tree takes exactly $2n$ energy units. It must be that $(\sum_{i=r}^{q-1} 2h_r) + w = 2n$ and thus we have

$$E \leq 4n \leq 2k \cdot OPT$$

where OPT is the energy cost of the optimal offline exploration. Indeed, $OPT \geq 2n/k$ since there are n edges and each has to be traversed twice by at least one robot.

Let $e_{\min} = min\{e_i : 1 \leq i \leq k\}$ and $e_{\max} = max\{e_i : 1 \leq i \leq k\}$ be respectively the minimal and the maximal energy used by the robots $R_{\min}$ and $R_{\max}$. One tour during a round takes at most $h_{q-1} + 3h_q \leq D + 3D = 4D$ energy units of an active robot. The active robot is chosen to be the one with the smallest energy used so far and therefore we have $e_{\max} - e_{\min} \leq 4D$. Certainly $OPT \geq 2D$, because there is at least one robot which has to reach the furtherest leaf at distance of D and return to v_0. This results in the upper bound

$$e_{\max} - e_{\min} \leq 2OPT .$$

Knowing the span of values of the elements of the sequence e_i we can use the following upper bound on the maximal value

$$e_{\max} \leq \frac{\sum_{i=1}^{k} e_i - (e_{\max} - e_{\min})}{k} + (e_{\max} - e_{\min})$$

and therefore we obtain

$$e_{\max} \leq \frac{2k \cdot OPT}{k} + (1 - 1/k) \cdot 2OPT \leq (4 - 2/k) \cdot OPT .$$

This analysis can be slightly improved (at the cost of readability) but it also can be proved that the competitive ratio of this algorithm asymptotically converges to 4.

Theorem 2. *The* `PushDfs` *algorithm explores an arbitrary tree and obtains the competitive ratio of at most* $4 - 2/k$ *for the online energy model.*

5 Conclusions

We have presented the lower bound of $\Omega\left(\log k / \log\log k\right)$ for a competitive ratio in the time model of the exploration. This is a significant improvement over the recent $2 - 1/k$ lower bound presented in [12]. The best algorithms for trees achieve a competitiveness of $O(k/\log k)$ and $O(\sqrt{D})$ (for sparse trees) which leaves a wide area for further research. Moreover, our result proves that the lack of a map is essentially harmful in the time related online graph exploration problem (and this remains even when we restrict ourself only to trees).

For the energy cost model there is an online algorithm with a constant competitive ratio for trees. Using a simple algorithm a team of k robots can explore a tree using only $4 - 2/k$ times the energy of the offline solution. This does not match yet the lower bound of $3/2$ presented in [14].

References

1. Betke, M., Rivest, R.L., Singh, M.: Piecemeal learning of an unknown environment. In: Proc. of the 6th Annual ACM Conference on Computational Learning Theory (COLT 1993), Association for Computing Machinery, pp. 277–286 (1993)
2. Bender, M., Slonim, D.: The power of team exploration: two robots can learn unlabeled directed graphs. In: FOCS 1994. Proc. of the 35th Annual Symposium on Foundations of Computer Science, pp. 75–85. IEEE Computer Society, Washington, DC (1994)
3. Bender, M.A., Fernández, A., Ron, D., Sahai, A., Vadhan, S.: The power of a pebble: Exploring and mapping directed graphs. Information and Computation 176, 1–21 (2005)
4. Fleischer, R., Trippen, G.: Exploring an unknown graph efficiently. In: Brodal, G.S., Leonardi, S. (eds.) Algorithms – ESA 2005. 13th Annual European Symposium, Palma de Mallorca, Spain, October 3-6, 2005. LNCS, vol. 3669, pp. 11–22. Springer, Heidelberg (2005)

5. Dessmark, A., Pelc, A.: Optimal graph exploration without good maps. Theoretical Computer Science 326, 343–362 (2004)
6. Gasieniec, L., Pelc, A., Radzik, T., Zhang, X.: Tree exploration with logarithmic memory. In: Proc. of ACM-SIAM Symp. on Discrete Algorithms (SODA 2007) (2007)
7. Dobrev, S., Flocchini, P., Kralovic, R., Ruzicka, P., Prencipe, G., Santoro, N.: Black hole search in common interconnection networks. Networks 47, 61–71 (2006)
8. Dobrev, S., Flocchini, P., Prencipe, G., Santoro, N.: Searching for a black hole in arbitrary networks: optimal mobile agent protocols. In: Proc. of the 21st Annual Symposium on Principles of Distributed Computing (PODC '02) pp.153–162 (2002)
9. Dessmark, A., Fraigniaud, P., Kowalski, D., Pelc, A.: Deterministic rendezvous in graphs. Algorithmica 46, 69–96 (2006)
10. Sugihara, K., Suzuki, I.: Distributed algorithms for formation of geometric patterns with many mobile robots. Journal of Robotic Systems 13, 127–139 (1996)
11. Das, S., Flocchini, P., Nayak, A., Santoro, N.: Distributed exploration of an unknown graph. In: Pelc, A., Raynal, M. (eds.) Structural Information and Communication Complexity. Proc. 12 International Colloquium, SIROCCO 2005, Mont Saint-Michel, France, May 24-26, 2005. LNCS, vol. 3499, pp. 99–114. Springer, Berlin Heidelberg New York (2005)
12. Fraigniaud, P., Gasieniec, L., Kowalski, D., Pelc, A.: Collective tree exploration. Networks 48, 166–177 (2006)
13. Dynia, M., Kutylowski, J., Heide, F.M.A.D, Schindelhauer, C.: Smart robot teams exploring sparse trees. In: Královič, R., Urzyczyn, P. (eds.) Mathematical Foundations of Computer Science 2006. Proc of the 31st International Symposium, MFCS 2006, Stará Lesná, Slovakia, August 28-September 1, 2006. LNCS, vol. 4162, pp. 327–338. Springer, Heidelberg (2006)
14. Dynia, M., Korzeniowski, M., Schindelhauer, C.: Power-aware collective tree exploration. In: Grass, W., Sick, B., Waldschmidt, K. (eds.) Architecture of Computing Systems - ARCS 2006. Proc of the 19th International Conference, Frankfurt/Main, Germany, March 13-16, 2006. LNCS, vol. 3894, pp. 341–351. Springer, Heidelberg (2006)
15. Hsiang, T., Arkin, E., Bender, M., Fekete, S., Mitchell, J.: Algorithms for rapidly dispersing robot swarms in unknown environments. In: Proc. of the 5th International Workshop on Algorithmic Foundations of Robotics, pp. 77–94. Springer, Heidelberg (2002)

Graph Searching with Advice

Nicolas Nisse* and David Soguet

LRI, Université Paris-Sud, Orsay, France
{nisse,soguet}@lri.fr

Abstract. Fraigniaud *et al.* (2006) introduced a new measure of difficulty for a distributed task in a network. The smallest *number of bits of advice* of a distributed problem is the smallest number of bits of information that has to be available to nodes in order to accomplish the task efficiently. Our paper deals with the number of bits of advice required to perform efficiently the graph searching problem in a distributed setting. In this variant of the problem, all searchers are initially placed at a particular node of the network. The aim of the team of searchers is to capture an invisible and arbitrarily fast fugitive in a monotone connected way, i.e., the cleared part of the graph is permanently connected, and never decreases while the search strategy is executed. We show that the minimum number of bits of advice permitting the monotone connected clearing of a network in a distributed setting is $O(n \log n)$, where n is the number of nodes of the network, and this bound is tight. More precisely, we first provide a labelling of the vertices of any graph G, using a total of $O(n \log n)$ bits, and a protocol using this labelling that enables clearing G in a monotone connected distributed way. Then, we show that this number of bits of advice is almost optimal: no protocol using an oracle providing $o(n \log n)$ bits of advice permits the monotone connected clearing of a network using the smallest number of searchers.

Keywords: Graph searching, Monotonicity, Bits of advice.

1 Introduction

The *search problem* has been widely used in the design of distributed protocols for clearing graphs in a decentralized manner [1,6,10,11]. In the search problem, the graph is regarded as a "contaminated" network that a team of *searchers* is aiming at clearing. Initially, the whole graph is *contaminated*. The searchers stand at some vertices of the graph and they are allowed to move along edges. An edge is *cleared* when it is traversed by a searcher. A clear edge e is preserved from recontamination if, for any path between e and a contaminated edge, a searcher is occupying a vertex of this path. The search problem deals with a sequence of moves of searchers, that satisfies: (1) initially all searchers stand at a particular vertex of the graph, *the homebase*, and (2) a searcher is allowed to move along an edge if it does not imply any recontamination. Such a sequence

---

* Additional supports from the project FRAGILE of the ACI Sécurité Informatique, and from the project GRAND LARGE of INRIA.

G. Prencipe and S. Zaks (Eds.): SIROCCO 2007, LNCS 4474, pp. 51–65, 2007.

of moves, or *steps*, is called a *search strategy*. Given a connected graph G and a vertex $v_0 \in V(G)$, the search problem consists in computing, in a distributed setting, a search strategy of G, with v_0 as the homebase, and using the fewest searchers as possible that results in all edges being simultaneously clear. The strategy is computed online by the searchers themselves. Note that, during the execution of a search strategy, the contaminated part of the graph never grows. The strategy is said *monotone* [5,15]. Moreover, the cleared part of the graph remains connected at any step. The strategy is said *connected* [1,2].

The main difference between the existing distributed protocols for clearing a graph is the amount of knowledge about the topology of the graph that searchers have *a priori*. In [1,10,11], the searchers know in advance the topology of the network in which they are launched, and clear the network in a polynomial time. Conversely, the protocol provided in [6] enables to clear any network without having any *a priori* information about its topology. However, the clearing of the network is connected but not monotone and it is performed in an exponential time. Thus, not surprisingly, it appears that there is a tradeoff between the amount of knowledge provided to the searchers and the performance of the search strategy.

In [12], Fraigniaud *et al.* propose a new framework for measuring the difficulty of a distributed task: the number of *bits of advice*. Given a distributed task, the minimum number of bits of advice for this problem represents the minimum total number of bits of information that has to be given to nodes or mobile agents to efficiently perform the task. This approach is quantitative, i.e. it considers the amount of knowledge without regarding what kind of knowledge is supplied. This paper addresses the problem of the minimum number of bits of advice permitting to solve the search problem.

1.1 Our Model

The searchers are modeled by synchronous autonomous mobile computing entities with distinct IDs. Otherwise searchers are all identical, run the same program and use at most $O(\log n)$ bits of memory, where n is the number of nodes of the network. A network is modeled by a synchronous undirected connected graph. *A priori*, the network is anonymous, that is, the nodes are not labelled. The $\deg(u)$ edges incident to any node u are labelled from 1 to $\deg(u)$, so that the searchers can distinguish the different edges incident to a node. These labels are called *port numbers*. Every node of the network has a zone of local memory, *whiteboard*, of size $O(\log n)$ bits in which searchers can read, erase, and write symbols. It is moreover assumed that searchers can access these whiteboards in fair mutual exclusion. An instance of the problem consists of a couple (G, v_0), where $G = (V, E)$ is a graph and $v_0 \in V$ is the homebase. An *oracle* [12,13] is a function $\mathcal{O}$ that maps any instance (G, v_0) to a function $f : V \rightarrow \{0,1\}^*$ assigning a binary string, called *advice*, to any node of the network. The size of the advice, i.e. the number of bits of advice, on a given instance, is the sum of the lengths of all the strings assigned to the nodes. Intuitively, the oracle provides additional knowledge to the nodes of the network.

The *search problem* consists in designing an oracle $\mathcal{O}$ and a protocol $\mathcal{P}$ using $\mathcal{O}$, with the following characteristics. For any instance $(G = (V, E), v_0)$, any vertex $v \in V$ is provided with the string $f(v)$, $f = \mathcal{O}(G, v_0)$. Protocol $\mathcal{P}$ must enable the optimal number of searchers to clear G starting from v_0. Moreover, the search strategy performed by searchers is computed locally. That is, the decision of the searcher at a vertex v (moving via some specific port number, switching its state, writing on the whiteboard) only depends on (1) the current state of the searcher, (2) the label $f(v)$ of the current vertex (3) the content of the current node's whiteboard (plus possibly the incoming port number if the searcher just entered the node). In particular, the searchers do not know in advance in which graph they are launched. The only information about the graph is the bit strings available locally at each node.

1.2 Our Results

We show that the minimum number of bits of advice permitting the clearing of any n-node graph, in a distributed setting, is $O(n \log n)$, and this bound is tight. More precisely, on one hand, we define an oracle $\mathcal{O}$ and a distributed protocol `Cleaner` that allow to solve the search problem for any connected n-node graph G starting from any vertex $v_0 \in V(G)$. Moreover, the clearing of G is performed in time $O(n^3)$. The searchers are modeled by automata with $O(\log n)$ bits of memory. The nodes' whiteboards have size $O(\log n)$. Actually, our protocol ensures that the whiteboard will only be used in order to allow two searchers present at the same node to exchange their states and IDs. Finally, the number of bits of advice provided by $\mathcal{O}$ is $O(n \log n)$ for any n-node graph. On the other hand, we show that this number of bits of advice is almost optimal: no protocol using an oracle providing $o(n \log n)$ bits of advice permits to solve the search problem.

1.3 Related Work

In many areas of distributed computing, the quality of algorithmic solutions for a given network problem often depends on the amount of knowledge given to the nodes of the network (see [9] for a survey). The comparison of two algorithms with knowledge appears however to be not obvious when they are provided with qualitatively different informations: upper bound on the size of the network [3], the entire topology of the network [8], etc. In [12], Fraigniaud et al. introduce the notion of bits of advice as a way to quantitatively measure the difficulty of a distributed task. As an example, Fraigniaud et al. [12] study the amount of knowledge that must be distributed on the vertices of the graph in order to perform broadcast and wake-up efficiently (i.e., using a minimum number of messages). They prove that the minimum number of bits of advice permitting to perform wake-up (resp., broadcast) with a linear number of messages is $\Theta(n \log n)$ (resp., $O(n)$) bits. This quantitatively differentiate the difficulty of broadcast and wake-up. Fraigniaud *et al.* [13] also study the minimum number of bits of advice that allows to efficiently explore a tree, i.e., with a better competitive ratio than a Depth First Search.

Introduced by Parson [18], *graph searching* looks for the smallest number of searchers required to clear a graph. However, in graph searching, the strategies are not constrained to be connected nor monotone (see [4] for a survey). The *search number* of the graph G, denoted $\mathbf{s}(G)$, is the minimum k such that there is a search strategy for G (not necessarily monotone nor connected) using at most k searchers that results in all edges being simultaneously clear. The graph searching problem has been extensively studied for its practical applications and for the close relationship between its several variants (edge-search, node-search, mixed-search [4]) and standard graph parameters like treewidth [19] and pathwidth [4]. The problem of finding the search number of a graph has been proved to be NP-hard [16]. According to the important Lapaugh's result [5,15], "recontamination does not help". That is, for any graph G, there is a monotone search strategy for G using at most $\mathbf{s}(G)$ searchers. Monotonicity plays a crucial role in graph searching, since a monotone search strategy ensures a clearing of the graph in a polynomial number of steps. It implies that the graph searching problem is in NP. This result is not valid anymore, as soon as the search strategy is constrained to be connected [20]. Several practical applications (decontamination of polluted pipes [18], speleological rescue [7], network security...) require the search strategy to be connected to ensure safe communications between searchers. Barrière *et al.* [2] prove that, clearing a tree T in a connected way requires at most $2\,\mathbf{s}(T) - 2$ searchers and that this bound is tight. The better bound known in the case of an arbitrary n-node graph G is $\mathbf{s}(G)(\log n + 1)$ [14].

Several protocols for clearing a network in distributed setting have been proposed in the literature. It has been proved that any distributed protocol clearing an asynchronous network in a monotone connected way requires at most one searcher more than in the synchronous case [11]. Moreover, this result remains valid even if the topology of the network is known in advance. In [6], Blin *et al.* proposed a distributed protocol that enables the optimal number of searchers to clear any network G in a fully decentralized manner. The strategy is computed online by the searchers themselves. The distributed computation must not require knowing the topology of the network in advance. Roughly speaking, their protocol ensures that searchers try every possible connected monotone partial search strategy. Thus, whilst the search strategy eventually computed by the searchers is monotone, failing search strategies investigated before lead to withdrawals, and therefore to recontamination. Flocchini *et al.* proposed protocols that address the graph searching problem in specific topologies (trees [1], hypercubes [11], tori and chordal rings [10], etc.). For each of these classes of graphs, the authors propose a protocol using the optimal number of searchers for clearing G in a monotone connected way with $O(\log n)$ bits of memory and whiteboards of $O(\log n)$ bits, that clears the graph in a polynomial time. Note that, encoding the entire topology requires $\Omega(n^2)$ bits.

2   Distributed Search Strategy Using Little Information

This section is devoted to prove the following theorem.

Theorem 1. *The search problem can be solved using $O(n \log n)$ bits of advice.*

To prove it, we describe an oracle $\mathcal{O}$ which provides an advice of size $O(n \log n)$, and a distributed protocol Cleaner that solve the search problem in a synchronous decentralized manner. Protocol Cleaner is divided in n phases, each one being divided in two stages of $O(n^2)$ rounds. Thus, the clearing of G is performed in a time $O(n^3)$.

2.1   The Oracle

In this section, we describe the oracle $\mathcal{O}$. For any instance $(G = (V, E), v_0)$ of the search problem, we consider a strategy S that is solution of the problem. The function $f = \mathcal{O}(G, v_0)$ is defined from S. Roughly speaking, the bits of advice supplied by $\mathcal{O}$ enable searchers using protocol Cleaner, to clear the vertex-set in the same order as S. Moreover, they allow the searchers to circulate in the cleared part of the graph avoiding recontamination. Let us define some notations.

Let $n = |V|$ and $m = |E|$. The strategy S can be defined by the order in which S clears the edges. Let $(e_1, \cdots, e_m)$ be this order. An edge e_i is smaller than an edge e_j, denoted by $e_i \preceq e_j$, if $i \leq j$. S also induces an order on the vertices of G. For any $v, w \in V$, we say that v is smaller than w, denoted $v \preceq w$, if the first cleared edge incident to v is smaller than the first cleared edge incident to w. Let $(v_0, \cdots, v_{n-1})$ be this order, i.e., $v_i \preceq v_j$ if and only if $i \leq j$.

For any $0 \leq i \leq n - 1$, let $f_i \in E$ be the first cleared edge incident to v_i. By definition, $f_0 = f_1 \prec f_2 \cdots \prec f_{n-1}$. For any $1 \leq i \leq n - 1$, the *parent* of v_i, denoted by $parent(v_i)$, is defined as the neighbour v of v_i such that, $\{v, v_i\} = f_i$. Note that $parent(v_i) \prec v_i$, and for any neighbour w of v_i, $f_i = \{parent(v_i), v_i\} \preceq \{w, v_i\}$. Intuitively, for any $1 \leq i \leq n - 1$, $f_i = \{parent(v_i), v_i\}$ is the edge by which a searcher has arrived to clear v_i. Conversely, the children of $v \in V$ are the vertices w such that $v = parent(w)$. For any $0 \leq i \leq n - 1$, let T_i be the subgraph of G whose vertex-set is $\{v_0, \cdots, v_i\}$, and the edge-set is $\{f_1, \cdots, f_i\}$. For any $0 \leq i \leq n - 1$, T_i is a spanning tree of $G[v_0, \cdots, v_i]$, which denotes the subgraph of G induced by $\{v_0, \cdots, v_i\}$. Intuitively, at the phase i of the execution of Protocol Cleaner, T_{i-1} is a spanning tree of the clear part of the graph. It is used to allow the searchers to move in the clear part, performing a DFS of T_{i-1}.

We now define a local labelling $\mathcal{L}(S)$ of the vertices of G. Again, this labelling depends on the strategy S that is considered. Let $v \in V(G)$. The label of a vertex v consists of the following local variables: a boolean TYPE_v, four integers TCU_v, TTL_v, LASTPORT_v, PARENT_v and a list CHILD_v of ordered pairs of integers. The index will be omitted whenever this omission does not cause any confusion. Intuitively, PARENT_v and CHILD_v enable the searchers to perform a DFS of a subtree spanning the cleared part. To avoid recontamination, the searchers must know which ports they can take or not, and the moment when such a move is possible, i.e. the phase of the protocol when a searcher can take some port. The informations about the ports are carried by $\mathtt{PARENT}_v$, $\mathtt{CHILD}_v$, and $\mathtt{LASTPORT}_v$. $\mathtt{CHILD}_v$, $\mathtt{TCU}_v$ and $\mathtt{TTL}_v$ carry information about phases. Moreover, if a searcher preserves a node from recontamination, we say that this searcher *guards* the node, otherwise the searcher is said *free*. A searcher which guards a node v will leave v by its largest edge. Such a move will not induce any recontamination because any other edges incident to v will have been previously cleared by free searchers. For this task, we need to distinguish two types of node with $\mathtt{TYPE}_v$.

In the following we will say that a port number p of a vertex v (resp., the edge incident to v, corresponding to p) is *labelled* if either there exists $\ell \leq n-1$ such that $(p, \ell) \in \mathtt{CHILD}_v$, or $p = \mathtt{LASTPORT}_v$, or $p = \mathtt{PARENT}_v$. Note that an edge may have two different labels, or may be unlabelled at one of its ends, and labelled at the other, or unlabelled at both ends. Let $0 \leq i \leq n-1$ be the integer such that $v = v_i$. Let e be the largest edge incident to v that is not in $E(T_{n-1})$, and let f be the largest edge incident to v that is not in $(T_{n-1}) \cup \{e\}$.

- $\mathtt{PARENT}_v$ is the port number of v leading to $parent(v)$ through an edge of $E(T_{n-1})$ (we set $\mathtt{PARENT}_{v_0} = -1$).
- $\mathtt{CHILD}_v$ is a list of ordered pairs of integers. Let $1 \leq p \leq \deg(v)$ and $0 < j \leq n-1$. $(p, j) \in \mathtt{CHILD}_v$ if and only if $v = parent(v_j)$ and p is the port number of v leading to v_j. In the following, $\mathtt{CHILD}_v(j)$ denotes the port number p of v such that $(p, j) \in \mathtt{CHILD}_v$.
- $\mathtt{TYPE}_v$ is a boolean variable. It equals 0 if the largest edge incident to v belongs to T_{n-1}. Otherwise, the variable $\mathtt{TYPE}_v$ equals 1. In the following we will say that a vertex is of type 0 (resp., type 1) if $\mathtt{TYPE}_v = 0$ (resp., $\mathtt{TYPE}_v = 1$). Roughly, a vertex is of type 0 if, in S, the searcher cleared the last uncleared incident edge to v, in order to reach a new vertex which was still uncleared.

  the last incident edge to reach a new vertex that was not occupied yet.
- $\mathtt{LASTPORT}_v = -1$ if $\mathtt{TYPE}_v = 0$, else $\mathtt{LASTPORT}_v$ is the port number corresponding to e.
- $\mathtt{TCU}_v$ (*Time to Clean Unlabelled port*), represents the phase when the free searchers must clear all the unlabelled ports of v. Case $\mathtt{TYPE}_v = 0$: if e does not exist, then $\mathtt{TCU}_v = -1$, else $\mathtt{TCU}_v$ is the largest k such that $f_{k-1} \preceq e$. Case $\mathtt{TYPE}_v = 1$: if f does not exist, then $\mathtt{TCU}_v = -1$, else $\mathtt{TCU}_v$ is the largest k such that $f_{k-1} \preceq f$.
- $\mathtt{TTL}_v$ (*Time To Leave*), represents the phase when, a searcher that guards v will leave v. Case $\mathtt{TYPE}_v = 0$: $\mathtt{TTL}_v = j$ such that v_j is the largest child of v. Case $\mathtt{TYPE}_v = 1$: $\mathtt{TTL}_v$ is the largest k such that $f_{k-1} \leq e$.

We now define the bits of advice $\mathcal{O}(G, v_0)$ provided by oracle $\mathcal{O}$ to G, using the labelling $\mathcal{L}(S)$. For any $0 \leq i \leq n-1$,
$\mathcal{O}(G, v_0)(v_i) = (i, n, \mathtt{TYPE}_{v_i}, \mathtt{PARENT}_{v_i}, \mathtt{LASTPORT}_{v_i}, \mathtt{TCU}_{v_i}, \mathtt{TTL}_{v_i}, \mathtt{CHILD}_{v_i})$.
The following lemma follows obviously from the definition of the oracle.

Lemma 1. *For any n-node graph, $\mathcal{O}$ provides $O(n \log n)$ bits of advice.*

2.2   The Protocol Cleaner

In this section, we define a distributed protocol Cleaner using the oracle $\mathcal{O}$, that enables to clear any n-node synchronous network G starting from the homebase v_0. Protocol Cleaner is formally described in Figures 1 and 2.

Let us roughly describe our protocol. Our searchers can be in seven different states: DFS_TEST, DFS_BACK, CLEAR_UNLABELLED, CLEAR_UNLABELLED_BACK, CLEAR, WAIT, GUARD. Initially, all searchers stand at v_0. Each of them reads n on the label $\mathcal{O}(G, v_0)(v_0)$ of v_0 to initialize their counters. Then the searcher with the largest Id is elected to guard v_0 and switches to state GUARD, the other searchers become free and switch to state DFS_TEST. After the phase $1 \leq i \leq n-1$, our protocol ensures the following. (1) A subgraph G' of $G[v_0, \cdots, v_i]$ containing T_i as a subgraph is cleared. (2) For any vertex v of the border of G', i.e. v is incident to an edge in $E(G')$ and an edge of $E(G) \setminus E(G')$, one searcher is guarding v (in state GUARD). (3) Any other searcher is free and stand at a vertex of G'.

During the first stage of the phase $i+1$, the free searchers are aiming at clearing the unlabelled edges of those vertices v of $V(G[v_0, \cdots, v_i])$ such that the largest unlabelled edge e incident to v satisfies $f_i \prec e \prec f_{i+1}$. Note that such an edge e belongs to $E(G[v_0, \cdots, v_i])$. For this purpose, any free searcher performs a DFS of T_i thanks to the labels PARENT and CHILD. The searcher is in state DFS_TEST if it goes down in the tree, in state DFS_BACK otherwise.

During this DFS, if the searcher meets a vertex v_j ($j \leq i$) labelled in such a way that $\text{TCU}_{v_j} = i+1$ (recall that TCU means *Time to Clear Unlabelled edges*), then the searcher clears all unlabelled edges of v_j and then it carries on the DFS. To clear the unlabelled edges of v_j, the searcher take successively, in the order of the port numbers, all the unlabelled ports. It takes each unlabelled port back and forth, in state CLEAR_UNLABELLED for the first direction, and CLEAR_UNLABELLED_BACK for the second direction.

Moreover, during this stage, any searcher that is guarding a vertex labelled in such a way that (TYPE = 1 and TCU < TTL = $i+1$) is aiming at clearing, in state CLEAR, the edge corresponding to the port number LASTPORT of the considered vertex (recall that TTL means *Time To Leave*). Protocol Cleaner ensures that the corresponding port number corresponds to the single contaminated edge incident to the considered vertex at this stage.

Before the first round of the second stage of phase $i+1$, the two following properties are satisfied: (1) if there exists a vertex v such that v is labelled with (TYPE$_v$ = 0 and TTL$_v$ = $i+1$), f_{i+1} is the only contaminated edge incident to $v = parent(v_{i+1})$, and (2) for any vertex v labelled in such a way that (TYPE$_v$ = 1 and TCU$_v$ = TTL$_v$ = $i+1$), the edge corresponding to LASTPORT$_v$ is the only contaminated edge incident to v.

During the second stage of the phase $i+1$, Protocol Cleaner performs the clearing of f_{i+1} (incident to $parent(v_{i+1}) \in V(G')$) and the clearing of any edge corresponding to port number LASTPORT$_{v_j}$ of a vertex v_j ($j \leq i$) labelled in such a way that (TYPE$_{v_j}$ = 1 and TCU$_{v_j}$ = TTL$_{v_j}$ = $i+1$). For this purpose, any free searcher performs a DFS of T_i.

Program of searcher A.

Initialisation: /* all searchers start at v_0 */
    Read n on $\mathcal{O}(G, v_0)$ to initialize the counter;
    if A is the searcher with the largest ID at v_0 **then**
        Switch to the state GUARD;
    else
        At the first round on the second stage of phase 1,
        Switch to the state DFS_TEST;
    endif

Program of searcher A at any round of stage $s \in \{0, 1\}$ of phase $1 \leq i \leq n$.

/* Searcher A arrives at node v_j, coming by port number p_ℓ of v_j */
(corresponding to the edge $\{v_\ell, v_j\}$).

Let p_{first} be the smallest unlabelled port number of v_j.
$p_{first} = -1$ if there are no such edges.
Let p_{next} be the smallest unlabelled port number p of v_j, such that $p > p_\ell$.
$p_{next} = -1$ if there are no such edges.

Let $p_{firstChild}$ be the port number p of v_j such that it exists $1 \leq k \leq n-1$ with p being labelled CHILD(k), and for any $1 \leq k' < k$, no port numbers of v_j are labelled CHILD(k'). $p_{firstChild} = -1$ if there are no such edges.
If $p_{firstChild} \neq -1$, let $firstChild$ denote the corresponding neighbour of v_j.

Let $p_{nextChild}$ be the port number p of v_j such that it exists $\ell < k \leq n-1$ with p being labelled CHILD(k), and for any $\ell \leq k' < k$, no port numbers of v_j are labelled CHILD(k'). If $p_{nextChild} \neq -1$, $nextChild$ denotes the corresponding neighbour of v_j.

Case:
    state $=$ DFS_TEST
        if $s = 1$ and there is a port p labelled CHILD(i) **then**
            Take port p in state CLEAR;
        else if $s = 0$ and TCU $= i$ **then**
            Take port p_{first} in state CLEAR_UNLABELLED;
        else if $p_{firstChild} \neq -1$ and $firstChild \preceq v_{i-1}$ **then**
            Take port $p_{firstChild}$ in state DFS_TEST;
        else Take port labelled PARENT in state DFS_BACK;
        endif

    state $=$ CLEAR_UNLABELLED_BACK
        if $p_{next} \neq -1$ **then**
            Take port p_{next} in state CLEAR_UNLABELLED;
        else if $p_{firstChild} \neq -1$ and $firstChild \preceq v_{i-1}$ **then**
            Take port $p_{firstChild}$ in state DFS_TEST;
        else Take port labelled PARENT in state DFS_BACK;
        endif

Fig. 1. Protocol Cleaner (1/2)

```
    state = CLEAR_UNLABELLED
        Take port pℓ in state CLEAR_UNLABELLED_BACK;

    state = DFS_BACK
        if s = 1 and there is a port p labelled CHILD(i) then
            Take port p in state CLEAR;
        else if p_nextChild ≠ −1 and nextChild ⪯ v_{i−1} then
            Take port p_nextChild in state DFS_TEST;
        else if PARENT ≠ −1 then
            Take port labelled PARENT in state DFS_BACK;
        else Take port CHILD(1) in state DFS_TEST;
        endif

    state = CLEAR
        if v_j ≺ v_i or deg(v_j) = 1 then
            if j > 0 then
                Take port labelled PARENT in state DFS_BACK;
            else Take port labelled CHILD(1) in state DFS_TEST;
            endif
        else Switch to the state WAIT;
        endif

/* Searcher that stands at node v_j */

    state = GUARD
        if TYPE = 1 then
            if TCU = TTL then
                At the first round of the second stage of phase TTL,
                Take port labelled LASTPORT in state CLEAR;
            else At the first round of the first stage of phase TTL,
                take port labelled LASTPORT in state CLEAR;
            endif
        else At the first round of the second stage of phase TTL
            take port CHILD(TTL) in state CLEAR;
        endif

    state = WAIT
        At the last round of this phase:
        if A is the searcher with the greatest ID at v_j then
            Switch to the state GUARD;
        else Take port labelled PARENT in state DFS_BACK;
        endif

end
```

Fig. 2. Protocol Cleaner (2/2)

When the searcher meets the vertex $parent(v_{i+1})$ whose a port number is labelled CHILD($i + 1$), it takes the corresponding edge in state CLEAR. Moreover, any searcher that is guarding the vertex $parent(v_{i+1})$ also takes the edge corresponding to CHILD($i + 1$) in state CLEAR if (TTL $= i + 1$ and TYPE $= 0$). Finally, any searcher that is guarding a vertex labelled in such a way that (TYPE $=$ 1 and TCU $=$ TTC $= i+1$), takes the edge corresponding to port number LASTPORT in state CLEAR. During this stage, any searcher arriving at v_{i+1} waits (in state WAIT) the last round of the stage if $\deg(v_{i+1}) > 1$, else it becomes free. During this last round, if $\deg(v_{i+1}) > 1$, the searcher with largest Id that stands at v_{i+1} is elected to guard v_{i+1} while other searchers are free and take the port labelled PARENT in state DFS_BACK.

2.3   Sketch of Proof of Cleaner

In order to prove the correctness of our protocol we need the following notations. A searcher is called free if it is not in state GUARD nor WAIT. For any $0 \le i \le n-1$, let $M_i = \{v \in V(G) \mid \text{for any edge } e \text{ incident to } v, e \preceq f_i\}$. $M_i \subseteq V(T_i)$ is the set of the vertices whose all incident edge, but f_i, have been cleared by S before the step corresponding to the clearing of f_i. Moreover, we set $M_n = V$. Thus, after the step corresponding to the clearing of f_i, no vertices in M_i need to be guarded in the strategy S. Note that, for any $0 \le j \le n - 1$, the set $M_j \setminus M_{j-1}$ is exactly the set of vertices v such that TTL $= j$. The proof of theorem 1 easily follows from the following lemma.

Lemma 2. *Let G be a connected graph and $v_0 \in V(G)$. Let S be a strategy that clears the graph G, starting from v_0, and using the smallest number of searchers. Let $\mathcal{O}(G, v_0)$ be the labelling of the vertices of G, using $\mathcal{L}(S)$. After the last round of the phase $i \ge 1$ of the execution of Protocol Cleaner, the cleared part of the graph G satisfies the following:*

1. *any edge in $\{f_0, \cdots, f_i\}$ is clear,*
2. *any edge incident to vertex in M_i is clear,*
3. *there is exactly one searcher in state GUARD at any vertex of $V(T_i) \setminus M_i$,*
4. *any other searcher is free and stands at a vertex of T_i,*
5. *for any vertex v with TCU $\le i$, any unlabelled edge of v is clear.*

Roughly speaking, this lemma implies that, after the phase i, those vertices that have been all their incident edges cleared by S (the strategy from which the oracle is defined) before the clearing of f_{i+1} are cleared by Protocol Cleaner as well. Moreover, there is a searcher in state GUARD at any vertex of the border of the clear part of the graph, which avoids any recontamination. The proof of the lemma is by induction on $1 \le i \le n$. One can easily check that the case $i = 1$ holds. Let us assume that the result holds for $1 \le i \le n - 1$. We prove that it still holds after the last round of the phase $i + 1$. We consider two cases according whether there is a free searcher or not. Due to lack of space, the proof of this lemma is omitted, and can be found in [17].

3   Lower Bound

In this section, we show that the upper bound proved in the previous section is almost optimal. More precisely, we prove that:

Theorem 2. *The search problem cannot be solved using only $o(n \log n)$ bits of advice.*

To prove the theorem, we build a $4n + 4$-node graph $\mathcal{G}_n$. Then, we prove that any distributed protocol requires $\Omega(n \log n)$ bits of advice to clear $\mathcal{G}_n$ in a monotone connected way starting from $v_0 \in V(\mathcal{G}_n)$, and using the fewest number of searchers.

Let $n \geq 4$. Let $t = 2n + 7$. Let $P = \{v_1, \cdots, v_t\}$ be a path and let K_{n-2}, resp. K_n, be a $(n - 2)$-clique, resp. a n-clique. We obtain the graph $\mathcal{G}_n$ by adding all edges between v_i and the vertices of K_{n-2}, for any $1 \leq i \leq t$. Then, let the node v_t coincide with a vertex of K_n. Finally, let us choose one vertex of K_{n-2} and denote it by v_0.

We now enumerate some technical lemmas that describe how any search strategy clears $\mathcal{G}_n$ using the fewest number of searchers.

Lemma 3. *The smallest number of searchers sufficient to clear $\mathcal{G}_n$ is n.*

Proof. Since $\mathcal{G}_n$ admits K_n as a minor, we get that the smallest number of searcher required to clear $\mathcal{G}_n$ is at least n. We now describe a strategy that clears $\mathcal{G}_n$ using n searchers. Starting from v_0, move one searcher to guard any vertex of K_{n-2}. Use the two remaining searchers to clear any edge of $E(K_{n-2})$. Then, move one remaining searcher to v_1. The second remaining searcher clears any edge between v_1 and K_{n-2}. Then, the searcher at v_1 move to v_2 and the second remaining searcher clears any edge between v_2 and K_{n-2}. And so on, until any vertex of P has been cleared. At this step, there are one searcher at any vertex of K_{n-2} and one searcher at v_t. Finally, let us use all the searchers to clear K_n. □

Lemma 4. *For any optimal search strategy that clears $\mathcal{G}_n$, the last vertex of $\mathcal{G}_n$ to have all its incident edges clear belongs to $V(K_n)$.*

Proof. During the clearing of K_n, the n searchers must stand at vertices of K_n. Thus, v_0 is not occupied by a searcher anymore. To avoid recontamination, any vertex of P and K_{n-2} must have all its incident edges clear. □

Due to lack of space, the proof of the following lemma is omitted, and can be found in [17].

Lemma 5. *For any optimal search strategy that clears $\mathcal{G}_n$, the first vertex of $\mathcal{G}_n$ to have all its incident edges clear is v_1 or v_2. Moreover, at this step, any vertex of K_{n-2} is occupied by a searcher, and no vertices of $\{v_4, \cdots, v_t\}$ have been occupied.*

The following lemma aims at proving that any strategy clearing $\mathcal{G}_n$ using n searchers and starting from v_0 is strongly constrained.

Lemma 6. *Let S be an optimal connected search strategy that clears $\mathcal{G}_n$ starting from v_0. For any $5 \leq i \leq t - 2$, at the first step of S when a searcher reaches v_i, the following is satisfied:*

- *any vertex in $V(K_n) \cup \{v_{i+1}, \cdots, v_t\}$ has all its incident edges contaminated;*
- *there is one searcher at any vertex of K_{n-2};*
- *any vertex in $\{v_1, \cdots, v_{i-2}\}$ has all its incident edges clear;*
- *either v_{i-1} has all its incident edges clear, or there is a searcher at v_{i-1} and v_{i-1} has only one incident edge that is still contaminated. In the latter case, the next move consists in moving a searcher along the last contaminated edge incident to v_{i-1}.*

Proof. Let s be the first step of the strategy such that, after this step, a searcher is occupying v_i. Let us consider the situation just before this step. Since $i \geq 5$, by Lemma 5, just before step s, v_1 or v_2 has all its incident edges clear, and there are one searcher at any vertex from K_{n-2} to preserve them from recontamination. Moreover, there is a vertex on the path between v_1 and v_i in P, that is occupied by a searcher for preserving v_1 or v_2 from recontamination. Let j, $1 < j < i$, be the minimum index such that a searcher is standing at v_j. Note that, for any k, $1 \leq k < j$, v_k has all its incident edge clear.

First, let us show that for any $\ell > i$, v_ℓ is not occupied before step s. For purpose of contradiction, let us assume v_ℓ is occupied. Since v_i has all its incident edges contaminated, for any k, $j < k < \ell$, v_k has all its incident edges contaminated. By Lemma 4 a vertex of K_n has at least one contaminated incident edge. Thus, for any k, $\ell < k \leq t$, v_k has all its incident edges contaminated, since there are no searchers on the path between v_k and K_n. Thus, there exits $k \neq i$ such that v_k has all its incident edges contaminated. Thus, the searchers at K_{n-2} cannot move, because they preserve recontamination from v_i and v_k. The searcher at v_ℓ cannot move because it preserves recontamination from v_i and K_n. The searcher at v_j may move at v_{j+1}, but then could not move anymore. Then the strategy fails, a contradiction. This proves the first item of the lemma.

Thus, before step s, there are one searcher at any vertex of K_{n-2}. These searchers preserve recontamination from v_i and v_t. Therefore, they cannot move. This proves the second item of the lemma.

According to the first item of the lemma, v_{i-1} has been reached before v_i. Since the strategy is monotone, just before the step s, a searcher is occupying v_{i-1}. Two cases must be considered:

- If s consists in moving a searcher occupying v_{i-1} along the edge $\{v_{i-1}, v_i\}$, the monotonicity of the strategy implies that either all edges incident to v_{i-1} are clear, or just before step s two searchers were occupying v_{i-1}. In the first case, the lemma is valid. Thus, let us assume that at least one edge incident to v_{i-1} is still contaminated after step s. Since $i \leq t - 2$, any vertex in $V(K_{n-2}) \cup \{v_i\}$ is occupied by a searcher, and incident to at least two contaminated edges: all edges incident to v_{i+1} and v_{i+2} are contaminated. If more than one edge incident to v_{i-1} is contaminated, the strategy fails. Therefore, at most one edge incident to v_{i-1} is contaminated, and the single possible move consists in moving the searcher at v_{i-1} along this edge.
- Else, the step s consists in moving a searcher along an edge between a vertex u of K_{n-2} and v_i. Since $i \leq t - 2$, there must be two searchers at u just before step s. Again, just after step s, any vertex in $V(K_{n-2}) \cup \{v_i\}$ is

occupied by a searcher, and incident to at least two contaminated edges: all edges incident to v_{i+1} and v_{i+2} are contaminated. Moreover, a searcher is occupying v_{i-1} and $\{v_{i-1}, v_i\}$ is contaminated. If another edge incident to v_{i-1} is contaminated, the strategy fails. Hence, at most one edge incident to v_{i-1} is contaminated, and the single possible move consists in moving the searcher at v_{i-1} along this edge.

This concludes the proof of the lemma.   □

A *local orientation* of a graph is a mapping from the incidence of the graph (between a vertex and an edge) into the port number of the graph. An instance of the problem consists of a graph, a vertex of this graph (the homebase) and a local orientation for this graph. Let $\mathcal{C}$ be the set of the following instances $\{(\mathcal{G}, v_0, \ell o) \mid \ell o \text{ is a local orientation of } \mathcal{G}\}$. Let $\mathcal{I} = |\mathcal{C}|$. The following lemma proves that any distributed protocol, using an arbitrary string of bits of advice, can clear only some amount of the instances of $\mathcal{C}$.

Lemma 7. *Let $\mathcal{P}$ be a distributed protocol for solving the search problem. Let f be a binary string of bits of advice provided by an oracle. Using f, $\mathcal{P}$ can clear at most $\mathcal{I} * (\frac{1}{n-2})^n$ instances of $\mathcal{C}$.*

Proof. Let $\mathcal{I}_{k,j}$ be the number of instances such that $(\mathcal{P}, f)$ allows to a searcher to clear j edges between v_k and K_{n-2}. We prove that, for $5 \leq k \leq n+5$ and any $1 \leq j \leq n-3$, $\mathcal{I}_{k,j} \leq \mathcal{I}_{k,j-1} \frac{n-j-1}{n-j}$.

Let us consider the last step such that exactly $0 \leq j \leq n-3$ edges between v_k and K_{n-2} are clear. By the lemma above, at this step, there is a searcher at v_k and a searcher at any vertex of K_{n-2}. Moreover, the remaining searcher cannot move to a vertex of $\{v_{k+1}, \cdots, v_t\}$. Let v be the vertex where this searcher stands. Using f, protocol $\mathcal{P}$ chooses a port number p that the remaining searcher must take. There are two cases according whether the remaining searcher stands at v_k or at a vertex of K_{n-2}.

- If the remaining searcher stands at v_k, it remains $n-j-1$ contaminated edges incident to this vertex and the strategy fails if p leads to v_{k+1}. Thus, the strategy fails in at least $\mathcal{I}_{k,j} \frac{1}{n-j-1}$ instances. Therefore, $\mathcal{I}_{k,j+1} \leq \mathcal{I}_{k,j} \frac{n-j-2}{n-j-1}$.
- If the remaining searcher stands at a vertex of K_{n-2}, it remains at most $n-3+t-k+1$ contaminated edges incident to this vertex and the strategy fails if p leads to one vertex in $\{v_{k+1}, \cdots, v_t\}$. Thus, the strategy fails in at least $\mathcal{I}_{k,j-1}(\frac{t-k}{n-3+t-k+1})$ instances. Hence, $\mathcal{I}_{k,j} \leq \mathcal{I}_{k,j-1} \frac{n-2}{t+n-2-k}$. To conclude, it is sufficient to remark that, since $n \geq 4$, $t = 2n+7$, $1 \leq j \leq n-3$ and $5 \leq k \leq n-5$, we have $\frac{n-2}{t+n-2-k} \leq \frac{n-2}{2n}$ and $\frac{n-j-2}{n-j-1} \geq \frac{n-3}{2}$. Thus, $\frac{n-2}{t+n-2-k} \leq \frac{n-j-2}{n-j-1}$.

Hence, $\mathcal{I}_{k,n-2} \leq \mathcal{I}_{k-1,n-2} \prod_{j=1..n-3}(\frac{n-j-2}{n-j-1}) = \mathcal{I}_{k-1,n-2}(\frac{1}{n-2})$. Using f, $\mathcal{P}$ can clear at most $\mathcal{I}_{n-5,n-2} \leq \mathcal{I}_{5,n-2}(\frac{1}{n-2})^n$. Since, $\mathcal{I}_{5,n-2} \leq \mathcal{I}$, the lemma holds.   □

Proof. of the Theorem 2. Let $N = |V(\mathcal{G})| = 4n+4$. To prove the theorem, it is sufficient to prove that for any $\alpha < 1/4$, and for any oracle that provides less than $q = \alpha N \log N$ bits of advice, no distributed protocol using $\mathcal{O}$ permit to clear all instances of $\mathcal{C}$. Let $\mathcal{O}$ be such an oracle. The number of functions f that the oracle $\mathcal{O}$ can output for $\mathcal{G}_n$ is at most $(q+1)2^q\binom{N+q}{N}$ [12]. Thus, there exists a set $\mathcal{S} \subseteq \mathcal{C}$ of at least $B = \frac{\mathcal{I}}{(q+1)2^q\binom{N+q}{N}}$ instances of $\mathcal{C}$ for which $\mathcal{O}$ returns the same string of bits of advice.

Let $\mathcal{P}$ be a distributed protocol that uses the oracle $\mathcal{O}$ for solving the search problem. By Lemma 7, $\mathcal{P}$ cannot clear more than $\mathcal{I} * (\frac{1}{n-2})^n$ instances of $\mathcal{C}$ using the same string of bits of advice.

To conclude, it remains to prove that $B > \mathcal{I} * (\frac{1}{n-2})^n$. Indeed,

$$B * (\frac{(n-2)^n}{\mathcal{I}}) = \frac{(n-2)^n}{(q+1)2^q\binom{N+q}{N}}$$

Using the Stirling formula we get that for n large enough,

$$B * (\frac{(n-2)^n}{\mathcal{I}}) \sim \frac{(n-2)^n}{2^{\alpha N \log N}(1+\alpha \log N)^N} * (\frac{\alpha \log N}{1+\alpha \log N})^{\alpha N \log N}$$

Since $N = 4n + 4$, we obtain:

$$\log[B * (\frac{(n-2)^n}{\mathcal{I}})] \sim (1 - 4\alpha)n \log n$$

Since $\alpha < 1/4$, we get that $B > \mathcal{I} * (\frac{1}{n-2})^n$. Thus, the result holds. ☐

References

1. Barrière, L., Flocchini, P., Fraigniaud, P., Santoro, N.: Capture of an intruder by mobile agents. In: 14th ACM Symp. on Parallel Algorithms and Architectures (SPAA), pp. 200–209 (2002)
2. Barrière, L., Fraigniaud, P., Santoro, N., Thilikos, D.: Connected and Internal Graph Searching. In: Bodlaender, H.L. (ed.) WG 2003. LNCS, vol. 2880, pp. 34–45. Springer, Heidelberg (2003)
3. Bender, M.A., Fernandez, A., Ron, D., Sahai, A., Vadhan, S.: The power of a pebble: Exploring and mapping directed graphs. Information and Computation 176, 1–21 (2002)
4. Bienstock, D.: Graph searching, path-width, tree-width and related problems (a survey). DIMACS Ser. in Discrete Mathematics and Theoretical Computer Science 5, 33–49 (1991)
5. Bienstock, D., Seymour, P.: Monotonicity in graph searching. Journal of Algorithms 12, 239–245 (1991)
6. Blin, L., Fraigniaud, P., Nisse, N., Vial, S.: Distributing Chasing of Network Intruders. In: Flocchini, P., Gąsieniec, L. (eds.) SIROCCO 2006. LNCS, vol. 4056, pp. 70–84. Springer, Heidelberg (2006)
7. Breisch, R.: An intuitive approach to speleotopology. Southwestern Cavers VI(5), 72–78 (1967)
8. Clementi, A.E.F., Monti, A., Silvestri, R.: Selective families, superimposed codes, and broadcasting on unknown radio networks. In: 12th Ann. ACM-SIAM Symp. on Discrete Algorithms (SODA) pp. 709–718 (2001)

9. Fich, F., Ruppert, E.: Hundreds of impossibility results for distributed computing. Distributed Computing 16, 121–163 (2003)

10. Flocchini, P., Luccio, F.L., Song, L.: Decontamination of chordal rings and tori. In: Proc. of 8th Workshop on Advances in Parallel and Distributed Computational Models (APDCM) (2006)

11. Flocchini, P., Huang, M.J., Luccio, F.L.: Contiguous search in the hypercube for capturing an intruder. In: Proc. of 18th IEEE Int. Parallel and Distributed Processing Symp. (IPDPS) (2005)

12. Fraigniaud, P., Ilcinkas, D., Pelc, A.: Oracle Size: a New Measure of Difficulty for Communication Tasks. In: 25th Annual ACM Symp. on Principles of Distributed Computing (PODC), pp. 179–187 (2006)

13. Fraigniaud, P., Ilcinkas, D., Pelc, A.: Tree Exploration with an Oracle. In: Královič, R., Urzyczyn, P. (eds.) MFCS 2006. LNCS, vol. 4162, pp. 24–37. Springer, Heidelberg (2006)

14. Fraigniaud, P., Nisse, N.: Connected Treewidth and Connected Graph Searching. In: Correa, J.R., Hevia, A., Kiwi, M. (eds.) LATIN 2006. LNCS, vol. 3887, pp. 470–490. Springer, Heidelberg (2006)

15. LaPaugh, A.: Recontamination does not help to search a graph. Journal of the ACM 40(2), 224–245 (1993)

16. Megiddo, N., Hakimi, S., Garey, M., Johnson, D., Papadimitriou, C.: The complexity of searching a graph. Journal of the ACM 35(1), 18–44 (1988)

17. Nisse, N., Soguet, D.: Graph searching with advice.Technical Report LRI-1469, University Paris-Sud, France (March 2007)

18. Parson, T.: Pursuit-evasion in a graph. In: Theory and Applications of Graphs. Lecture Notes in Mathematics, pp. 426–441. Springer, Heidelberg (1976)

19. Seymour, P., Thomas, R.: Graph searching and a min-max theorem for tree-width. J. Combin. Theory Ser. B. 58, 22–33 (1993)

20. Yang, B., Dyer, D., Alspach, B.: Sweeping Graphs with Large Clique Number. In: Fleischer, R., Trippen, G. (eds.) ISAAC 2004. LNCS, vol. 3341, pp. 908–920. Springer, Heidelberg (2004)

From Renaming to Set Agreement

Achour Mostefaoui, Michel Raynal, and Corentin Travers

IRISA, Université de Rennes, 35042 Rennes, France
{achour,raynal,ctravers}@irisa.fr

Abstract. The M-renaming problem consists in providing the processes with a new name taken from a new name space of size M. A renaming algorithm is adaptive if the size M depends on the number of processes that want to acquire a new name (and not on the total number n of processes). Assuming each process proposes a value, the k-set agreement problem allows each process to decide a proposed value in such a way that at most k different values are decided. In an asynchronous system prone to up to t process crash failures, and where processes can cooperate by accessing atomic read/write registers only, the best that can be done is a renaming space of size $M = p + t$ where p is the number of processes that participate in the renaming. In the same setting, the k-set agreement problem cannot be solved for $t \geq k$.

This paper focuses on the way a solution to the renaming problem can help solving the k-set agreement problem when $k \leq t$. It has several contributions. The first is a t-resilient algorithm ($1 \leq t < n$) that solves the k-set agreement problem from any adaptive $(n + k - 1)$-renaming algorithm, when $k = t$. The second contribution is a lower bound that shows that there is no wait-free k-set algorithm based on an $(n+k-1)$-renaming algorithm that works for any value of n, when $k < t$. This bound shows that, while a solution to the $(n + k - 1)$-renaming problem allows solving the k-set agreement problem despite $t = k$ failures, such an additional power is useless when $k < t$. In that sense, in an asynchronous system made up of atomic registers, $(n + k - 1)$-renaming allows progressing from $k > t$ to $k = t$, but does not allow bypassing that frontier. The last contribution of the paper is a wait-free algorithm that constructs an adaptive $(n + k - 1)$-renaming algorithm, for any value of the pair (t, k), from a failure detector of the class Ω_*^k (this last algorithm is a simple adaptation of an existing renaming algorithm).

1 Introduction

Asynchronous Computability. Renaming and set agreement are among the basic problems that lie at the core of computability in asynchronous systems prone to process crashes. The *renaming* problem (introduced in [3]) consists in designing an algorithm that allows processes (that do not crash) to obtain new names from a new name space that is as small as possible. In the following M denotes the size of the new name space, and a corresponding algorithm is called an M-renaming algorithm.

G. Prencipe and S. Zaks (Eds.): SIROCCO 2007, LNCS 4474, pp. 66–80, 2007.
© Springer-Verlag Berlin Heidelberg 2007

A *wait-free* algorithm is an algorithm that allows each process that does not crash to terminate in a finite number of computation steps, whatever the behavior of the other processes (i.e., despite the fact that all the other processes are extremely slow, or even have crashed) [12]. It has been shown that, in a system of n processes that can communicate through atomic read/write registers only, the smallest new name space that a wait-free renaming algorithm can produce is lower bounded by $M = 2n - 1$ [15]. More generally, in an asynchronous system where up to t processes may crash, the smallest value of M is $n + t$ (the wait-free case corresponds to $t = n - 1$).

A renaming algorithm is *adaptive* if the size of the new name space depends only on the number of processes that ask for a new name (and not on the total number of processes). Let p be the number of processes that *participate* in the renaming. Several adaptive algorithms have been designed such that the size of the new name space is $M = 2p - 1$ (e.g., [2,5]). These adaptive algorithms are consequently optimal with respect to the size of the new name space.

Recently, with the aim of circumventing the $M = 2p - 1$ lower bound, researchers have investigated the use of base objects stronger than atomic registers in order to solve the renaming problem. Following this line of research, it has been shown in [19] that, as soon as k-test&set objects can be used, the renaming problem can be wait-free solved with a new name space the size of which is $M = 2p - \lceil \frac{p}{k} \rceil$ [1]. Among the processes that access it, a k-test&set object ensures that at least one and at most k processes obtain the value 1 (they win), while all the other processes obtain the value 0 (they lose)[2]. It has also been shown in [10] that the renaming problem can be wait-free solved with a new name space of size $M = p + k - 1$ as soon as k-set agreement objects can be used. According to the base objects they use, respectively, both algorithms are optimal with respect to the size of their new name space.

The k-set agreement problem (sometimes abbreviated k-set), has been introduced in [8]. It is a paradigm of coordination problems encountered in distributed computing and is defined as follows. Each process is assumed to propose a value. The problem consists in designing an algorithm such that (1) each process that does not crash decides a value (termination), (2) a decided value is a proposed value (validity), and (3) no more than k different values are decided (agreement). (The well-known consensus problem is nothing else than the 1-set agreement problem.) The parameter k can be seen as the coordination degree (or the difficulty) associated with the corresponding instance of the problem. The smaller k is, the more coordination among the processes is imposed: $k = 1$ means the strongest possible coordination, while $k = n$ means no coordination.

It has been shown in [6,15,22] that, in an asynchronous system made up of processes that communicate through atomic registers only, and where up to t processes may crash, there is no wait-free k-set agreement algorithm for $k \leq t$.

---

[1] The renaming algorithm presented in [19] is actually based on k-set agreement objects. But, as observed by E. Gafni, these objects can be trivially replaced by k-test&set objects without affecting the behavior of the renaming algorithm.

[2] The usual test&set object is a 1-test&set object.

Differently, when $k > t$ the problem can be trivially solved (a predefined set of k processes write their proposal, and a process decides the first proposal it reads).

Randomized or failure detector-based algorithms have been proposed to circumvent the previous impossibility result [13,17,18]. An algorithm that wait-free solves the $(n-1)$-set agreement in a system of n crash-prone asynchronous processes from $(2n-2)$-renaming objects is described in [9].

Content of the Paper. The paper has three contributions. The first is motivated by the computability power of the renaming problem with respect to the set agreement problem. More specifically, the paper considers systems made up of n processes. In such a system, an algorithm is *t-resilient* if it always preserves its safety and liveness properties when no more than t processes commit failures. (The notion of t-resilience boils down to the wait-free notion when $t = n-1$.) The first contribution investigates the t-resilience notion to solve the k-set agreement problem from renaming objects. It presents a t-resilient algorithm that solves the k-set problem from an adaptive $(n+k-1)$-renaming object when $k = t$. Interestingly, this result generalizes a previous result presented in [9] that also considers $k = t$, but only for the wait-free case (i.e., $t = n-1$). So, the algorithm presented in the paper works for any value of t. When we consider the constructions relating renaming and set agreement that are known, we obtain the transformations described in Figure 1. Interestingly, it follows from the proposed algorithm (that considers $k = t$) that, in asynchronous shared memory systems prone to a single process crash ($t = 1$), a solution to the renaming problem allows solving the consensus problem (and vice-versa).

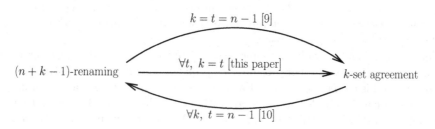

Fig. 1. Piecing together the transformations

The second contribution of the paper is a lower bound. While, in an asynchronous shared memory system made up of atomic registers only, the k-set agreement problem can be (trivially) solved when $k > t$, and is impossible to solve when $k \leq t$, the previous algorithm shows that enriching the system with an adaptive $(n+k-1)$-renaming algorithm allows progressing from $k > t$ to $k = t$. So, an important question is the following: does an $(n+k-1)$-renaming algorithm allows bypassing the $k = t$ frontier? The second contribution of the paper shows that such a renaming algorithm is not powerful enough to do it. More precisely, it shows that, in an asynchronous shared memory system made up of atomic registers and $(n+k-1)$-renaming, there are values of n for which it is not possible to solve the k-set agreement problem when $k < t$. (Showing that this is true for any value of n remains an open problem.)

The last contribution is a wait-free algorithm that builds a $(p+k-1)$-renaming object from an oracle of the class Ω_*^k. Such an oracle class has been introduced in [21]. It generalizes the "leader" oracles (failure detectors) classes introduced in [7,11,19,20]. Basically, such an oracle provides the processes with a primitive leader() that always returns a set of at most k processes, and after some unknown but finite time, returns always the same set that contains at least one correct participating process. Interestingly, that algorithm is a simple generalization of an $(n + t)$-renaming algorithm described in [4] (that is in turn an adaptation to the shared memory setting of an $(n + t)$-renaming algorithm designed for message-passing systems [3]).

Roadmap. The paper is made up of 5 sections. Section 2 describes the computation model. Section 3 presents a t-resilient algorithm that solves the k-set problem from a single $(n+k-1)$-renaming object. Section 4 shows that $(n+k-1)$-renaming does not allow solving the k-set agreement problem when $k < t$, for any value of n. Then, Section 5 presents a wait-free construction from Ω_*^k to an adaptive $(p + k - 1)$-renaming object.

2 Basic Computation Model

Process Model. The system is made up of n asynchronous processes $p_1, \ldots, p_n$. The integer i is the index of p_i while its identity is kept in id_i. Π denotes the set of indexes, i.e., $\Pi = \{1, \ldots, n\}$. *Asynchronous* means that there is no bound on the time it takes for a process to execute a computation step. A process may crash (halt prematurely). After it has crashed a process executes no step. A process executes correctly its algorithm until it possibly crashes. The integer t, $0 \leq t < n$, denotes an upper bound on the number of processes that may crash; t is known by the processes. A process that does not crash in a run is *correct* in that run; otherwise, it is *faulty* in that run.

Communication Model. The processes cooperate by accessing atomic read/write registers. *Atomic* means that each read or write operation appears as if it has been executed instantaneously at some time between its begin and end events [16]. Each atomic register is a one-writer/multi-readers (1WnR) register. This means that a single process (statically determined) can write it. Moreover such a register is a write-once register (the writing process writes it at most once). Atomic registers are denoted with uppercase letters. The atomic registers are structured into arrays. $X[1..n]$ being such an array, $X[i]$ denotes the register of that array that p_i only is allowed to write. A process can have local registers. Such registers are denoted with lowercase letters with the process index appearing as a subscript (e.g., $winner_i$ is a local register of p_i).

The processes are provided with an atomic snapshot operation [1] denoted snapshot(X), where $X[1..n]$ is an array of atomic registers. It allows a process p_i to atomically read the whole array. This means that the execution of a snapshot()

operation appears as it has been executed instantaneously at some point in time between its begin and end events. Such an operation can be built from 1WnR atomic registers [1].

The value $\bot$ denotes a default value that can appear only in the algorithms described in the paper. It always remains everywhere else unknown to the processes.

Notions of t-resilience and Wait-freeness. An algorithm is t-resilient if it copes with up to t process failures. In our context, this means that it satisfies its safety and liveness (termination) properties despite up to t process crashes. A wait-free algorithm is an $(n-1)$-resilient algorithm.

Notion of Adaptive Renaming. In the renaming problem, each process p_i has an initial name denoted id_i (that it is the only to know). These names are from a very large name space, i.e., $\max(id_1, \dots, id_n) >> n$. A renaming algorithm is adaptive with respect to the size of its new name space, if that size depends on the number of processes that actually participate in the renaming algorithm. A process participates in an algorithm as soon as it has written an atomic register used by that algorithm. Let us remark that an adaptive renaming algorithm cannot systematically assign the new name i to p_i. This is because, if only p_n wants to acquire a new name, the new name space is $[1..n]$, which depends on the number of processes instead of depending on the number of participating processes (here a single process). To rule out this type of ineffective solution, the following symmetry requirement is usually considered for the renaming problem [4]: the code executed by p_i with name id is the same as the code executed by process p_j with name id. This means that the process indexes can be used only for addressing purposes.

As indicated in the introduction, if p processes participate in a renaming algorithm based on atomic registers only, the best that can be done is an adaptive name space of size $M = 2p - 1$. This means that if "today" p' processes acquire new names, their new names belong to the interval $[1..2p' - 1]$. If "tomorrow" p'' additional processes acquire new names, these processes will have their new names in the interval $[1..2p - 1]$ where $p = p' + p''$.

3   From Adaptive $(p + k - 1)$-Renaming to k-Set Agreement

Considering an asynchronous system made up of n processes, where up to t $(1 \leq t < n)$ may crash and where the processes can cooperate through 1WnR write-once atomic registers, plus an adaptive $(p + t - 1)$-renaming object (where $p \leq n$ is the number of participating processes), this section presents and proves correct an algorithm that builds a t-set agreement object.

3.1   Principles and Description of the t-Resilient Algorithm

The principle of the transformation algorithm rests on two simple ideas.

1. First, use the underlying adaptive renaming object to partition the participating processes into two groups: the processes the name of which is smaller
   or equal to t (the winners); and the processes the name of which is greater
   than t (the losers). So, there are at most t winners.
2. Then, direct a process p_i to decide a value proposed by a winner. If p_i does
   not see winner processes, direct it to decide the value proposed by a process
   that has proposed a value but not yet obtained a new name.

To make operational these ideas, the shared memory is composed of two arrays
of 1WnR write-once atomic registers.

- The array $PROP[1..n]$, initialized to $[\bot, \ldots, \bot]$, is such that $PROP[i]$ will
  contain the value (denoted v_i) proposed by p_i to the set agreement problem.
  A process p_i becomes *participating* as soon as $PROP[i] \neq \bot$.
- The aim of the array $RENAMED[1..n]$, also initialized to $[\bot, \ldots, \bot]$, is to
  allow the processes to benefit from the renaming object. When a process p_i
  has obtained a new name, $RENAMED[i]$ is set to 1 if its new name is smaller
  or equal to t (p_i is then a winner), while $RENAMED[i]$ is set to 0 if p_i is a
  loser. It trivially follows that $RENAMED[i] \neq \bot$ means that p_i has acquired
  a new name.

The behavior of a process p_i is described in Figure 2. A process p_i invokes
kset_propose$_t(v_i)$ where v_i is the value it proposes to the k-set agreement problem.
It decides a value when it executes the return(v) statement (line 09) where v is
the value it decides. The way it implements the previous design ideas can be
decomposed in two stages.

1. The first stage is composed of the lines 01-04. After it has deposited its proposal (line 01), obtained a new name (line 02), and updated $RENAMED[i]$
   accordingly (line 03), a process p_i atomically reads the array $RENAMED$
   (using the snapshot() operation) until it sees that at least $n - t$ processes
   have acquired new names (line 04).
2. The second stage, composed of the lines 05-09, is the decision stage. It p_i
   sees a winner, it decides the value proposed by that winner process (lines 05,
   06 and 09). If p_i sees no winner, it decides the value proposed by a process
   that (from its point of view) has not yet obtained a new name. The proof
   will show that this is a consistent rule for deciding a value.

3.2 Proof of the Algorithm

The proof considers that (1) $k = t$, i.e., the size of the new name space of the
underlying adaptive renaming is $M = p + t - 1$ when p processes participate, and
(2) at least $(n - t)$ correct processes participate in the k-set agreement problem.

Lemma 1. *The number of values that are decided is at most t, and a decided
value is a proposed value.*

```
operation kset_propose(v_i):
(1)    PROP[i] ← v_i;
(2)    new_name_i ← rename(id_i);
(3)    RENAMED[i] ← 1 if new_name_i ≤ t, 0 otherwise;
(4)    repeat renamed_i ← snapshot(RENAMED)
             until |{j : renamed_i[j] ≠ ⊥}| ≥ (n − t);
(5)    let winners_i = {j : renamed_i[j] = 1};
(6)    if winners_i ≠ ∅ then ℓ_i ← any value ∈ winners_i
(7)           else let set_i = {j : PROP[j] ≠ ⊥ ∧ renamed_i[j] = ⊥};
(8)                ℓ_i ← any value ∈ set_i
(9)    end if;
(10)   return(PROP[ℓ_i])
```

Fig. 2. From $(n + k − 1)$-renaming to k-set, for $k = t$, $\forall t$ (code for p_i)

Proof. Let RENAMED$_i$ be the last value of $renamed_i$ when p_i exits the **repeat** loop at line 04. As a process p_x writes $RENAMED[x]$ at most once, we have RENAMED$_i[x] \neq \bot \wedge$ RENAMED$_j[x] \neq \bot \Rightarrow$ RENAMED$_i[x]=$RENAMED$_j[x]$. Let us define RENAMED$_i \leq$ RENAMED$_j$ as $\forall x :$ RENAMED$_i[x] \neq \bot \Rightarrow$ RENAMED$_i[x] =$ RENAMED$_j[x]$. Due to the atomicity property of the snapshot() operation (line 04) we have $\forall i, j:$ RENAMED$_i \leq$ RENAMED$_j \vee$ RENAMED$_j \leq$ RENAMED$_i$ (this is sometimes called the *containment* property provided by the snapshot() operation).

If no process ever executes line 05, the agreement and validity property are trivially satisfied. So, let us assume that at least one process executes line 05. Moreover, let RENAMED be the smallest array value obtained by a process when it exits the **repeat** loop at line 04. We consider two cases.

– $\exists x:$ RENAMED$[x] = 1$.
  In that case there is at least one winner, namely, p_x. Due to the containment property, RENAMED$_i[x] = 1$ for any process p_i that decides. It follows from that observation and the lines 05-06 that any process that decides, does decide the value proposed by a winner process. As at most t processes can obtain a new name comprised between 1 and t (lines 02-03), it follows that there are at most t winners. Consequently, no more than t different values can be decided.
– $\forall x:$ RENAMED$[x] \neq 1$.
  In that case, let $R = \{x :$ RENAMED$[x] = 0\}$ (hence, all other entries of RENAMED are equal to $\bot$). Due to the exit condition of the **repeat** loop (line 04), we have $|R| \geq n - t$, from which it follows that $|\Pi \setminus R| \leq t$. We claim (claim $C1$) that any process p_i that decides, decides a value proposed by a process p_y such that $y \in \Pi \setminus R$. Combining this claim with $|\Pi \setminus R| \leq t$, we conclude that at most t different values can be decided.
  Proof of the claim $C1$. Let p_i be a process that decides. It decides the value in $PROP[y]$ where y has been determined at line 06 or line 08.
    • p_i selects y at line 06. In that case, p_i decides the value proposed by a process p_y such that RENAMED$_i[y] = 1$. As RENAMED $\leq$ RENAMED$_i$

(snapshot containment property), and RENAMED does not contain the value 1, we conclude that $y \notin R$, and the claim $C1$ follows.

- p_i selects y at line 08. In that case, p_i decides a value proposed by a process p_y such that $\text{RENAMED}_i[y] = \perp$. We claim (claim $C2$) that $set_i \neq \emptyset$, i.e., p_y does exist. As $\text{RENAMED}_i[y] = \perp$ and $\text{RENAMED} \leq \text{RENAMED}_i$, we conclude from the definition of R that $y \notin R$, which proves the claim $C1$.

Proof of the claim $C2$ ($set_i \neq \emptyset$). Let p_i be a process that executes line 07. That process is such that $\forall x \in \Pi$: $\text{RENAMED}_i[x] = \perp$ or 0. Let $R_i = \{x : \text{RENAMED}_i[x] = 0\}$, and $\alpha = |R_i|$. Moreover, let $r = |\{x : PROP[x] \neq \perp\}|$ where the value of $PROP[x]$ is the value read by p_i at line 07. (See Figure 3, where the time instants are such that $\tau_0 < \tau_2 < \tau_3 < \tau_4$). We show that $\alpha < r$, from which the claim follows (namely, there is a process p_y such that $PROP[y] \neq \perp \wedge \text{RENAMED}_i[y] = \perp$ when p_i executes line 07).

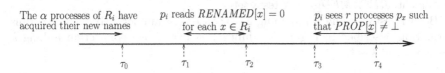

The α processes of R_i have acquired their new names

p_i reads $RENAMED[x] = 0$ for each $x \in R_i$

p_i sees r processes p_x such that $PROP[x] \neq \perp$

$\tau_0 \qquad \tau_1 \qquad \tau_2 \qquad \tau_3 \qquad \tau_4$

Fig. 3. Timing scenario

1. Let us first consider the processes p_x of the set R_i (i.e., the processes p_x such that $\text{RENAMED}_i[x] = 0$). These processes have obtained new names in a name space $[1..M]$ before time τ_0. We can conclude from the text of the algorithm that the new name obtained by each of these processes p_x (a loser) is such that $new\_name_x > t$ (lines 02 and 03). As there are α such processes we have $t + \alpha \leq M$.

2. Let ρ be the number of processes that started participating in the renaming before τ_0. We have seen (item 1) that M is the greatest name obtained by a process of R_i and that name has been obtained before τ_0. As the algorithm is adaptive, we have $M \leq \rho + t - 1$.

3. As the ρ processes started participating in the renaming before τ_0, they updated their entry in $PROP$ to a non-$\perp$ value before τ_0, and consequently we have $\rho \leq r$.

4. It follows from the previous items that $t + \alpha \leq M \leq \rho + t - 1 \leq r + t - 1$, from which we conclude $\alpha < r$, that terminates the proof of the claim $C2$.
$$\square_{Lemma\ 1}$$

Lemma 2. *Each correct process decides a value.*

Proof. As there are at least $n - t$ correct process that participate in the set agreement problem, no process can block forever at line 04. Moreover, as the set set_i of a process p_i that executes line 07 is not empty (see the claim $C2$ in the proof of Lemma 1), the entry ℓ_i from which p_i decides is well-defined (it does exist). It follows that each correct process decides. $\square_{Lemma\ 2}$

Theorem 1. *The algorithm described in Figure 2 is a t-resilient t-set agreement algorithm.*

Proof. The proof follows directly from Lemma 1 and Lemma 2. $\square_{Theorem\ 1}$

3.3   From k-Test&Set to k-Set

In the k-test&set problem, the processes invoke an operation k_test&set() and obtains the value 1 (winner), or the value 0 (loser). The values returned to the processes satisfy the following property: there are at least one and at most k winners.

In a very interesting way, the algorithm described in Figure 2 allows solving the k-set problem from any solution to the k-test&set problem, when $k = t$, $\forall t$. The only "modification" consists in replacing the lines 02-03 by the following statement: $RENAMED[i] \leftarrow$ k_test&set().

Both 1-test&set and n-renaming have consensus number 2 [10,19]. The transformation described in Figure 2 exhibits another strong connection linking k-test&set and k-set.

4   An Impossibility Result

Theorem 2. *The k-set agreement problem cannot be solved in asynchronous systems made up of atomic registers and a solution to the adaptive $(n + k - 1)$-renaming problem, for any value of n, $k < t$ and $t = n - 1$.*

Proof. The proof uses the following notations:

- f_k: the function $p \rightarrow 2p - \lceil \frac{p}{k} \rceil$.
- g_k: the function $p \rightarrow p + k - 1$.
- (n, k)-TS: the k-tes&set problem with up to n processes. (At least one and most k processes are winners.)
- (n, k)-SA: the k-set agreement problem with up to n processes.
- (n, f_k)-AR: the adaptive M-renaming problem with $M = f_k(p)$ (where $p \leq n$ is the number of processes that participate in the renaming).
- (n, g_k)-AR: the adaptive M-renaming problem with $M = g_k(p)$ (where $p \leq n$ is the number of processes that participate in the renaming).
- Any solution to the (n, ℓ)-XX problem (where XX is TS, SA, or AR, and ℓ is k, f_k or g_k) defines a corresponding (n, ℓ)-XX object.

Let us first observe that $\forall p$, $\forall k$, we have $f_1(p) = g_1(p) \leq g_k(p)$. This means that any solution to (n, f_1)-AR is a solution to (n, g_k)-AR.

The proof consists in showing the following: $\forall k$, $\forall n \geq 2k + 1$: there is no algorithm that solves (n, k)-SA from (n, g_k)-AR. The proof is by contradiction. Let us assume that there is an algorithm $\mathcal{A}$ that, for $t = n - 1$, solves (n, k)-SA from (n, g_k)-AR with $n \geq 2k + 1$. The $(2, 1)$-SA problem plays a key role in proving the contradiction.

1. On one side.
   - The $(2,1)$-TS problem and the $(2,1)$-SA problem are equivalent [9].
   - There is a wait-free construction of (n,k)-TS from $(2,1)$-TS objects [9].
   - The (n,f_1)-AR problem can be wait-free solved from $(n,1)$-TS objects [19].
   - For any $k \geq 1$, the (n,g_k)-AR problem can be wait-free solved from (n,f_1)-AR objects (previous observation).
   - Due to the assumption, the algorithm $\mathcal{A}$ solves the (n,k)-SA problem from (n,g_k)-AR objects with $n \geq 2k+1$, when $t = n-1$.
   - It follows that, when $t = n-1$, it is possible to solve the (n,k)-SA problem from $(2,1)$-SA objects for $n \geq 2k+1$.
2. On the other side.
   - It is shown in [14] that $k \geq j\lfloor\frac{t+1}{m}\rfloor + \min\left(j, (t+1) \bmod m\right)$ is a necessary requirement for having a t-resilient k-set agreement algorithm for n processes, when these processes share atomic registers and (m,j)-SA objects (objects that allow solving j-set agreement among m processes).
   - Let us consider the case where the (m,j)-SA objects are $(2,1)$-SA objects. Let us recall $t = n-1$. We have then: $k \geq \lfloor\frac{t+1}{2}\rfloor + \min\left(1, (t+1) \bmod 2\right)$, from which we obtain the necessary requirement $k \geq \lfloor\frac{n}{2}\rfloor$.
   - It follows that, for $t = n-1$, $k \geq \lfloor\frac{n}{2}\rfloor$ (i.e., $2k \geq n$) is a necessary requirement for solving the (n,k)-SA problem from $(2,1)$-SA objects and atomic registers.
3. The previous items 1 and 2 contradict each other. It follows that the initial assumption $\mathcal{A}$ cannot hold, which proves the theorem. $\square_{Theorem\ 2}$

5   From Ω_*^k to $(p+k-1)$-Renaming

This section enriches the picture by proposing a wait-free algorithm that solves the adaptive M-renaming problem with $M = \min(2p-1, p+k-1)$, p being the number of processes that participate in the algorithm. In addition to 1WnR atomic registers, this algorithm uses an oracle of the class Ω_*^k. Interestingly, when all the correct processes participate and the oracle has no additional power (i.e., $k \geq t+1$), this algorithm boils down to a t-resilient algorithm described in [4] that solves the $(n+t)$-renaming problem.

5.1   The Class of Oracles Ω_*^k

This class has been defined in [21]. An oracle of the class Ω_*^k provides the processes with an operation denoted leader(). (As indicated in the introduction, this definition is based on the leader oracle classes introduced in [11,19,20].) When a process p_i invokes that operation, it provides it with an input parameter, namely a set X of processes, and obtains a set of process identities as a result[3].

---

[3] The definition of Ω_*^k is not expressed in the framework introduced by Chandra and Toueg to define failure detector classes. More precisely, in their framework, the failure detector operation that a process can issue has no input parameter. It would be possible to express Ω_*^k in their framework. We don't do it in order to keep the presentation simpler.

The semantics of Ω_*^k is based on a notion of time, whose domain is the set of integers. It is important to notice that this notion of time is not accessible to the processes. An invocation of leader(X) by a process p_i is *meaningful* if $i \in X$. If $i \notin X$, it is *meaningless*. The primitive leader() is defined by the following properties where L_X denotes the set of processes returned by an invocation leader(X).

- Termination (wait-free). Any invocation of leader() by a correct process always terminates (whatever the behavior of the other processes).
- Bounded size leadership. Whatever X, the set L_X returned by a leader(X) invocation is such that $|L_X| \le k$.
- Triviality. A meaningless invocation can return any set (of size k) of processes.
- Eventual multi-leadership for each input set X: For any $X \subseteq \Pi$, such that $X \cap \textit{Correct} \neq \emptyset$, there is a time τ_X such that, $\forall \tau \ge \tau_X$, all the meaningful leader(X) invocations (that terminate) return the same set L_X and this set is such that $L_X \cap X \cap \textit{Correct} \neq \emptyset$.

The intuition that underlies this definition is the following. The set X passed as input parameter by the invoking process p_i is the set of all the processes that p_i considers as being currently *participating* in the computation. (This also motivates the notion of meaningful and meaningless invocations: an invoking process is trivially participating).

Given a set X of participating processes that invoke leader(X), the eventual multi-leadership property states that there is a time after which these processes obtain the same set L_X of at most k leaders, and at least one of them is a correct process of X. Let us observe that the (at most $k - 1$) other processes of L_X can be any subset of processes (correct or not, participating or not).

It is important to notice that the time τ_X from which this property occurs is not known by the processes. Moreover, before that time, there is an anarchy period during which each process, as far as its leader(X) invocations are concerned, can obtain different sets of any number of leaders. Let us also observe that if a process p_i issues two meaningful invocations leader($X1$) and leader($X2$) with $X1 \neq X2$, there is no relation linking L_{X1} and L_{X2}, whatever the values of $X1$ and $X2$ (e.g., the fact that $X1 \subset X2$ imposes no particular constraint on L_{X1} and L_{X2}).

Let us consider an execution in which all the invocations leader(X) are such that $X = \Pi$ (the whole set of processes are always considered as participating). In that case, Ω_*^k boils down to the failure detector class denoted Ω^k introduced in [20]. If additionally, $k = 1$, we obtain the classical leader failure detector Ω introduced in [7].

When $X \subseteq \Pi$ and $k = 1$, Ω_*^k boils down to the failure detector class introduced in [11]. It is shown in [11] that Ω is weaker than Ω_*^1 that in turn is weaker than $\Diamond \mathcal{P}$ (the class of eventually perfect failure detectors: after some finite but unknown time, an eventually perfect failure detector suspects all the crashed processes and only them).

5.2 An Adaptive $\min(2p-1, p+k-1)$-Renaming Algorithm

As previously mentioned, the adaptive renaming algorithm that is now presented is inspired from a t-resilient renaming algorithm designed for read/write registers only, described in [4].

Atomic Registers. The algorithm uses an array of 1WnR atomic registers, denoted $STATE[1..n]$. Each register $STATE[i]$ contains three fields. The first field, denoted $STATE[i].old$, is for the initial name of p_i. The second field, denoted $STATE[i].prop$, is for the new name that p_i is currently trying to acquire. Finally, the third field, denoted $STATE[i].done$, is set to true once p_i has obtained a new name ($STATE[i].prop$ contains then the new name of p_i). Initially, each atomic register $STATE[i]$ is initialized to $< \bot, \bot, false >$.

Process Behavior. A process starts the renaming algorithm by setting a local flag denoted $done_i$ to *false*, and its current proposal for a new name to $\bot$ (line 01). Then, it enters a **repeat** loop and leaves it only when it has acquired a new name (line 15).

In the loop body, a process p_i first writes its current state in $STATE[i]$ to inform the other processes about its current progress, and then atomically reads $STATE$ (using the snapshot() operation) to obtain a consistent view of the global state. If it has not yet determined a name proposal or there is another process that has chosen the same name proposal (line 05), p_i enters the lines 06-11 to determine another name proposal. Differently, if its current name proposal is not proposed by another process (the test of line 05 is then negative), p_i commits its last proposal that becomes its new name (line 12), informs the other processes (line 13), and decides that new name (line 15).

To determine a name proposal, a process p_i proceeds as follows. It first determines the processes that are competing to have a new name. Those are the processes p_j that, from p_i's point of view, are participating in the renaming (namely, the processes p_j such that $state_i[j].old \neq \bot$) and have not yet obtained a new name (i.e., such that $\neg(state_i[j].done)$). Before starting the next execution of the loop body, some processes have to change their new name proposal (otherwise, it could be possible that they loop forever). So, a process p_i does the following.

- According to the set of processes perceived as competing with it, p_i computes a current set of leaders (line 07).
- If it does not appear in the set of leaders, p_i starts directly another execution of the loop body. Let us notice that, in that case, p_i's new name proposal is not modified.
- Differently, if it appears in the set of leaders (line 08), p_i determines a new name proposal before starting another execution of the loop body. This determination (done exactly as in [4]) consists for p_i in first computing its rank within the leader set, and then taking as its new name proposal the first integer not yet used by the other processes (lines 09-10).

5.3   Proof of the Algorithm

Lemma 3. *Let p be the number of processes that participate in the renaming. The size of the new name space is $M = \min(2p - 1, p + k - 1)$.*

Proof. Let us consider a run in which p processes participate. Let p_i be a process that returns a new name (line 15). The new name obtained by p_i is the last name it has proposed (at line 10 during the previous iteration). When p_i defined its last name proposal, at most $p - 1$ other processes have previously defined a name proposal, i.e., $|\{j : (j \neq i) \wedge (state_i[j].prop \neq \perp)\}| \leq p - 1$ (O1). Moreover, due to the definition of Ω_*^k, when it defines its last name proposal, the rank of p_i in $leaders_i$ is at most $\min(p, k)$ (O2). It follows from (O1) and (O2) that the last name proposal computed by p_i is upper bounded by $(p - 1) + \min(p, k)$, i.e., $M = \min(2p - 1, p - 1 + k)$. □$_{Lemma\ 3}$

Lemma 4. *No two processes decide the same new name.*

Proof. [Preliminary Remark. This proof is verbatim the same as the corresponding proof in [4]. We give it only for completeness purpose. As noticed in [4], this follows from the fact that this proof does not depend on the way the new names are chosen. It is based only on the structure of the algorithm and the containment property of the the snapshot() operation.]

The proof is by contradiction. Let us assume that p_i and p_j obtain the same new name a. Let STATE$_i$ (resp., STATE$_j$) be the last snapshot value obtained by p_i (resp., p_j) before returning its new name a. Due to the sequence of the lines 10, 02 and 04 executed by p_i (resp., p_j) before deciding its new name, we have STATE$_i[i].prop = a$ (resp., STATE$_j[j] = a$). Moreover, after having written its last new name proposal, a process does not change its entry of $STATE.prop$.

Due to the containment property of the snapshot($STATE$) operation, we have STATE$_i \leq$STATE$_j$ or STATE$_j \leq$STATE$_i$. Let us assume without loss of generality that STATE$_i \leq$STATE$_j$. It follows from the containment property that STATE$_j[i].prop=$STATE$_i[i].prop = a$. According to the test of line 05, p_j proceeds to lines 06-11 to select a new name proposal distinct from STATE$_j[i].prop = a$, which proves the lemma. □$_{Lemma\ 4}$

Lemma 5. *Each correct process that participates obtains a new name.*

Proof. As in [4], the proof is by contradiction. Let us assume that a process takes infinitely many steps without obtaining a new name. Let $CORRECT$ be the set of correct processes, and NT the subset of correct processes that do not terminate. Let τ be a time such that:

1. Each (correct or not) participating process p_j has written its initial name id_j in $STATE[j].old$ before $\tau_1 < \tau$.
2. Each (correct or not) process p_j that decides, has set $STATE[j].done$ to $true$ before $\tau_2 < \tau$.

3. Each process $p_i \in NT$ has taken at least one snapshot of $STATE$ between $\max(\tau_1, \tau_2)$ and τ.

Due to the containment property provided by the snapshot() primitive, it follows that, after τ, each process $p_i \in NT$ sees the same set of participating processes and the same set of processes that have decided.

4. Let $\tau_3 < \tau$ be the time from which the multi-leadership property of Ω^k_* remains forever satisfied.

Let $contending_x[\tau']$ be the value, at time τ', of the set $\{j : (state_x[j].old \neq \perp) \wedge \neg(state_x[j].done)\}$. Let p_i be a process of NT, and $CTD = contending_i[\tau]$. Let us observe that, at any time $\tau' \geq \tau$, and for each process $p_j \in NT$, we have $contending_j[\tau'] = CTD$. Moreover, $NT \subseteq CTD$. It follows from the properties of Ω^k_*, that there is a set $leaders$ such that, after τ, each time a process $p_j \in NT$ invokes leader(CTD), it obtains $leaders$. Since $CTD \setminus NT$ contains only faulty processes, and (due to the definition of τ) $leaders \cap CTD \cap CORRECT \neq \emptyset$, the set $leaders \cap CTD$ is not empty and contains at least one correct process.

As $|leaders| \leq k$, all the correct processes in $leaders \cap CTD$ select a new name proposal when they execute the lines 09-11, and these new name proposals are all different (this follows from the fact that they select their rank from the same set $leaders$). It follows that they decide their new name. A contradiction with the assumption that the processes of NT do not terminate. $\square_{Lemma\ 5}$

```
operation rename(id_i):
(1)   prop_name_i ← ⊥; done_i ← false;
(2)   repeat
(3)     STATE[i] ←< id_i, prop_name_i, done_i >;
(4)     state_i ← snapshot(STATE);
(5)     if (prop_name_i = ⊥) ∨ (∃j : (j ≠ i) ∧ (state_i[j].prop = prop_name_i))
(6)       then contending_i ← {j : (state_i[j].old ≠ ⊥) ∧ ¬(state_i[j].done)};
(7)            leaders_i ← leader(contending_i);
(8)            if id_i ∈ leaders_i then
(9)              let r_i = rank of id_i in leaders_i;
(10)             prop_name_i ← r_i-th integer ∉ X where
(11)                X = {state_i[j].prop : (j ≠ i) ∧ (state_i[j].prop ≠ ⊥)} end if
(12)       else new_name_i ← prop_name_i; done_i ← true;
(13)            STATE[i] ←< id_i, prop_name_i, done_i > end if
(14)  until done_i;
(15)  return(prop_name_i)
```

Fig. 4. From Ω^k_* to adaptive M-renaming with $M = \min(2p - 1, p + k - 1)$ (p_i's code)

Theorem 3. *The algorithm described in Figure 4 is an adaptive wait-free M-renaming algorithm with $M = \min(2p - 1, p + k - 1)$.*

Proof. The theorem follows from Lemma 3, Lemma 4 and Lemma 5. $\square_{Theorem\ 3}$

References

1. Afek, Y., Attiya, H., Dolev, D., Gafni, E., Merritt, M., Shavit, N.: Atomic Snapshots of Shared Memory. Journal of the ACM 40(4), 873–890 (1993)
2. Afek, Y., Merritt, M.: Fast, Wait-Free $(2k-1)$-Renaming. 18th ACM Symposium on Principles of Distributed Computing (PODC'99), pp. 105–112 (1999)
3. Attiya, H., Bar-Noy, A., Dolev, D., Peleg, D., Reischuk, R.: Renaming in an Asynchronous Environment. Journal of the ACM 37(3), 524–548 (1990)
4. Attiya, H., Welch, J.P.: Distributed Computing: Fundamentals, Simulations and Advanced Topics, 2nd edn. p. 414. Wiley-Interscience, New York (2004)
5. Borowsky, E., Gafni, E.: Immediate Atomic Snapshots and Fast Renaming. 12th ACM Symp on Principles of Distributed Computing (PODC'93), pp. 41–51 (1993)
6. Borowsky, E., Gafni, E.: Generalized FLP Impossibility Results for t-Resilient Asynchronous Computations. 25th ACM Symposium on Theory of Distributed Computing (STOC'93), pp. 91–100 (1993)
7. Chandra, T., Hadzilacos, V., Toueg, S.: The Weakest Failure Detector for Solving Consensus. Journal of the ACM 43(4), 685–722 (1996)
8. Chaudhuri, S.: More Choices Allow More Faults: Set Consensus Problems in Totally Asynchronous Systems. Information and Computation 105, 132–158 (1993)
9. Gafni, E.: Read/Write Reductions. DISC/GODEL presentation given as introduction to the 18th Int'l Symposium on Distributed Computing (DISC'04) (2004)
10. Gafni, E.: Renaming with k-set Consensus: an Optimal Algorithm in $n+k-1$ Slots. In: Shvartsman, A.A. (ed.) OPODIS 2006. LNCS, vol. 4305, pp. 36–44. Springer, Heidelberg (2006)
11. Guerraoui, R., Kapałka, M., Kouznetsov, P.: The Weakest Failure Detectors to Boost Obstruction-Freedom. In: Dolev, S. (ed.) DISC 2006. LNCS, vol. 4167, pp. 376–390. Springer, Heidelberg (2006)
12. Herlihy, M.P.: Wait-Free Synchronization. ACM Transactions on Programming Languages and Systems 13(1), 124–149 (1991)
13. Herlihy, M.P., Penso, L.D.: Tight Bounds for k-Set Agreement with Limited Scope Accuracy Failure Detectors. Distributed Computing 18(2), 157–166 (2005)
14. Herlihy, M.P., Rajsbaum, S.: Algebraic Spans. Mathematical Structures in Computer Science 10(4), 549–573 (2000)
15. Herlihy, M.P., Shavit, N.: The Topological Structure of Asynchronous Computability. Journal of the ACM 46(6), 858–923 (1999)
16. Herlihy, M.P., Wing, J.M.: Linearizability: a Correctness Condition for Concurrent Objects. ACM TOPLAS 12(3), 463–492 (1990)
17. Mostéfaoui, A., Raynal, M.: k-Set Agreement with Limited Accuracy Failure Detectors. 19th ACM Symp. on Principles of Distr. Comp. pp. 143–152 (2000)
18. Mostéfaoui, A., Raynal, M.: Randomized Set Agreement. 13th ACM Symposium on Parallel Algorithms and Architectures (SPAA'01), pp. 291–297 (2001)
19. Mostéfaoui, A., Raynal, M., Travers, C.: Exploring Gafni's reduction land: from Ω^k to wait-free adaptive $(2p\text{-}[p/k])$-renaming via k-set agreement. In: Dolev, S. (ed.) DISC 2006. LNCS, vol. 4167, Springer, Heidelberg (2006)
20. Neiger, G.: Failure Detectors and the Wait-free Hierarchy. In: Proc. 14th ACM Symposium on Principles of Distributed Computing (PODC'95), pp. 100–109 (1995)
21. Raynal, M., Travers, C.: In search of the holy grail: looking for the weakest failure detector for wait-free set agreement. In: Shvartsman, A.A. (ed.) OPODIS 2006. LNCS, vol. 4305, Springer, Heidelberg (2006)
22. Saks, M., Zaharoglou, F.: Wait-Free k-Set Agreement is Impossible: The Topology of Public Knowledge. SIAM Journal on Computing 29(5), 1449–1483 (2000)

A Self-stabilizing Algorithm for the Median Problem in Partial Rectangular Grids and Their Relatives*

Victor Chepoi, Tristan Fevat, Emmanuel Godard, and Yann Vaxès

LIF-Laboratoire d'Informatique Fondamentale de Marseille, UMR 6166,
Marseille, France
{chepoi,fevat,godard,vaxes}@lif.univ-mrs.fr

Abstract. Given a graph $G = (V, E)$, a vertex v of G is a *median vertex* if it minimizes the sum of the distances to all other vertices of G. The median problem consists in finding the set of all median vertices of G. In this note, we present a self-stabilizing algorithm for the median problem in partial rectangular grids. Our algorithm is based on the fact that partial rectangular grids can be isometrically embedded into the Cartesian product of two trees, to which we apply the algorithm proposed by Antonoiu, Srimani (1999) and Bruell, Ghosh, Karaata, Pemmaraju (1999) for computing the medians in trees. Then we extend our approach from partial rectangular grids to plane quadrangulations.

1 Introduction

Given a connected graph G one is sometimes interested in finding the vertices minimizing the total distance $\sum_u d(u, x)$ to the vertices u of G, where $d(u, x)$ is the distance between u and x. A vertex x minimizing this expression is called a *median* (vertex) of G. The median problem consists in finding the set of all median vertices. The median problem arises with one of the basic models in discrete facility location [29] and with majority consensus in classification and data analysis [6,8]. This is also a classical topic in graph theory [9,29]. Linear time algorithms for computing medians are known for several classes of graphs: among them are trees, planar quadrangulations and triangulations with degree-constraints, partial rectangular grids [16], and a few other classes of graphs. A distributed algorithm for computing medians in graphs is given in [27].

In distributed systems, the median is a suitable location for information exchange and communication. Indeed, to place a common resource at a median site minimizes the cost of sharing the resource with other sites. Note also that [23] shows that, given a tree-network, choosing a median and then routing all the information through it minimizes the number of messages sent during any execution of any distributed sorting algorithm. Moreover, partial rectangular grids

---

* The first and the fourth authors were partly supported by the ANR grant BLAN06-1-138894 (projet OPTICOMB). The second and the third authors were supported by the ACI grant "Jeunes Chercheurs"(TAGADA project).

G. Prencipe and S. Zaks (Eds.): SIROCCO 2007, LNCS 4474, pp. 81–95, 2007.
© Springer-Verlag Berlin Heidelberg 2007

and trees are among the most used topologies in the design of microprocessors and distributed architectures. It is therefore of important practical interest to solve the median problem in a distributed setting on such a topology.

A distributed system can be defined as a set of processors exchanging information between neighbors. The system state, called *global state* or *configuration*, is the union of all the local states. A processor has only a local knowledge of the system, this knowledge varying according to the system connectivity. It is often desirable to maintain the system in a certain set of states, *the legitimate states*. An algorithm running in a system is said to be *self-stabilizing* if any execution has a suffix in the set of the legitimate states [19]. Self-stabilization is very desirable and useful in distributed systems because it provides immunity to transient failures and can even make possible in some cases a dynamical and transparent modification of the system topology. Besides self-stabilizing algorithms are often elegant and simple. The self-stabilizing paradigm was introduced by Dijkstra in 1973 [18]. He gave three self-stabilizing mutual exclusion algorithms on rings, opening a field of research still extremely dynamic today in distributed calculus. Self-stabilizing algorithms have been conceived to answer problems of routing [17], synchronization [5,26], leader election [20], spanning tree construction [2,4], maximum flow [24], mutual exclusion [28], and some other problems.

Antonoiu, Srimani [3] and Bruell et al. [11] proposed a strikingly simple and nice self-stabilizing algorithm for computing the median set of a tree T. The state of each node is an integer s. At each step, the algorithm updates the s-value of the currently active vertex v: it sets $s(v) = 1$ if v is a leaf, otherwise it computes the sum of the s-values of all neighbors of v minus the largest s-value of a neighbor and then adds 1 (denoted by $1 + \sum(N_s^-(v))$). If the current s-value of v is different from $1 + \sum(N_s^-(v))$, then this value becomes the new s-value of v. The algorithm terminates when there are no more s-values to modify. Interestingly, this happens to be a "valid" global state as the medians of T are the vertices with maximum s-values. The authors of [3,11] establish that the algorithm stabilizes in a polynomial number of steps. See also [3,11,10] for self-stabilizing algorithms solving other facility location problems on tree-networks.

In this note, we propose self-stabilizing median computation algorithms for two classes of plane graphs: partial rectangular grid and even squaregraphs. Our algorithms are based on the fact that such graphs isometrically embed into the Cartesian product of two trees and that the median in the initial graph G can be derived from the medians in two spanning trees of G closely related to the two tree-factors. Using the sense of direction in the grid, the algorithm computes the tree-factors and apply the algorithm of [3,11] to compute the medians of both spanning trees. This computation is performed anonymously. For an even squaregraph G, the algorithm first computes a spanning tree of G using the self-stabilizing spanning tree algorithm of Afek, Kutten, and Yung [1]. This algorithm needs unique identities to be available for every node of the network. Then the algorithm "repairs" this spanning tree in order to produce the two tree-factors and their spanning trees relatives, to which the median computation algorithm of [3,11] is applied. The algorithms have a round complexity of

$O(n)$ and $O(n^2)$ respectively. By self-stabilization, our algorithms are resilient to transient failures. Concerning non-transient failures like the permanent crash of a link or a processor, if the resulting topology is still a partial grid or an even squaregraph, then the algorithm will dynamically adjust to the changes. If not, and the crash creates a hole in the partial grid, then it is of course likely that our algorithm will not find the correct medians or even will not converge, because the computed factors could then contain cycles.

The article is organized as follows. In the first part, we investigate the properties of partial grids, squaregraphs and their medians which are used in the algorithm. In the second part, we describe the algorithms for computing their medians and give a proof of the correctness as well as an upper bound on the time and round complexities. This is, to our knowledge, the first self-stabilizing algorithm for location problems on non-tree networks.

2 Partial Grids, Squaregraphs and Their Medians

2.1 Preliminaries

In a graph $G = (V, E)$, the *length* of a path from a vertex x to a vertex y is the number of edges in the path. The *distance* $d_G(x, y)$ (or $d(x, y)$ if G is obvious from the context) between x and y is the length of a shortest path connecting x and y. The interval $I(u, v)$ between two vertices u, v of G is the set $I(u, v) = \{x \in V : d(u, v) = d(u, x) + d(x, v)\}$. A subset $S \subseteq V$ is called *gated* [25] if for each $v \notin S$ there exists a unique vertex $v' \in S$ (the gate of v in S) such that $v' \in I(v, u)$ for every $u \in S$. For every edge uv of G, define $W(u, v) = \{x \in V : d(u, x) < d(v, x)\}$. Given two connected graphs $G = (V(G), E(G))$ and $H = (V(H), E(H))$, we say that G admits an *isometric embedding* into H if there exists a *mapping* $\alpha : V(G) \to V(H)$ such that $d_H(\alpha(x), \alpha(y)) = d_G(x, y)$ for all vertices $x, y \in V(G)$. The *Cartesian product* $H = H_1 \times H_2$ of two connected graphs H_1, H_2 is defined upon the Cartesian product of the vertex sets of the corresponding graphs (called *factors*), i.e., $V(H) = \{u = (u_1, u_2) : u_1 \in V(H_1), u_2 \in V(H_2)\}$. Two vertices $u = (u_1, u_2)$ and $v = (v_1, v_2)$ are adjacent in H if and only if the vectors u and v coincide except at one position i, in which we have two vertices u_i and v_i adjacent in H_i. The *distance* $d_H(x, y)$ between two vertices $x = (x_1, x_2)$ and $y = (y_1, y_2)$ of H is $d_{H_1}(x_1, y_1) + d_{H_2}(x_2, y_2)$.

2.2 Medians

In the following, we consider graphs with weighted vertices. A *weight function* is any mapping π from the vertex set to the positive real numbers. The total weighted distance of a vertex x in G is given by $M_\pi(x) = \sum_u \pi(u) d(u, x)$. A vertex x minimizing this expression is a *median* (vertex) of G with respect to π, and the set of all medians is the *median set* $\mathrm{Med}_\pi(G)$. For a subset of vertices $S \subseteq V$, denote by $\pi(S) = \sum_{s \in S} \pi(s)$ the weight of S. We continue with the following property of median functions:

Lemma 1. *For each edge uv in a graph G, $M_\pi(u) - M_\pi(v) = \pi(W(v,u)) - \pi(W(u,v))$.*

Goldman and Witzgall [25] established that if the weight of a gated set S of G is larger than one half of the total weight, then $\mathrm{Med}_\pi(G) \subseteq S$. In trees, in partial rectangular grids, in squaregraphs, and, more generally, in all median graphs, for each edge uv, the sets $W(u,v)$ and $W(v,u)$ are gated, and constitute a partition of G. Recall that G is a *median graph* if for each triplet u, v, w the intersection $I(u,v) \cap I(v,w) \cap I(w,u)$ consists of a single vertex. We can then infer that these graphs satisfy the following *majority rule* (which is a folklore for trees):

Lemma 2. *[6] If G is a median graph, then $u \in \mathrm{Med}_\pi(G)$ iff $\pi(W(u,v)) \geq \pi(W(v,u))$ for each neighbor v of u. If T is a tree, then $\pi(W(u,v)) = \pi(W(v,u))$ if and only if $\mathrm{Med}_\pi(T) = \{u, v\}$.*

2.3 Partial Grids and Squaregraphs

A *rectangular system* or a *partial rectangular grid* is the subgraph of the regular rectangular grid which is formed by the vertices and the edges of the grid lying either on a simple circuit of the grid (with possibly some vertices visited more than once) or inside the region bounded by this circuit. Every partial rectangular grid is a connected plane graph with inner faces of length four and inner vertices of degree four (the converse in general is not true). More generally, a *squaregraph* [15] is a plane graph with inner faces of length four and inner vertices of degree at least four. An *even squaregraph* is a squaregraph in which all inner vertices have even degrees (see Fig. 1). Squaregraphs constitute a particular subclass of median graphs. Median graphs arise in several areas of discrete mathematics, geometry, and theoretical computer science.

2.4 Isometric Embedding into Products of Two Trees

Now, we will describe the isometric embedding of partial rectangular grids and even squaregraphs $G = (V, E)$ into the Cartesian product of two trees. For this we will use the notations in [7,15]. For a squaregraph or a partial grid G denote by ∂G the bounding cycle of G.

First, let $G = (V, E)$ be a partial rectangular grid bounded by the cycle ∂G. Denote by E_1 the set of vertical edges of G and consider the graph $G_1 = (V, E_1)$. It is clear that the connected components of G_1 are paths of G with end-vertices on ∂G. Define the graph $T_1 = (V(T_1), E(T_1))$ whose vertices are the connected components of G_1 and two components P' and P'' are adjacent if and only if there exists an edge of G with one end in P' and another one in P''. In the same way we can define the set E_2 of horizontal edges, the graph G_2, and the tree $T_2 = (V(T_2), E(T_2))$. We obtain the following canonical embedding α of G into the Cartesian product $T_1 \times T_2$. For any vertex v of G, we set $\alpha_1(v)$ (resp. $\alpha_2(v)$) to be the connected component of v in G_1 (resp. G_2). The embedding is

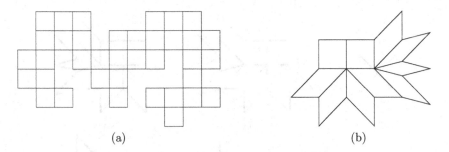

Fig. 1. A rectangular grid (a) and an even squaregraph (b)

defined by $\alpha(v) = (\alpha_1(v), \alpha_2(v))$. It can be verified that α provides an isometric embedding of G into $T_1 \times T_2$. For all vertices x, y of G, we have $d_G(x, y) = d_{T_1}(\alpha_1(x), \alpha_1(y)) + d_{T_2}(\alpha_2(x), \alpha_2(y))$. From now on, we will identify a vertex of G with the couple of vertical and horizontal paths to which it belongs. We call such a path P a *fiber*, as P is equal to the subgraph induced by $\alpha_1^{-1}(P)$.

This canonical embedding of partial grids can be generalized to all graphs isometrically embeddable into Cartesian products of two trees. It was established in [7] that a graph G can be embedded into the Cartesian product of two trees if and only if G is a cube-free median graph without odd bipartite wheels. In particular, from this characterization follows that even squaregraphs admit such embeddings. To derive the embedding, the edges of an even squaregraph $G = (V, E)$ are divided into two sets E_1 and E_2 subject to the constraint that two incident edges e_1 and e_2 of a common inner face of G belong to different edge-sets. Equivalently, if we define the *side-graph* of G as the graph having the edges of G as the vertex-set and two edges e_1, e_2 of G are adjacent in the side-graph if and only if e_1 and e_2 are incident sides of some inner face of G, then the side-graph is bipartite. Note that the bipartition $\{E_1, E_2\}$ of E satisfies the following two conditions: (i) all "parallel" edges of G, i.e., edges which belong to the same equivalence class of the transitive closure of the binary relation "to be opposite edges of a common inner face of G" all belong to the same color-class, and (ii) if we consider all edges incident to an inner vertex v of G, and number them counterclockwise, starting with an arbitrary edge, then the edges having numbers of the same parity all belong to the same color-class E_1 or E_2. Note also that the set of all edges parallel to a given edge e (i.e. all edges that belong to the equivalence class of e) constitutes a cut-set of the graph G.

Analogously to the case of partial grids, the connected components of the graphs $G_1 = (V, E_1)$ (resp. $G_2 = (V, E_2)$) are called the fibers of G_1 (resp. G_2). They are (gated) trees. Define the graphs $T_i = (V(T_i), E(T_i)), i = 1, 2$, whose vertices are the fibers of G_i and two fibers F' and F'' are adjacent if and only if there exists an edge of G with one end in F' and another one in F''. Then T_1 and T_2 are trees. We obtain an isometric embedding α of G into the Cartesian product $T_1 \times T_2$ of the two trees, so that for any vertex v of G, $\alpha(v) = (\alpha_1(v), \alpha_2(v)) = (P, Q)$, where $\alpha_1(v) = P$ and $\alpha_2(v) = Q$ are the fibers of the graphs G_1 and G_2 sharing the vertex v [7].

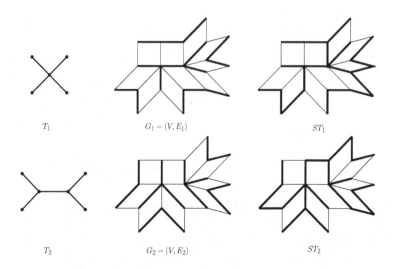

Fig. 2. An even squaregraph and its trees T_i and ST_i

This isometric embedding into the Cartesian product of two trees is used to establish one of the two properties on which our algorithms are based. Each vertex F_i of T_i is given the weight $\pi_i(F_i)$ equal to the total weight of vertices of G located in F_i.

Proposition 1. *Let G be an even squaregraph or a partial grid. A vertex $u = (\alpha_1(u), \alpha_2(u))$ is a median vertex of G if and only if $F_1 = \alpha_1(u)$ and $F_2 = \alpha_2(u)$ are median vertices of the trees T_1 and T_2 endowed with the weight functions π_1 and π_2 respectively.*

Proof. Denote by F_i a fiber of the graph G_i which is a median vertex of the tree T_i, $i = 1, 2$. We assert that $F_1 \cap F_2 \neq \emptyset$. Suppose not and let $F_1 \cap F_2 = \emptyset$. From the definition of F_1 and F_2 we immediately conclude that F_2 is completely contained in the same connected component of the graph $G \setminus F_1$ obtained from G by removing all vertices of F_1. Thus all vertices x of F_2 have their image $\alpha_1(x)$ in the same connected component of $T_1 \setminus F_1$. Denote by F_1' the neighbor of F_1 in the subtree of $T_1 \setminus F_1$ which contains F_2. From Lemma 2 and the fact that F_1 is median in T_1, we deduce that $\pi_1(W_{T_1}(F_1, F_1')) \geq \frac{1}{2}\pi_1(V(T_1)) = \frac{1}{2}\pi(V)$. In the same way, defining F_2' to be the neighbor of F_2 in the connected component of $T_2 \setminus F_2$ which contains F_1, we obtain $\pi_2(W_{T_2}(F_2, F_2')) \geq \frac{1}{2}\pi(V)$. The sets of vertices of G whose images are respectively in $W_{T_1}(F_1, F_1')$ and $W_{T_2}(F_2, F_2')$ are disjoint and do not entirely cover the graph G. Since $\pi(x) > 0$ for any vertex x of G, we obtain a contradiction with Lemma 2. Thus $F_1 \cap F_2 \neq \emptyset$, whence there exists a vertex m of G such that $\alpha_1(m) = F_1$ and $\alpha_2(m) = F_2$ are medians in T_1 and T_2 respectively. Thanks to the isometric embedding, we obtain

$$M_\pi(m) = \sum_{x \in V} \pi(x)d_G(m, x) = \sum_{x \in V} \pi(x)d_{T_1}(F_1, \alpha_1(x)) + \sum_{x \in V} \pi(x)d_{T_2}(F_2, \alpha_2(x))$$

$$= \sum_{R \in V(T_1)} \pi_1(R)d_{T_1}(F_1, R) + \sum_{Q \in V(T_2)} \pi_2(Q)d_{T_2}(F_2, Q).$$

Writing up a similar expression for any other vertex v of G and using the fact that F_1 and F_2 are medians of T_1 and T_2, respectively, we conclude that $M_\pi(m) \leq M_\pi(v)$, thus m is a median vertex of G. Conversely, the previous equality also shows that any median vertex of $\mathrm{Med}_\pi(G)$ can be expressed as the intersection of two median paths, one of T_1 and another of T_2. $\qquad\square$

Before proving the second property of medians of partial grids and even square-graphs, we define two particular spanning trees ST_1 and ST_2 of an even square-graph G. ST_1 contains all edges of E_1 plus exactly one edge running between each pair of incident fibers of the graph $G_1 = (V, E_1)$. The choice of this edge is arbitrary (we can also select an edge belonging to the bounding cycle of G). We call such extra-edges the *switch edges* of ST_1 and denote them by E_1'. Clearly, since all fibers of G_1 are trees, the graph $ST_1 = (V, E_1 \cup E_1')$ is indeed a spanning tree of G. Analogously, we define the spanning tree $ST_2 = (V, E_2 \cup E_2')$.

Proposition 2. *A vertex F_i of T_i ($i = 1, 2$) is a median vertex with respect to the weight function π_i if and only if F_i contains a median vertex m of the tree ST_i with respect to the weight function π.*

Proof. First suppose that m is a median vertex of ST_i (i.e., $m \in \mathrm{Med}_\pi(ST_i)$) and let F_i be the vertex of T_i such that $\alpha_i(m) = F_i$. In other words, F_i is the fiber of G_i containing m. Suppose that F_i is not a median of T_i (i.e., $F_i \notin \mathrm{Med}_{\pi_i}(T_i)$). Lemma 2 yields that $M_{\pi_i}(F') < M_{\pi_i}(F_i)$ for some vertex F' of T_i adjacent to F_i. By Lemma 1 and the definition of T_i we conclude that $M_{\pi_i}(F_i) - M_{\pi_i}(F') = \pi(W(x', x)) - \pi(W(x, x')) > 0$, where $x'x$ is any edge running between F' and F_i. If m has a neighbor m' in F' and mm' is a switch edge, then it can easily be seen from the definition of ST_i that all vertices of $W(x', x)$ (this set is defined in G) are closer to x' than to x in ST_i. This implies $M_\pi(m) - M_\pi(m') \geq \pi(W(x', x)) - \pi(W(x, x')) > 0$, contrary to the assumption that m is a median of ST_i. On the other hand, if the switch between F_i and F' is the edge $p'p$ with $p' \in F'$ and $p \in F_i$, and we denote by m' the neighbor of m on the unique path connecting m with p in the tree-fiber F_i, then again, in the tree ST_i all vertices of $W(p', p)$ are closer to m' than to m. Since $\pi(W(p', p)) > \frac{1}{2}\pi(V)$, Lemma 2 yields that m is not a median vertex of ST_i, a contradiction.

Conversely, suppose that F_i is a median vertex of the tree T_i. We assert that $F_i \cap \mathrm{Med}_\pi(ST_i) \neq \emptyset$. Remove from T_i all edges incident to F_i and denote by $S_1, \ldots, S_k$ the resulting subtrees of T_i not containing F_i. Then Lemmas 1 and 2 imply that $\pi_i(S_j) \leq \frac{1}{2}\pi(V)$ for any subtree S_j. Now, if we pick the switch edge $x_j m_j$ running between S_j and F_i with $x_j \in S_j$ and $m_j \in F_i$, then from the definition of the spanning tree ST_i we infer that S_j coincides with the set of all vertices which are closer to x_j than to m_j in ST_i. The majority rule for trees implies that $M_\pi(m_j) \le M_\pi(x_j)$ holds in ST_i. Now, the median function M_π on trees is convex [29]. Since $M_\pi(m_j) \le M_\pi(x_j)$, this implies that $M_\pi(x_j) \le M_\pi(y_j)$ for any vertex $y_j \in S_j \setminus \{x_j\}$. Since any vertex z outside F_i is located in some subtree S_j, we conclude that $M_\pi(m_j) \le M_\pi(x_j) \le M_\pi(z)$ holds in ST_i. This shows that indeed F_i must contain at least one median vertex of the tree ST_i. $\square$

We obtain the following corollary as a direct consequence of the two previous properties:

Corollary 1. $m \in Med_\pi(G)$ if and only if $\alpha_i(m) \in Med_\pi(ST_i)$, for $i = 1, 2$.

3 Algorithms for the Median Problem

In the introduction, we outlined the self-stabilizing algorithm for the median problem in a tree proposed in [3,11]. We continue with a more detailed account of the model used by this and our algorithms. Then we present the algorithm of [3,11] and our algorithms for partial grids and even squaregraphs.

3.1 Computational Model

The nodes of the graph $G = (V, E)$ are seen as processors executing the same algorithm. Each processor $v \in V$ has a memory whose value (its *state*) can be read by its neighbors, but can only be changed by v itself. A distributed algorithm is a set of rules (a pair of *precondition* and *command*) that describe how a processor has to change its current state (the command) according to the state of all its neighbors (the precondition or guard). We say that a rule R is *activable* at a processor v if the neighborhood of v satisfies the precondition of R. In this case, the node v is also said to be *activable*. If a rule R is activable in v, an *atomic move* for v consists in reading the states of all its neighbors, computing a new value of its state according to the command of R, and writing this value to the local memory. An *execution* is a sequence of moves. This is indeed an interleaved (central deamon) asynchronous model of computation. (Implicit) termination or stabilization is reached when there are no more activable rules. The described model is a standard model for distributed computing originally introduced by Dijkstra [18] and used in many following papers. In particular, it was extensively studied in [12]. It was used in [11], where the algorithms are expressed in the language of "guarded commands". In his thorough study of computational models, it is coined by Chalopin as the "interleaved cellular model" [13, chapter 5]. In [1], it appears as the "local detection paradigm".

A distributed algorithm is said to be *self-stabilizing* if an execution starting from any arbitrary global state has a suffix belonging to the set of legitimate states. For our purposes, we additionally suppose that the state variable of each node has a specific bit named the *median flag*. Then a global state is *legitimate* if the median flag of a node v is set up if and only if v is a median vertex of G. The *time complexity* of a self-stabilizing algorithm is the maximum number of moves that are performed until stabilization. A *round* [19] is a sequence of moves such that each node activable at the beginning of the round is activated at least once. The *round complexity* of a self-stabilizing algorithm is the maximum number of rounds required by an execution to reach a legitimate state. Whenever we compose two or more self-stabilizing algorithms, then this will be done as a *fair composition* as defined in Subsection 2.7 of [19]. Additionally, the input-output binary dependency between our algorithms will be acyclic. Then Theorem 2.2 of [19] guarantees the self-stabilization of the resulting composite algorithm. In all three algorithms described below, only $O(\log n)$ bits are used to store the state of each node.

We specify now the structural information that is used by each of the three median computation algorithms. In the algorithm for trees, neither nodes nor edges have identifiers. In this case, the system is said to be *anonymous*. Whereas in the algorithm for partial rectangular grids, nodes are anonymous but edges are not: for each node, there exists a labeling of the outgoing edges that has the property of (weak) "sense of direction", allowing to compute the second neighborhood of each node in a self-stabilizing manner. Informally, a system represented by a directed graph is said to have *sense of direction* if it is possible to know, from the labels associated to the edges, whether different walks from any given node v end up in the same node or not. The use of sense of direction in a distributed system often leads to significant improvements on computability and complexity [21]. Finally, in the algorithm for even squaregraphs, each processor has a unique identifier (as a matter of fact, it is unclear for us whether this is a necessary structural information).

3.2 Trees

Let $T = (V, E)$ be a tree with n vertices and let π be a weight function on V. We need the following notations:

- $v.s$-value is the local value of the vertex v. It is also called the s-value of v;
- $\gamma_1(v) = \{u \in V : uv \in E\}$ is the set of neighbors of the vertex v in T;
- $N_s(v) = \{u.s\text{-value} : u \in \gamma_1(v)\}$ is a multiset:
- $N_s^-(v) = N_s(v) \backslash \{\max(N_s(v))\}$.

The main result of [3,11] is the proof of self-stabilization in polynomial time for the following algorithm (which we slightly modify to capture weighted medians as well). Algorithms are described by a list of rules, and to simplify the formulation of preconditions, we assume in all the following that a rule is not activable if its execution would not change the state of the node.

MEDIAN TREE		
(v is a leaf)	$\longrightarrow$	$v.s$-value$= \pi(v)$
(v is not a leaf)	$\longrightarrow$	$v.s$-value$= \pi(v) + \sum(N_s^-(v))$.

In a stabilized state, the median vertices are those vertices whose s-value is greater than the s-values of all their neighbors. It is shown in [11] that MEDI-ANTREE stabilizes in $O(e)$ rounds, where e is the maximum distance of a median to a leaf. Moreover, the algorithm makes $O(n^3 \cdot c_s)$ moves in the worst case, where c_s is the maximum initial s-value of any processor [11].

3.3   Partial Rectangular Grids

Let $G = (V, E)$ be a partial grid with n vertices bounded by the cycle ∂C. The first neighborhood $\gamma_1(v)$ of a vertex v of G is the set $V \cap \{v_N, v_S, v_E, v_W\}$, where v_N, v_S, v_E and v_W are the vertices of the square grid located at the North, the South, the East, and the West of v, respectively. The second neighborhood $\gamma_2(v)$ of v in G is the set $V \cap \{v_{NE}, v_{NW}, v_{SE}, v_{SW}\}$, where v_{NE}, v_{NW}, v_{SE} and v_{SW} are the vertices of the square grid which are located at the North East, the North West, the South East, and the South West of v, respectively (see Figure 3). In the following, with $D \in \{N, S, E, W\}$, we use the notation $Neighb(v, D)$ if $v_D \in \gamma_1(v)$.

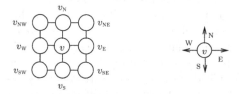

Fig. 3. The first and second neighborhood of a vertex

The model used by the algorithm MEDIANPARTIALGRID is similar to the one of MEDIANTREE except that the system has the "polar" sense of direction (that is, the knowledge of the North, East, South and West outgoing edges). The algorithm MEDIANPARTIALGRID consists of three phases.

Phase 1. In this phase, in order to compute the sets $\gamma_{ST_1}(v)$ and $\gamma_{ST_2}(v)$ of its neighbors in the spanning trees ST_1 and ST_2 respectively, a processor v has to know about the existence of edges between its first neighborhood $\gamma_1(v)$ and its second neighborhood $\gamma_2(v)$. For example, for the tree ST_1, this can be done by communicating with the first neighborhood $\gamma_1(v)$ and applying the following rule. For $D \in \{N, S, E, W\}$, $v_D \in \gamma_{ST_1}(v)$ if $Neighb(v, D)$ and

- either $D \in \{N, S\}$,
- or $D \in \{E, W\}$, and $\neg(Neighb(v, N) \land Neighb(v_N, D) \land Neighb(v_{ND}, S))$.

Phase 2. In this phase, each processor v runs the algorithm MEDIANTREE in parallel on each of the spanning trees ST_1 and ST_2. The variable $v.s_i$-value is the s-value of v for the tree ST_i. Then a median of ST_i can be identified by the fact that the s_i-value of the respective processor is maximum in its neighborhood on ST_i, $i = 1, 2$.

Phase 3. In this phase, a classical broadcasting self-stabilizing algorithm [19, chap.4, p 97] is replicated in both directions on every vertical (respectively, horizontal) path to compute two additional boolean variables $v.b_1$ and $v.b_2$. The variable $v.b_i$ will be set if the path $\alpha_i(v)$ is a median vertex of T_i.

Once the "vertical" and "horizontal" broadcasting algorithms stabilize, the median set of G is formed by all vertices v of the partial grid G for which $v.b_1 \wedge v.b_2$ is true.

Theorem 1. *The algorithm* MEDIANPARTIALGRID *computes the median set of a partial grid* G *with* n *vertices in* $O(n)$ *rounds and* $O(c_s n^3)$ *moves.*

Proof. First we show that the algorithm stabilizes. Indeed, by simulating the execution of MEDIANTREE on the two spanning trees ST_1 and ST_2, we obviously maintain the self-stabilization because the read and written variables are distinct (we use s_1 for the tree ST_1 and s_2 for ST_2). By composing these algorithms with the broadcasting algorithm, the self-stabilization is still maintained according to Theorem 2.2 of [19].

As to the time complexity, notice that the algorithm makes $O(n^3 \cdot c_s)$ moves in the worst case, where c_s is the maximum initial s-value of any processor. Since the time complexity of MEDIANTREE is greater than the time complexity of classical self-stabilizing broadcasting algorithms, the time complexity of ME-DIANPARTIALGRID is of the same order as the time complexity of MEDIANTREE which is given in [11], concluding the proof. In the same way the round complexity of MEDIANTREE is $O(e)$, where $e \leq n$ is the maximum distance from a median to a leaf in the spanning tree, yielding $O(n)$ round complexity for MEDIANPARTIALGRID.

Finally, we show the correctness of our algorithm, i.e. that in a global stabilized state the processors v that have their two boolean variables $v.b_1$ and $v.b_2$ set to true are medians of the partial grid G. Indeed, by Corollary 1 a vertex (processor) v is median in G if and only if it is on the vertical path of a median of ST_1 (thus variable $v.b_1$ set to true) and on the horizontal path of a median of ST_2 (variable $v.b_2$ set to true). Since the algorithm MEDIANTREE correctly computes the median set of each tree ST_1 and ST_2, we are done. □

3.4   Even Squaregraphs

Let $G = (V, E)$ be an even squaregraph. The "irregular" structure of G does not allow an easy use of the sense of direction as in the case of partial squaregrids. To obtain a bipartition of edges used in the construction of the trees T_1, T_2 and ST_1, ST_2, we will use the self-stabilizing algorithm for constructing a spanning BFS-tree of a graph designed by Afek, Kutten, and Yung [1] (for a survey on other related algorithms for this problem, see [22]). This algorithm requires that each vertex v has a unique identifier $v.\mathsf{Id}$. This extra-information allows to break the symmetry in order to select as the root of the spanning tree the vertex having the highest identifier.

Our median self-stabilizing algorithm MEDIANEVENSQUAREGRAPH consists of four phases. In each phase, we present the specific conditions which allow to test if the state of a vertex is legal or not for the current phase. If the respective condition is not satisfied, then we describe the modifications which must be undertaken in order to return the system to a legal state. We establish that after a finite number of activations, the corresponding conditions of the current phase are satisfied by all vertices, and thus the next phase can start.

Phase 1. In this phase, we construct a spanning tree using the algorithm of Afek and al [1]. When this phase terminates, each vertex v has computed the identifier v.Root of the root node of the resulting spanning BFS-tree, the identifier v.Parent of its father in this tree and an integer v.Distance which is the tree-distance between v and the root. This phase ends if the following condition holds in each node v:
Condition $st(v)$:
$$\{[(v.\text{Root} = v.\text{Id}) \wedge (v.\text{Parent} = v.\text{Id}) \wedge (v.\text{Distance} = 0)] \vee [(v.\text{Root} > v.\text{Id}) \wedge$$
$$(v.\text{Parent} \in v.\text{Edge-list}) \wedge (v.\text{Root} = v.\text{Parent.Root}) \wedge$$
$$(v.\text{Distance} = v.\text{Parent.Distance} + 1)]\} \wedge (v.\text{Root} \geq \max_{x \in v.\text{Edge-list}} x.\text{Root})$$

The algorithm, which is executed by each processor so that the system stabilizes in a state in which all these conditions are satisfied for each node, is described in details and analyzed in [1].

Phase 2. In this second phase of the algorithm, we aim to partition the edges of G into two subsets E_1 and E_2 so that two incident edges belonging to a common square-face are included in different sets E_1 and E_2. For a vertex v, we assume that the edges containing v as an end-vertex are numbered $0, \ldots, deg(v) - 1$ in the order in which they appear in the counterclockwise traversal, so that two edges incident to v and belonging to a common square have numbers of different parity. For each $v \in V$, the variable v.Color equals i if all edges incident to v which appear at even positions in the adjacency list of v belong to E_i and the remaining edges belong to E_{3-i}. Each vertex v runs the following algorithm:

$st(v) \wedge (v.\text{Id} = v.\text{Root})$	$\longrightarrow$ $v.\text{Color} := 1$
$st(v) \wedge (v.\text{Id} \neq v.\text{Root}) \wedge$ (the edge connecting the vertex v with v.Parent occurs with the same parity in the adjacency list of v as in the adjacency list of v.Parent)	$\longrightarrow$ $v.\text{Color} := v.\text{Parent.Color}$
$st(v) \wedge (v.\text{Id} \neq v.\text{Root}) \wedge$ (the edge connecting the vertex v to v.Parent occurs with different parities in the adjacency lists of v and v.Parent)	$\longrightarrow$ $v.\text{Color} := 3 - v.\text{Parent.Color}$

We say that the condition $col(v)$ *is satisfied by a vertex* $v \in V$ if none of the preconditions of the three previous actions is satisfied. Then Phase 2 terminates if the condition $col(v)$ is satisfied by all vertices $v \in V$ of G. At this time each vertex v knows the list v.Edge $-$ list$_i$ of its neighbors in G_i.

Phase 3. In this phase, the algorithm constructs the spanning trees ST_i, $i = 1, 2$. For sake of simplicity, these trees will be rooted at the same vertex as the root of the spanning BFS-tree computed in Phase 1. To encode the tree ST_i, we introduce for each $v \in V$ a new variable $v.\text{Parent}_i$ which will denote the father of the vertex v in the tree ST_i. The algorithm consists in "correcting" the spanning tree of Phase 1 by replacing $v.\text{Parent}$ with a vertex belonging to $v.\text{Edge-list}_i$ if this list contains a vertex located at the same distance to the root as the vertex $v.\text{Parent}$. Since all fibers of the graphs G_i are gated sets (in G), the tree obtained in this way has all the edges of E_i and exactly one switch edge of E_{3-i} running between each pair of neighboring fibers of G_i. Actually, if $v'v$ is a switch edge with $d_G(r, v') < d_G(r, v)$ and v belongs to the fiber F, then necessarily v is the gate of r in the fiber F and v' is the father of v in the BFS-tree. Indeed, the root r can be connected with any vertex u of F of G_i by a shortest path passing via the gate v of r in F, and thus the edge vv' will be included in the BFS-tree before the edge running from u to its father. Each processor $v \in V$ runs at this phase the following algorithm:

$col(v) \wedge (v.\text{Id} = v.\text{Root})$	$\longrightarrow \quad v.\text{Parent}_i := v.\text{Id}$
$col(v) \wedge (v.\text{Id} \neq v.\text{Root}) \wedge$ (the edge connecting v to $v.\text{Parent}$ belongs to E_i)	$\longrightarrow \quad v.\text{Parent}_i := v.\text{Parent}$
$col(v) \wedge (v.\text{Id} \neq v.\text{Root}) \wedge$ (the edge connecting the vertex v to $v.\text{Parent}$ belongs to E_{3-i}) $\wedge$ ($v.\text{Parent.Distance} < \min_{x \in v.\text{Edge-list}_i} x.\text{Distance}$)	$\longrightarrow \quad v.\text{Parent}_i := v.\text{Parent}$
$col(v) \wedge (v.\text{Id} \neq v.\text{Root}) \wedge$ (the edge connecting the vertex v with $v.\text{Parent}$ belongs to E_{3-i}) $\wedge$ ($v.\text{Parent.Distance} \geq \min_{x \in v.\text{Edge-list}_i} x.\text{Distance}$)	$\longrightarrow \quad v.\text{Parent}_i := \text{argmin}_{x \in v.\text{Edge-list}_i} x.\text{Distance}$

We say that the condition $st_i(v)$ *is satisfied by the vertex* v if none of the preconditions of the previous actions is satisfied. Since the spanning trees ST_1 and ST_2 are constructed at the same time, Phase 3 terminates if the condition $st_i(v)$ is satisfied at each vertex $v \in V$ and for each index $i \in \{1, 2\}$. At the end of this phase, each node $v \in V$ knows its father $v.\text{Parent}_i$ in the tree ST_i.

Phase 4. With trees ST_1 and ST_2 at hand, the algorithm is similar to that for partial grids. First, the algorithm MEDIANTREE is used to compute the medians of the trees ST_1 and ST_2. Then, using a self-stabilizing broadcasting algorithm, we set to "true" the variable $v.\text{b}_i$ of each vertex v belonging to the same connected component of the graph G_i as a median vertex of the tree ST_i. Once this broadcasting algorithm stabilizes, the median set $\text{Med}_\pi(G)$ of G is formed by all vertices v of G for which $v.\text{b}_1 \wedge v.\text{b}_2$ is true.

Theorem 2. *The algorithm* MEDIANEVENSQUAREGRAPH *computes the median set of an even squaregraph G with n vertices in $O(n^2)$ rounds.*

Proof. The algorithm given by Afek et al. [1] for constructing a spanning tree stabilizes in $O(n^2)$ rounds. This shows that Phase 1 stabilizes in $O(n^2)$ rounds. The Phase 2 needs $O(n)$ rounds in the worst case. By induction we can show that when a vertex v located at distance i from the root is activated during the round $i + 1$, the variable $w.\text{Color}$ of its father w, which is at distance $i - 1$

from the root, has already been correctly computed (by induction hypothesis). Thus the rules of Phase 2 correctly compute the value of v.Color using that of w.Color. Since two edges incident to the same vertex v and belonging to the same square have numbers of different parity in the degree list of v, they will be included in different sets E_1 and E_2 by the algorithm. It remains to show that any edge uv is inserted in the same edge-list by both vertices u and v. For this we proceed by induction on $k := d_G(v, r) < d_G(u, r)$. Our previous argument shows that the assertion holds when v is the father of u in the BFS-tree. So, suppose that v' is the father of v and u' is the father of u in the BFS-tree and that $u' \neq v$. It was shown in [14] for graphs more general than squaregraphs that, if u and v are adjacent in G, then their fathers u' and v' are also adjacent. Since $d_G(v', r) = k - 1$, by induction hypothesis we conclude that u' and v' inserted the edge $u'v'$ in the same set, say E_1. Now, since each vertex inserts two incident edges of a square in different edge-lists and the edge connecting a vertex with its father is set in the same list by the two ends, we conclude that the edges $u'u$ and $v'v$ are in the lists E_2 of their extremities. This implies that the edge uv will be put in the list E_1 by both vertices u and v, thus establishing our assertion.

The rules of Phase 3 depend only of the information computed at Phases 1 and 2, thus in order to correctly compute in Phase 3 the trees ST_1 and ST_2, it suffices that each vertex is activated at this phase once. As to the Phase 4, the algorithm MEDIANTREE stabilizes in $O(e)$ rounds, where e is the largest eccentricity of a median vertex of G [11], thus its complexity is the same as that of broadcasting. Summarizing, we conclude that the construction in $O(n^2)$ rounds of a spanning tree of G dominates the overall complexity of our algorithm. Most importantly, Proposition 2 establishes that MEDIANEVENSQUAREGRAPH correctly computes $\mathrm{Med}(G)$. □

Remark 1. Note that, given a spanning tree of an even squaregraph and unique identifiers, it would be possible to reconstruct locally a map representing the topology of the graph and then compute *internally* the median set. But it would require much more than $O(\log n)$ bits per node and therefore it is not a satisfactory solution.

References

1. Afek, Y., Kutten, S., Yung, M.: The Local Detection Paradigm and its Applications to Self-Stabilization. Theor. Comput. Sci. 186, 199–229 (1997)
2. Aggarwal, S., Kutten, S.: Time optimal self-stabilizing spanning tree algorithm. In: Shyamasundar, R.K. (ed.) Foundations of Software Technology and Theoretical Computer Science. LNCS, vol. 761, pp. 400–410. Springer, Heidelberg (1993)
3. Antonoiu, G., Srimani, P.K.: A self-stabilizing distributed algorithm to find the median of a tree graph. J. Comput. Syst. Sci. 58, 215–221 (1999)
4. Antonoiu, G., Srimani, P.K.: Distributed self-stabilizing algorithm for minimum spanning tree construction. In: Euro-Par'97, pp. 480–487 (1997)
5. Awerbuch, B., Kutten, S., Mansour, Y., Patt-Shamir, B., Varghese, G.: Time optimal self-stabilizing synchronization. In: STOC'93, pp. 652–661 (1993)

6. Bandelt, H.-J., Barthélemy, J.-P.: Medians in median graphs. Discr. Appl. Math. 8, 131–142 (1984)
7. Bandelt, H.-J, Chepoi, V., Eppstein, D.: Ramified rectilinear polygons (in preparation)
8. Barthélemy, J.-P., Monjardet, B.: The median procedure in cluster analysis and social choice theory. Math. Soc. Sci. 1, 235–268 (1981)
9. Buckley, F., Harary, F.: Distances in Graphs. Addison-Wesley, Redwood City, CA (1990)
10. Blair, J.R.S, Manne, F.: Efficient self-stabilizing algorithms for tree networks. Tech. Rept. 232, Univ. Bergen (2002)
11. Bruell, S.B., Ghosh, S., Karaata, M.H., Pemmaraju, S.V.: Self-stabilizing algorithms for finding centers and medians of trees. SIAM J. Computing 29, 600–614 (1999)
12. Boldi, P., Vigna, S.: An effective characterization of computability in anonymous networks. In: Welch, J.L. (ed.) DISC 2001. LNCS, vol. 2180, pp. 33–47. Springer, Heidelberg (2001)
13. Chalopin, J.: Algorithmique Distribuée, Calculs Locaux et Homomorphismes de Graphes, PhD Thesis Université Bordeaux I (2006)
14. Chepoi, V.: Graphs of some CAT(0) complexes. Adv. Appl. Math. 24, 125–179 (2000)
15. Chepoi, V., Dragan, F., Vaxès, Y.: Addressing, distances and routing in triangular systems with applications in cellular and sensor networks. Wireless Networks 12, 671–679 (2006)
16. Chepoi, V., Fanciullini, C., Vaxès, Y.: Median problem in some plane triangulations and quadrangulations. Comput. Geom. 27, 193–210 (2004)
17. Datta, A.K., Derby, J.L, Lawrence, J.E., Tixeuil, S.: Stabilizing hierarchical routing. J. Interconnexion Networks 1, 283–302 (2000)
18. Dijkstra, E.W.: Self-stabilizing systems in spite of distributed control. Comm. ACM 17, 643–644 (1974)
19. Dolev, S.: Self-stabilization. MIT Press, Cambridge (2000)
20. Dolev, S., Israeli, A., Moran, S.: Self-stabilization of dynamic systems assuming only read/write atomicity. In: PODC'90, pp. 103–118 (1990)
21. Flocchini, P., Mans, B., Santoro, N.: Sense of direction: formal definitions and properties. In: SIROCCO'95, pp. 9–34 (1995)
22. Gärtner, F.: A survey of self-stabilizing spanning-tree construction algorithms (2003)
23. Gerstel, O., Zaks, S.: A new characterization of tree medians with applications to distributed algorithms. Networks 24, 23–29 (1994)
24. Ghosh, S., Gupta, A., Pemmaraju, S.V.: A self-stabilizing algorithm for the maximum flow problem. Distributed Computing 10, 167–180 (1997)
25. Goldman, A.J., Witzgall, C.J.: A localization theorem for optimal facility placement. Transp. Sci. 4, 406–409 (1970)
26. Herman, T., Ghosh, S.: Stabilizing phase-clocks. Inf. Process. Lett. 54, 259–265 (1995)
27. Korach, E., Rotem, D., Santoro, N.: Distributed algorithms for finding centers and medians in networks. ACM Trans. Program. Lang. Syst. 6, 380–401 (1984)
28. Nesterenko, M., Mizuno, M.: A quorum-based self-stabilizing distributed mutual exclusion algorithm, J. J. Parallel Distrib. Comput. 62, 284–305 (2002)
29. Tansel, B.C, Francis, R.L, Lowe, T.J.: Location on networks: a survey. Management Sci. 29, 482–511 (1983)

A New Self-stabilizing Maximal Matching Algorithm

Fredrik Manne[1], Morten Mjelde[1], Laurence Pilard[2], and Sébastien Tixeuil[3,*]

[1] University of Bergen, Norway
{fredrikm,mortenm}@ii.uib.no
[2] University of Iowa, USA
laurence-pilard@uiowa.edu
[3] LRI-CNRS UMR 8623 & INRIA Grand Large, Université Paris Sud, France
tixeuil@lri.fr

Abstract. The maximal matching problem has received considerable attention in the self-stabilizing community. Previous work has given different self-stabilizing algorithms that solves the problem for both the adversarial and fair distributed daemon, the sequential adversarial daemon, as well as the synchronous daemon. In the following we present a single self-stabilizing algorithm for this problem that unites all of these algorithms in that it stabilizes in the same number of moves as the previous best algorithms for the sequential adversarial, the distributed fair, and the synchronous daemon. In addition, the algorithm improves the previous best moves complexities for the distributed adversarial daemon from $O(n^2)$ and $O(\delta m)$ to $O(m)$ where n is the number of processes, m is the number of edges, and δ is the maximum degree in the graph.

1 Introduction

A *matching* in an undirected graph is a subset of edges in which no pair of edges is adjacent. A matching M is *maximal* if no proper superset of M is also a matching. Matchings are typically used in distributed applications when pairs of neighboring nodes have to be set up (*e.g.* between a server and a client). As current distributed applications usually run continuously, it is expected that the system is dynamic (nodes may leave or join the network), so an algorithm for the distributed construction of a maximal matching should be able to reconfigure on the fly. *Self-stabilization* [3,4] is an elegant approach to forward recovery from transient faults as well as initializing a large-scale system. Informally, a self-stabilizing systems is able to recover from any transient fault in finite time, without restricting the nature or the span of those faults.

The environment of a self-stabilizing algorithm is modeled by the notion of a *daemon*. There are two main characteristics for the daemon: it can be either *sequential* (or central, meaning that exactly one eligible process is scheduled for

---

* Support for this work was given by the Aurora program for collaboration between France and Norway.

G. Prencipe and S. Zaks (Eds.): SIROCCO 2007, LNCS 4474, pp. 96–108, 2007.
© Springer-Verlag Berlin Heidelberg 2007

execution at a given time) or *distributed* (meaning that any subset of eligible processes can be scheduled for execution at a given time), and in an orthogonal way, it can be *fair* (meaning that in any execution, every eligible processor is eventually scheduled for execution) or *adversarial* (meaning that the daemon only guarantees global progress, *i.e.* some eligible process is eventually scheduled for execution). An extreme case of a fair daemon is the *synchronous* daemon, where all eligible processes are scheduled for execution at every step. Of course, any algorithm that can cope with the distributed daemon can cope with the sequential daemon or the synchronous daemon, and any algorithm that can handle the adversarial daemon can be used with a fair or a synchronous daemon, but the converse is not true in either case.

There exists several self-stabilizing algorithms for computing a maximal matching in an unweighted general graph. Hsu and Huang [10] gave the first such algorithm and proved a bound of $O(n^3)$ on the number of steps assuming an adversarial daemon. This analysis was later improved to $O(n^2)$ by Tel [12] and finally to $O(m)$ by Hedetniemi et al. [9]. The original algorithm assumes an anonymous network and operates therefore under the sequential daemon in order to achieve symmetry breaking. Indeed, one can show that in some symmetric networks there exists no deterministic self-stabilizing solution to the maximal matching problem.

By using randomization, Gradinariu and Johnen [7] proposed a scheme to give processes a local name that is unique within distance 2, and used this scheme to run Hsu and Huang's algorithm under an adversarial distributed daemon. However, only a finite stabilization time was proved. Using the same technique of randomized local symmetry breaking, Chattopadhyay et al. [2] later gave a maximal matching algorithm with $O(n)$ round complexity (in their model, this is tantamount to $O(n^2)$ steps), but assuming the weaker fair distributed daemon.

In [5] Goddard *et al.* describe a synchronous version of Hsu and Huang's algorithm and show that it stabilizes in $O(n)$ rounds. Although not explicitly proved in the paper, it can be shown that their algorithm also copes with the adversarial distributed daemon using $\theta(n^2)$ steps. Here, symmetry is broken using unique identifiers at every process. In [8], Gradinariu and Tixeuil provide a general scheme to transform an algorithm using the sequential adversarial daemon into an algorithm that copes with the distributed adversarial daemon. Using this scheme with Hsu and Huang's algorithm yields a step complexity of $O(\delta m)$, where δ denotes the maximum degree of the network.

Our contribution is a new self-stabilizing algorithm that stabilizes after $O(m)$ steps both under the sequential and under the distributed adversarial daemon. Under a distributed fair daemon the algorithm stabilizes after $O(n)$ rounds. Thus, this algorithm unifies the moves complexities of the previous best algorithms both for the sequential and for the distributed fair daemon and also improves the previous best moves complexity for the distributed adversarial daemon. As a side effect, we improve the best known algorithm for the adversarial daemon by lowering the environment requirements (distributed *vs.* sequential). To break symmetry, we assume that node identifiers are unique within distance

2 (this can be done using the scheme of [2,7]). The following table compares features of the aforementioned algorithms and ours (best feature for each category is presented in boldface).

Reference	Daemon	Step complexity	Round complexity	Asymmetry
[9,10,12]	sequential adversarial	**O(m)**		**anonymous**
[7]	**distributed adversarial**	finite		distance 2
[2]	distributed fair	$O(n^2)$	**O(n)**	distance 2
[5]	synchronous	$O(n^2)$	**O(n)**	unique ID
[8]	**distributed adversarial**	$O(\delta m)$		unique ID
This paper	**distributed adversarial**	**O(m)**	**O(n)**	distance 2

The rest of this paper is organized as follows. In Section 2 we give a short introduction to self-stabilizing algorithms and the computational environment we use. In Section 3 we describe our algorithm and prove its correctness and speed of convergence in Section 4. Finally, in Section 5 we conclude.

2   Model

A system consists of a set of processes where two adjacent processes can communicate with each other. The communication relation is typically represented by a graph $G = (V, E)$ where each process is represented by a node in V and two processes i and j are adjacent if and only if $(i,j) \in E$. The set of neighbors of a node $i \in V$ is denoted by $N(i)$. The neighbors of a set of processes $A \subseteq V$ is defined as follows $N(A) = \{j \in V - A, \exists i \in A \text{ s.t. } (i,j) \in E\}$. A process maintains a set of variables. Each variable ranges over a fixed domain of values. An action has the form $\langle name \rangle : \langle guard \rangle \longrightarrow \langle command \rangle$. A *guard* is a boolean predicate over the variables of both the process and those of its neighbors. A *command* is a sequence of statements assigning new values to the variables of the process.

A *configuration* of the system is the assignment of a value to every variable of each process from its corresponding domain. Each process contains a set of actions. An action is *enabled* in some configuration if its guard is **true** at this configuration. A process is *eligible* if it has at least one enabled action. A *computation* is a maximal sequence of configurations such that for each configuration s_i, the next configuration s_{i+1} is obtained by executing the command of at least one action that is enabled in s_i (a process that executes such an action makes a *move* or a *step*). Maximality of a computation means that the computation is infinite or it terminates in a configuration where none of the actions are enabled.

A *daemon* is a predicate on executions. We distinguish several kinds of daemons: the *sequential* daemon make the system move from one configuration to the next by executing exactly one enabled action, the *synchronous* daemon makes the system move from one configuration to the next one by executing all enabled actions, the *distributed* daemon makes the system move from one configuration to the next one by executing any non empty subset of enabled actions.

Note that the sequential and synchronous daemons are instances of the more general (*i.e.* less constrained) distributed daemon. Also, a daemon is *fair* if any action that is continuously enabled is eventually executed, and *adversarial* if it may execute *any* enabled action at every step. Again, the adversarial daemon is more general than the fair daemon.

A system is self-stabilizing for a given specification, if it automatically converges to a configuration that conforms to this specification, independently of its initial configuration and without external intervention.

We consider two measures for evaluating complexity of self-stabilizing programs. The *step* complexity investigates the maximum number of process moves that are needed to reach a configuration that conforms to the specification (*i.e.* a *legitimate* configuration), for all possible starting configurations. The *round* complexity considers that executions are observed in rounds: a round is the smallest sequence of an execution in which every process that was eligible at the beginning of the round either makes a move or has its guard(s) disabled since the beginning of the round.

3   The Algorithm

In the following we present and motivate our algorithm for computing a maximal matching. The algorithm is self-stabilizing and does not make any assumptions on the network topology. A set of edges $M \subseteq E$ is a *matching* if and only if $x, y \in M$ implies that x and y do not share a common end point. A matching M is *maximal* if no proper superset of M is also a matching.

Each process i has a variable p_i pointing to one of its neighbors or to *null*. We say that processes i and j *are married* to each other if and only if i and j are neighbors and their p-values point to each other. In this case we will also refer to i as being married without specifying j. However, we note that in this case j is unique. A process which is not married is *unmarried*.

We also use a variable m_i to let neighboring processes of i know if process i is married or not. To determine the value of m_i we use a predicate $PRmarried(i)$ which evaluates to true if and only if i is married. Thus predicate $PRmarried(i)$ allows process i to know if it is currently married and the variable m_i allows neighbors of i to know if i is married. Note that the value of m_i is not necessarily equal to $PRmarried(i)$.

Our self-stabilizing scheme is given in Algorithm 1. It is composed of four mutual exclusive guarded rules as described below.

The *Update* rule updates the value of m_i if it is necessary, while the three other rules can only be executed if the value of m_i is correct. In the *Marriage* rule, an unmarried process that is currently being pointed to by a neighbor j tries to marry j by setting $p_i = j$. In the *Seduction* rule, an unmarried process that is not being pointed to by any neighbor, point to an unmarried neighbor with the objective of marriage. Note that the identifier of the chosen neighbor has to be larger than that of the current process. This is enforced to avoid the creation of cycles of pointer values. In the *Abandonment* rule, a process i resets

---

Algorithm 1. A self-stabilizing maximal matching algorithm

Variables of process i:
  $m_i \in \{\text{true, false}\}$
  $p_i \in \{null\} \cup N(i)$

Predicate:
  $PRmarried(i) \equiv \exists j \in N(i) : (p_i = j \text{ and } p_j = i)$

Rules:
  Update:
    if $m_i \neq PRmarried(i)$
    then $m_i := PRmarried(i)$

  Marriage:
    if $m_i = PRmarried(i)$ **and** $p_i = null$ **and** $\exists j \in N(i) : p_j = i$
    then $p_i := j$

  Seduction:
    if $m_i = PRmarried(i)$ **and** $p_i = null$ **and** $\forall k \in N(i) : p_k \neq i$
      and $\exists j \in N(i) : (p_j = null$ **and** $j > i$ **and** $\neg m_j)$
    then $p_i := Max\{j \in N(i) : (p_j = null$ **and** $j > i$ **and** $\neg m_j)\}$

  Abandonment:
    if $m_i = PRmarried(i)$ **and** $p_i = j$ **and** $p_j \neq i$ **and** $(m_j$ **or** $j \leq i)$
    then $p_i := null$

---

its p_i value to *null*. This is done if the process j which it is pointing to does not point back at i and if either *(i)* j is married, or *(ii)* j has a lower identifier than i. Condition *(i)* allows a process to stop waiting for an already married process while the purpose of Condition *(ii)* is to break a possible initial cycle of p-values.

We note that if *PRmarried(i)* holds at some point of time then from then on it will remain true throughout the execution of the algorithm. Moreover, the algorithm will never actively create a cycle of pointing values since the *Seduction* rule enforces that $j > i$ before process i will point to process j. Also, all initial cycles are eventually broken since the guard of the *Abandonment* rule requires that $j \leq i$.

Figure 1 gives a short example of the execution of the algorithm. The initial configuration is as shown in Figure 1a, where $id_i > id_j > id_k$. Here both processes j and k attempt to become married to i. In Figure 1b process i has executed a *Marriage* move, and i and j are now married. In Figure 1c both i and j execute an *Update* move, setting their m-values to *true*. And finally, in Figure 1d process k executes an *Abandonment* move.

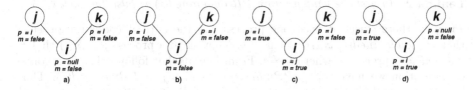

Fig. 1. Example

4   Proof of Correctness

In the following we will first show that when Algorithm 1. has reached a stable configuration it also defines a maximal matching. We will then bound the number of steps the algorithm needs to stabilize both for the adversarial and fair distributed daemon. Note that the sequential daemon is a subset of the distributed one, thus any result for the latter also applies to the former.

4.1   Correct Stabilization

We say that a configuration is *stable* if and only if no process can execute a move in this configuration. We now proceed to show that if Algorithm 1. reaches a stable configuration then the p and m-values will define a maximal matching M where $(i, j) \in M$ if and only if $(i, j) \in E, p_i = j$, and $p_j = i$ while both m_i and m_j are true. In order to perform the proof, we define the following five mutual exclusive predicates:

$PRmarried(i) \quad \equiv \exists j \in N(i) : (p_i = j \text{ and } p_j = i)$
$PRwaiting(i) \quad \equiv \exists j \in N(i) : (p_i = j \text{ and } p_j \neq i \text{ and } \neg PRmarried(j))$
$PRcondemned(i) \equiv \exists j \in N(i) : (p_i = j \text{ and } p_j \neq i \text{ and } PRmarried(j))$
$PRdead(i) \quad\quad \equiv (p_i = null) \text{ and } (\forall j \in N(i) : PRmarried(j))$
$PRfree(i) \quad\quad \equiv (p_i = null) \text{ and } (\exists j \in N(i) : \neg PRmarried(j))$

Note first that each process will evaluate exactly one of these predicates to true. Moreover, also note that *PRmarried(i)* is the same as in Algorithm 1.

   We now show that in a stable configuration each process i evaluates either *PRmarried(i)* or *PRdead(i)* to true, and when this is the case, the p-values define a maximal matching. To do so, we first note that in any stable configuration the m-values reflects the current status of the process.

Lemma 1. *In a stable configuration we have* $m_i = PRmarried(i)$ *for each* $i \in V$.

Proof. This follows directly since if $m_i \neq PRmarried(i)$ then i is eligible to execute the *Update(i)* rule.   □

We next show in the following three lemmas that no process will evaluate either *PRwaiting(i)*, *PRcondemned(i)*, or *PRfree(i)* to true in a stable configuration.

Lemma 2. *In a stable configuration $PRcondemned(i)$ is false for each $i \in V$.*

Proof. If there exists at least one process i in the current configuration such that $PRcondemned(i)$ is true then p_i is pointing to a process $j \in N(i)$ that is married to a process k where $k \neq i$. From Lemma 1 it follows that in a stable configuration we have $m_i = PRmarried(i)$ and $m_j = PRmarried(j)$. Thus in a stable configuration the predicate ($m_i = PRmarried(i)$ **and** $p_i = j$ **and** $p_j \neq i$ **and** m_j) evaluates to true. But then process i is eligible to execute the *Abandonment* rule contradicting that the current configuration is stable. □

Lemma 3. *In a stable configuration $PRwaiting(i)$ is false for each $i \in V$.*

Proof. Assume that the current configuration is stable and that there exists at least one process i such that $PRwaiting(i)$ is true. Then it follows that p_i is pointing to a process $j \in N(i)$ such that $p_j \neq i$ and j is unmarried. Note first that if $p_j = null$ then process j is eligible to execute a *Marriage* move. Also, if $j < i$ then process i can execute an *Abandonment* move.

Assume therefore that $p_j \neq null$ and that $j > i$. It then follows from Lemma 2 that
$\neg PRcondemned(j)$ is true and since j is not married we also have $\neg PRmarried(j)$. Thus $PRwaiting(j)$ must be true. Repeating the same argument for j as we just did for i it follows that if both i and j are ineligible for a move then there must exist a process k such that $p_j = k$, $k > j$, and $PRwaiting(k)$ also evaluates to true. This sequence of processes cannot be extended indefinitely since each process must have a higher id than the preceding one. Thus there must exist some process in V that is eligible for a move and the assumption that the current configuration is stable is incorrect. □

Lemma 4. *In a stable configuration $PRfree(i)$ is false for each $i \in V$.*

Proof. Assume that the current configuration is stable and that there exists at least one process i such that $PRfree(i)$ is true. Then it follows that $p_i = null$ and that there exists at least one process $j \in N(i)$ such that j is not married.

Next, we look at the value of the different predicates for the process j. Since j is not married it follows that $PRmarried(j)$ evaluates to false. Also, from lemmas 2 and 3 we have that both $PRwaiting(j)$ and $PRcondemned(j)$ must evaluate to false. Finally, since i is not married we cannot have $PRdead(j)$. Thus we must have $PRfree(j)$. But then the process with the smaller id of i and j is eligible to propose to the other, contradicting the fact that the current configuration is stable. □

From lemmas 2 through 4 we immediately get the following corollary.

Corollary 1. *In a stable configuration either $PRmarried(i)$ or $PRdead(i)$ holds for every $i \in V$.*

We can now show that a stable configuration also defines a maximal matching.

Theorem 1. *In any stable configuration the m and p-values define a maximal matching.*

Proof. From Corollary 1 we know that either $PRmarried(i)$ or $PRdead(i)$ holds for every $i \in V$ in a stable configuration. Also, from Lemma 1 it follows that m_i is true if and only if i is married. It is then straightforward to see that the p-values define a matching.

To see that this matching is maximal assume to the contrary that it is possible to add one more edge (i, j) to the matching so that it still remains a legal matching. To be able to do so we must have $p_i = null$ and $p_j = null$. Thus we have $\neg PRmarried(i)$ and $\neg PRmarried(j)$ which again implies that both $PRdead(i)$ and $PRdead(j)$ evaluates to true. But according to the $PRdead$ predicate two adjacent processes cannot be dead at the same time. It follows that the current matching is maximal. $\square$

4.2   Convergence for the Distributed Adversarial Daemon

In the following we will show that Algorithm 1. will reach a stable configuration after at most $3 \cdot n + 2 \cdot m$ steps under the distributed adversarial daemon.

First we note that as soon as two processes are married they will remain so for the rest of the execution of the algorithm.

Lemma 5. *If processes i and j are married in a configuration C ($p_i = j$ and $p_j = i$) then they will remain married in any ensuing configuration C'.*

Proof. Assume that $p_i = j$ and $p_j = i$ in some configuration C. Then process i cannot execute neither the *Marriage* nor the *Seduction* rule since these require that $p_i = null$. Similarly, i cannot execute the *Abandonment* rule since this requires that $p_j \neq i$. The exact same argument for process j shows that j also cannot execute any of the three rules *Marriage*, *Seduction*, and *Abandonment*. Thus the only rule that processes i and j can execute is *Update* but this will not change the values of p_i or p_j. $\square$

A process discovers that it is married through executing the *Update* rule. Thus this is the last rule a married process will execute in the algorithm. This is reflected in the following.

Corollary 2. *If a process i executes an* Update *move and sets $m_i = true$ then i will not move again.*

Proof. From the predicate of the *Update* rule it follows that when process i sets $m_i = true$ there must exist a process $j \in N(i)$ such that $p_i = j$ and $p_j = i$. Thus from Lemma 5 the only move i can make is an *Update* move. But since the m_i value is correct and p_i and p_j will not change again this will not happen. $\square$

Since a married process cannot become "unmarried" we also have the following restriction on the number of times the *Update* rule can be executed by any process.

Corollary 3. *Any process executes at most two* Update *moves.*

We will now bound the number of moves from the set {*Marriage, Seduction, Abandonment*}. Each such move is performed by a process i in relation to one of its neighbors j. We will call any such move made by either i or j with respect to the other as an i, j-*move*.

Lemma 6. *For any edge* $(i, j) \in E$, *there can at most be three steps in which an* i, j-*move is performed.*

Proof. Let $(i, j) \in E$ be an edge such that $i < j$. We then consider four different cases depending on the initial values of p_i and p_j at the start of the algorithm. Note from Algorithm 1. that the only values that p_i and p_j can take on are $p_i \in \{null\} \cup N(i)$ and $p_j \in \{null\} \cup N(j)$. For each case we will show that there can at most be three steps in which i, j-moves occur.

Case (i): $p_i \neq j$ and $p_j \neq i$. Since $i < j$ the first i, j-move cannot be process j executing a *Seduction* move. Also, as long as $p_i \neq j$, process j cannot execute a *Marriage* move. Thus process j cannot execute an i, j-move until after process i has first made an i, j-move. It follows that the first possible i, j-move is that i executes a *Seduction* move simultaneously as j makes no move. Note that at the starting configuration of this move we must have $\neg m_j$. If the next i, j-move is performed by j simultaneously as i performs no move then this must be a *Marriage* move which results in $p_i = j$ and $p_j = i$. Then by Lemma 5 there will be no more i, j-moves. If process i makes the next i, j-move (independently of what process j does) then this must be an *Abandonment* move. But this requires that the value of m_j has changed from false to true. Then by Corollary 2 process j will not make any more i, j-moves and since $p_j \neq null$ and $p_j \neq i$ for the rest of the algorithm it follows that process i cannot execute any future i, j-move. Thus there can at most be two steps in which i, j-moves are performed.

Case (ii): $p_i = j$ and $p_j \neq i$. If the first i, j-move only involves process j then this must be a *Marriage* move resulting in $p_i = j$ and $p_j = i$ and from Lemma 5 neither i nor j will make any future i, j-moves. If the first i, j-move involves process i then it must make an *Abandonment* move. Thus in the configuration prior to this move we must have $m_j = \text{true}$. It follows that either $m_j \neq PRmarried(j)$ or $p_j \neq null$. In both cases process j cannot make an i, j-move simultaneously as i makes its move. Thus following the *Abandonment* move by process i we are at Case *(i)* and there can at most be two more i, j-moves. Hence, there can at most be a total of three steps with i, j-moves.

Case (iii): $p_i \neq j$ and $p_j = i$. If the first i, j-move only involves process i then this must be a *Marriage* move resulting in $p_i = j$ and $p_j = i$ and from Lemma 5 neither i nor j will make any future i, j-moves. If the first i, j-move involves process j then this must be an *Abandonment* move. If process i does not make a simultaneous i, j-move then this will result in configuration i) and there can at most be two more steps with i, j-moves for a total of three steps containing i, j-moves.

If process i does make a simultaneous i, j-move then this must be a *Marriage* move. We are now at a similar configuration as Case *(ii)* but with $\neg m_j$. If the second i, j-move involves process i then this must be an *Abandonment* move implying that m_j has changed to true. It then follows from Corollary 2 that process j (and therefore also process i) will not make any future i, j-move leaving a total of two steps containing i, j-moves. If the second i, j-move does not involve i then this must be a *Marriage* move performed by process j and resulting in $p_i = j$ and $p_j = i$ and from Lemma 5 neither i nor j will make any future i, j-moves.

Case (iv): $p_i = j$ and $p_j = i$. In this case it follows from Lemma 5 that neither process i nor process j will make any future i, j-moves.   □

It should be noted in the proof of Lemma 6 that only an edge (i, j) where we initially have either $p_i = j$ or $p_j = i$ (but not both) can result in three i, j-moves, otherwise the limit is two i, j-moves per edge. When we have three (i, j)-moves across an edge (i, j) we can charge these moves to the processor that was initially pointing to the other. In this way each process will at most be incident on one edge which it is charged three moves for. From this observation we can now give the following bound on the total number of steps needed to obtain a stable solution.

Theorem 2. *Algorithm 1. will stabilize after at most* $3 \cdot n + 2 \cdot m$ *steps under the distributed adversarial daemon.*

Proof. From Corollary 2 we know that there can be at most $2n$ *Update* moves, each which can occur in a separate step. From Lemma 6 it follows that there can at most be three i, j-moves per edge. But as observed, there is at most one such edge incident on each process i for which process i is charged for, otherwise the limit is two i, j-moves. Thus the total number of i, j-moves is at most $n + 2 \cdot m$ and the result follows.   □

From Theorem 2 it follows that Algorithm 1. will use $O(m)$ moves on any connected system when assuming a distributed daemon. Since the distributed daemon encompasses the sequential daemon this result also holds for the sequential daemon.

4.3   Convergence for the Distributed Fair Daemon

Next we consider the number of rounds used by Algorithm 1. when operated under the distributed fair daemon. Note that one round may encompass several steps, and we only require that every process eligible at the start of a round either executes at least one rule during the round or becomes ineligible to do so. This also implies that moves made in the same round may or may not be simultaneous. Since the fair distributed daemon is a subset of the adversarial distributed daemon any results that were shown in Section 4.2 also applies here. We will now show that Algorithm 1. converges after at most $2 \cdot n + 1$ rounds for this daemon.

We define that a process $i \in V$ is *active* if either $PRmarried(i)$ or $PRdead(i)$ is false. A process that is not active is *inactive*. From Corollary 2 it follows that any process $i \in V$ where $PRmarried(i)$ is true will not become active again for the rest of the algorithm. This also implies that if $PRdead(i)$ is true in some configuration then it will remain so for the rest of the algorithm.

Lemma 7. *Let $A \subseteq V$ be a maximal connected set of active processes in some configuration of the algorithm. If $|A| > 2$ then after at most four more rounds the size of A has decreased by at least 2.*

Proof. We first note that the size of A cannot increase during the execution of the algorithm. Assume now that no processes in A gets married during the next four rounds. We will show that this leads to a contradiction.

After the first round every process $j \in N(A)$ must have $m_i = true$. This follows since any process $j \in N(A)$ must have $PRdead(j) = $ false (by definition) and will therefore have $PRmarried(j) = true$. Thus if m_j is initially false for a process $j \in N(A)$ then after the first round m_j will be set to true. Similarly, if a node $i \in A$ has $m_i = true$ then m_i will be set to false after the first round. According to the assumption that no processes in A gets married, the m-values will not change during the next three rounds.

Next, consider any $i \in A$ that either initially or after the first round satisfies $p_i = j$ such that either $j \in N(A)$ or $j < i$ (or both). It follows that if $j \in N(A)$ then $m_j = true$ after the first round, and if $j < i$ then i will be eligible for an *Abandonment* move before j can execute a *Marriage* move (otherwise they get married). Thus in either case, process i is eligible for an *Abandonment* move no later than after the first round. Also note that the situation where $p_i = j$ and $j < i$ cannot occur again after the first round. This is because prior to this configuration we must have $p_j = i$ and $m_i = true$, which is not possible if $i \in A$.

Thus after the second round a process $i \in A$ cannot execute an *Abandonment* move since this requires that either $m_{p_i} = true$ or that $i > p_i$. Since no process can execute an *Abandonment* move it also follows that no process can execute a *Marriage* move since this would lead to two processes getting married. Thus at this stage a process can only execute a *Seduction* move and a process that is not eligible for a *Seduction* move at this point will not become eligible for a *Seduction* move after the third round since no m-value is changed and no p-value is set to *null* during the third round.

Hence, at the start of the third round we have that for every $i \in A$ either *(i)* $p_i = null$ or *(ii)* $p_i = j$ where $j \in N(j) \cap A$. If Case *(i)* is true for every process in A, then since $|A| \geq 2$ then at least the process with the lowest id in A is eligible for a *Seduction* move. Therefore no later than after the third round there exists at least one process $i_1 \in A$ where $p_{i_1} = i_2$ such that $i_2 \in N(j) \cap A$. Further, let $\{i_1, i_2, ..., i_k\}$ be a path of maximal length such that $i_{x+1} \in N(i_x) \cap A$ and $p_{i_x} = i_{x+1}$, $1 \leq x < k$. Note that while the *Seduction* moves made by the processes during the third round may be performed in different steps, no process will become eligible for an *Update* or *Abandonment* move, since they must be preceded by a *Marriage* and *Update* move, respectively. It follows that each $i_x \in A$ and also that $i_x < i_{x+1}$. Since the length of the path is finite we have $p_{i_k} = null$.

The process i_k is now eligible for a *Marriage* move and therefore cannot be eligible for any other move. As noted, process $p_{i_{k-1}}$ cannot be eligible for an *Abandonment* move at this point since $i_{k-1} < i_k$ and m_k = false. Thus following the fourth round processes i_{k-1} and i_k will become married, contradicting our assumption and the result follows.   □

Note that if A in Lemma 7 only contains one node i then either $PRwaiting(i)$ or $PRcondemned(i)$ must be true initially. In either case, after at most two moves i will have updated m_i and executed an *Abandonment* move such that $PRdead(i)$ is true.

Obviously $|A| \leq |V|$, and from Lemma 5 we know that once married, a process will remain married for the rest of the algorithm. From this we get that at most $2 \cdot n$ rounds are needed to find the matching. However, after the matching has been found every married process may execute an *Update* move, and every unmarried process may execute an *Abandonment* move. Both of these can be done in the same round. Note that it is not necessary for a process i that is unmarried when the algorithm terminates to execute a final *Update* move as m_i = false after the first round and remains false throughout the algorithm. From this we get the following theorem.

Theorem 3. *Algorithm 1. will stabilize after at most $2 \cdot n + 1$ rounds when using a fair distributed daemon.*

5   Conclusion

We have presented a new self-stabilizing algorithm for the maximal matching problem that improves the time step complexity of the previous best algorithm for the distributed adversarial daemon, while at the same time as meeting the bounds of the previous best algorithms for the sequential and the distributed fair daemon.

It is well known that a maximal matching is a $\frac{1}{2}$-approximation to the *maximum* matching, where the maximum matching is a matching such that no other matching with strictly greater size exists in the network. In [6], Goddard *et al.* provide a $\frac{2}{3}$-approximation for a particular class of networks (trees and rings of size not divisible by 3). Also, in particular networks such as Trees in [11,1] or bipartite graphs in [2], self-stabilizing algorithms have been proposed for maximum matching. However, no self-stabilizing solution with a better approximation ratio than $\frac{1}{2}$ currently exists for general graphs. Thus it would be of interest to know if it is possible to create a self-stabilizing algorithm for general graphs that achieves a better approximation ratio than $\frac{1}{2}$, or even an optimal solution.

References

1. Blair, J., Manne, F.: Efficient self-stabilzing algorithms for tree networks. In: ICDS, pp. 20–26. IEEE Computer Society Press, Los Alamitos (2003)
2. Chattopadhyay, S., Higham, L., Seyffarth, K.: Dynamic and self-stabilizing distributed matching. In: PODC, pp. 290–297 (2002)

3. Dijkstra, E.W.: Self-stabilizing systems in spite of distributed control. Commun. ACM 17(11), 643–644 (1974)
4. Dolev, S.: Self Stabilization. MIT Press (March 2000)
5. Goddard, W., Hedetniemi, S.T., Jacobs, D.P., Srimani, P.K.: Self-stabilizing protocols for maximal matching and maximal independent sets for ad hoc networks. In: IPDPS, p. 162. IEEE Computer Society Press, Los Alamitos (2003)
6. Goddard, W., Hedetniemi, S.T., Shi, Z.: An anonymous self-stabilizing algorithm for 1-maximal matching in trees. In: Arabnia, H.R. (ed.) PDPTA, pp. 797–803. CSREA Press (2006)
7. Gradinariu, M., Johnen, C.: Self-stabilizing neighborhood unique naming under unfair scheduler. In: Sakellariou, R., Keane, J.A., Gurd, J.R., Freeman, L. (eds.) Euro-Par 2001. LNCS, vol. 2150, pp. 458–465. Springer, Heidelberg (2001)
8. Gradinariu, M., Tixeuil, S.: Conflict managers for self-stabilization without fairness assumption. Technical Report 1459, LRI, Université Paris Sud (September 2006)
9. Hedetniemi, S.T., Jacobs, D.P., Srimani, P.K.: Maximal matching stabilizes in time $o(m)$. Inf. Process. Lett. 80(5), 221–223 (2001)
10. Hsu, S-C., Huang, S.-T.: A self-stabilizing algorithm for maximal matching. Inf. Process. Lett. 43(2), 77–81 (1992)
11. Karaata, M.H., Saleh, K.A.: Distributed self-stabilizing algorithm for finding maximum matching. Comput. Syst. Sci Eng. 15(3), 175–180 (2000)
12. Tel, G.: Maximal matching stabilizes in quadratic time. Inf. Process. Lett 49(6), 271–272 (1994)

Labeling Schemes with Queries

Amos Korman* and Shay Kutten**

Information Systems Group, Faculty of IE&M, The Technion, Haifa, 32000 Israel
pandit@tx.technion.ac.il, kutten@ie.technion.ac.il

Abstract. Recently, quite a few papers studied methods for representing network properties by assigning *informative labels* to the vertices of a network. Consulting the labels given to any two vertices u and v for some function f (e.g. "distance(u, v)") one can compute the function (e.g. the graph distance between u and v). Some very involved lower bounds for the sizes of the labels were proven.

In this paper, we demonstrate that such lower bounds are very sensitive to the number of vertices consulted. That is, we show several almost trivial constructions of such labeling schemes that beat the lower bounds by large margins. The catch is that one needs to consult the labels of *three* vertices instead of two. We term our generalized model *labeling schemes with queries*.

Additional contributions are several extensions. In particular, we show that it is easy to extend our schemes for tree to work also in the dynamic scenario. We also demonstrate that the study of the queries model can help in designing a scheme for the traditional model too. Finally, we demonstrate extensions to the non-distributed environment. In particular, we show that one can preprocess a general weighted graph using almost linear space so that flow queries can be answered in almost constant time.

Keywords: Labeling schemes, routing schemes, distance queries.

1   Introduction

Background: Network representations play a major role in many domains of computer science, ranging from data structures, graph algorithms, and combinatorial optimization to databases, distributed computing, and communication networks. In most traditional network representations, the names or identifiers given to the vertices betray no useful information, and they serve only as pointers to entries in the data structure, which forms a *global* representation of the network. Recently, quite a few papers studied methods for representing network properties by assigning *informative labels* to the vertices of the network (see e.g., [40,9,29,36,45]).

Let f be a function on pairs of vertices (e.g., distance). Informally, the goal of an f-labeling scheme is to label the vertices of a graph G in such a way that for

---

 * Supported in part at the Technion by an Aly Kaufman fellowship.
** Supported in part by a grant from the Israel Science Foundation.

G. Prencipe and S. Zaks (Eds.): SIROCCO 2007, LNCS 4474, pp. 109–123, 2007.

every two vertices $u, v \in G$, the value $f(u, v)$ (e.g., the distance between u and v) can be inferred by merely inspecting the labels of u and v. Of course, this can be done trivially using labels that are large enough (e.g., every label includes the description of the whole graph). Therefore, the main focus of the research concerning labeling schemes is to minimize the amount of information (the sizes of the labels) required. Informally, an f-labeling scheme can be viewed as a way of distributing the graph structure information concerning f to the vertices of the graph, using small chunks of information per vertex.

Rather involved proofs were introduced to lower bound the sizes of such labeling schemes. The main contribution of this paper is the demonstration that these lower bounds are very sensitive to the model used. Intuitively, if the labels of three vertices can be consulted (rather than two, such as u and v above), it is very easy to reduce the sizes significantly, much below the previous lower bounds. Moreover, our query labeling schemes can be obtained using very simple methods, sometimes trivial.

Elaborating somewhat more (formal definitions appear in Section 2), this paper introduces the notion of f-*labeling schemes with queries* which generalizes the notion of f-labeling schemes. The idea is to distribute the global information (relevant to f) to the vertices, in such a way that $f(u, v)$ can be inferred by inspecting not only the labels of u and v but possibly the labels of additional vertices. We note that all the constructions given in this paper calculate $f(u, v)$ by inspecting the labels of three vertices (u and v above, and some w). That is, given the labels of u and v, we first find a vertex w and then consult its label to derive $f(u, v)$. However, in the concluding section we discuss generalizations to inspecting additional labels.

We also show several additional extensions. One extension is meant to demonstrate an advantage of the simplicity of the design of labeling schemes with queries. It helped us to simplify the design of a labeling scheme for the traditional model (with no queries). We did that in two steps: first we designed the simple scheme with queries, and then "simulated" this scheme in the old model (with no queries). (The "simulation" had some associated cost, which made the resulting scheme work for an approximation function f, rather than for the exact function we would have liked).

A second extension of the result is to dynamic scenarios. We note that most previous research concerning distributed network representations considered the *static* scenario, in which the topology of the underlying network is fixed. This is probably due to the fact that designing for the dynamic scenario is more complex. Some recent papers did tackle task of labeling dynamic networks in a distributed fashion. Such labeling methods should, of course, be dynamic too. Indeed, the designs in these recent paper tend to be harder and more complex. We show that the effect of introducing a query to the dynamic case is similar to the effect on the static case. That is- the sizes can be reduced considerably, and the construction of the schemes is rather easy (though somewhat more complex than in the static case). To do that, we modify the model translation methods of [35] and [38], and then use them to extend our static labeling schemes with queries on trees to the dynamic scenario. We then show that the sizes of the resulted schemes are smaller than those of the schemes for the older model. The reduction in the label size is similar to the reduction in the static case.

In a final extension, we show that our methods are also useful in the non-distributed environment.

Related work: In this subsection we mostly survey results concerning labeling schemes (with no queries). However, let us first mention an area of research (namely, overlay and Peer to Peer networks) that may serve as a practical motivation for our work, and for some other studies concerning labeling schemes. We stress, though, that the main motivation for this paper is theoretical.

When the third vertex w (mentioned above) is near by to u, it may be quite cheap for u to access the main memory at w, sometimes even cheaper than consulting the disk at u itself. See, for example [41,34]. Indeed, some of our schemes below are based on such a "near by" w. Even when w is remote, accessing it may be cheap in some overlay networks. The main overhead there is finding w (which can be done in our constructions by u using v's label) and creating the connection to it. Such models are presented explicitly, e.g. in [31,15,2,48], where such remote accesses are used to construct and to use overlay data structures. Famous overlay data structures that can fit such models appear for example in [49,55,42,23,43]. In some of these overlay networks, a vertex w is addressed by its contents. This may motivate common labeling schemes that assume content addressability.

Implicit labeling schemes were first introduced in [12,40]. Labeling schemes supporting the adjacency and ancestry functions on trees were investigated in [40,9,8].

Distance labeling schemes were studied in [44,29,52,17,4,39]. In particular, [44] showed that the family of n vertex weighted trees with integer edge capacity of at most W enjoys a scheme using $O(\log^2 n + \log n \log W)$-bit labels. This bound was proven in [29] to be asymptotically optimal.

Labeling schemes for routing on trees were investigated in a number of papers until finally optimized in [21,22,54]. For the *designer port* model, in which the designer of the scheme can freely enumerate the port numbers of the nodes, [21] shows how to construct a routing scheme using labels of $O(\log n)$ bits on n-node trees. In the *adversary port* model, in which the port numbers are fixed by an adversary, they show how to construct a routing scheme using labels of $O(\log^2 n / \log \log n)$ bits on n-node trees. In [22] they show that both label sizes are asymptotically optimal. Independently, a routing scheme for trees using $(1 + o(1)) \log n$-bit labels was introduced in [54] for the designer port model.

Two variants of labeling schemes supporting the nearest common ancestor (NCA) function in trees appear in the literature. In an id-NCA labeling schemes, the vertices of the input graph are assumed to have disjoint identifiers (using $O(\log n)$ bits) given by an adversary. The goal of an id-NCA labeling scheme is to label the vertices such that given the labels of any two vertices u and v, one can find the identifier of the NCA of u and v. Static labeling schemes on trees supporting the separation level and id-NCA functions were given in [45] using $\Theta(\log^2 n)$-bit labels. The second variant considered is the label-NCA labeling scheme, whose goal is to label the vertices such that given the labels of any two vertices u and v, one can find the label (and not the pre-given identifier) of the NCA of u and v. In [5] they present a label-NCA labeling scheme on trees enjoying $\Theta(\log n)$-bit labels.

In [36] they give a labeling scheme supporting the flow function on n-node general graphs using $\Theta(\log^2 n + \log n \log W)$-bit labels, where W is the maximum capacity of an edge. They also show a labeling scheme supporting the k-vertex-connectivity function on general graphs using $O(2^k \log n)$-bit labels. See [27] for a survey on (static) labeling schemes.

Most of the research concerning labeling schemes in the dynamic settings considered the following two dynamic models on tree topologies. In the *leaf-dynamic* tree model, the topological event that may occur is that a leaf is either added to or removed from the tree. In the *leaf-increasing* tree model, the only topological event that may occur is that a leaf joins the tree.

The study of dynamic distributed labeling schemes was initiated in [38,37]. In [38], a dynamic labeling scheme is presented for distances in the leaf-dynamic tree model with $O(\log^2 n)$ label size and $O(\log^2 n)$ amortized message complexity, where n is the current tree size. β-approximate distance labeling schemes (in which, given two labels, one can infer a β-approximation to the distance between the corresponding nodes) are presented [37]. Their schemes apply for dynamic models in which the tree topology is fixed but the edge weights may change.

Two general translation methods for extending static labeling schemes on trees to the dynamic setting are considered in the literature. Both approaches fit a number of natural functions on trees, such as ancestry, routing, label-NCA, id-NCA etc. Given a static labeling scheme on trees, in the leaf-increasing tree model, the resulting dynamic scheme in [38] incurs overheads (over the static scheme) of $O(\log n)$ in both the label size and the communication complexity. Moreover, if an upper bound n_f on the final number of vertices in the tree is known in advance, the resulting dynamic scheme in [38] incurs overheads (over the static scheme) of $O(\log^2 n_f / \log \log n_f)$ in the label size and only $O(\log n / \log \log n)$ in the communication complexity. In the leaf-dynamic tree model there is an extra additive factor of $O(\log^2 n)$ to the amortized message complexity of the resulted schemes.

In [35], it is shown how to construct for many functions $k(x)$, a dynamic labeling scheme in the leaf-increasing tree model extending a given static scheme, such that the resulting scheme incurs overheads (over the static scheme) of $O(\log_{k(n)} n)$ in the label size and $O(k(n) \log_{k(n)} n)$ in the communication complexity. As in [38], in the leaf-dynamic tree model there is an extra additive factor of $O(\log^2 n)$ to the amortized message complexity of the resulted schemes. In particular, by setting $k(n) = n^\epsilon$, dynamic labeling schemes are obtained with the same asymptotic label size as the corresponding static schemes and sublinear amortized message, namely, $O(n^\epsilon)$.

1.1 Our Contribution

We introduce the notion of f-labeling schemes with queries that is a natural generalization of the notion of f-labeling schemes. Using this notion we demonstrate that by increasing slightly the number of vertices whose labels are inspected, the size of the labels decreases considerably. Specifically, we inspect the labels of 3 vertices instead of 2, that is, we use a single *query*. In particular, we show that there exist simple labeling schemes with one query supporting the distance function on n-node trees as well as the flow function on n-node general graphs with label size $O(\log n + \log W)$, where W is the maximum (integral) capacity of an edge. (We note that the lower bound for labeling schemes without queries for each of these problems is $\Omega(\log^2 n + \log n \log W)$ [29,36].) We also show that there exists a labeling scheme with one query supporting the id-NCA function on n-node trees with label size $O(\log n)$. (The lower bound for schemes without queries is $\Omega(\log^2 n)$ [45].) In addition, we show a routing labeling scheme with one query in the fixed-port model using $O(\log n)$-bit labels. (The lower bound (see [22]) for the case of no queries is $\Omega(\frac{\log^2 n}{\log \log n})$.) We note that all the schemes we introduce have asymptotically optimal label size for schemes with one query. (The matching lower bound proofs are straightforward in most of the cases.) Moreover, most of the results are obtained by simple constructions, which strengthens the motivation for this model.

We then show several extensions that are somewhat more involved. In particular, we show that our labeling schemes with queries on trees can be extended to the dynamic scenario using model translation methods based on those of [38,35]. In order to save in the message complexity, we needed to make some adaptations to those methods, as well as to one of the static routing schemes of [21]. Second, we show that the study of the queries model can help with the traditional model too. That is, using ideas from our routing labeling scheme with one query, we show how to construct a 3-approximation routing scheme *without queries* for unweighted trees in the fixed-port model with $\Theta(\log n)$-bit labels.

Finally, we turn to a non-distributed environment and demonstrate similar constructions. That is, first, we show a simple method to transform previous results on NCA queries on static and dynamic trees in order to support also distance queries. Then, we show that one can preprocess a general weighted graph using almost linear space so that flow queries can be answered in almost constant time.

2 Preliminaries

Let T be a tree and let v be a vertex in T. Let $deg(v)$ denote the degree of v. For a non-root vertex $v \in T$, let $p(v)$ denote the parent of v in T. In the case where the tree T is weighted (respectively, unweighted), the *depth* of a vertex is defined as its weighted (resp., unweighted) distance to the root. The *nearest common ancestor* of u and w, $NCA(u,w)$, is the common ancestor of both u and w of maximum depth. Let $\mathcal{T}(n)$ denote the family of all n-node unweighted trees. Let

$T(n, W)$ (respectively, $\mathcal{G}(n, W)$) denote the family of all n-node weighted trees (resp., connected graphs) with (integral) edge weights bounded above by W.

Incoming and outgoing links from every node are identified by so called *port-numbers*. When considering routing schemes, we distinguish between the following two variants of port models. In the *designer port* model the designer of the scheme can freely assign the port numbers of each vertex (as long as these port numbers are unique), and in the *fixed-port* model the port numbers at each vertex are assigned by an adversary. We assume that each port number is encoded using $O(\log n)$ bits.

We consider the following functions which are applied on pairs of vertices u and v in a graph $G = \langle V, E \rangle$. (1) **flow** (maximum legal flow between u and v), (2) **distance** (either weighted, or unweighted), (3) **routing** (the port in u to the next vertex towards v). If the graph is a tree T then we consider also the following functions: (4) **separation level** (depth of $NCA(u, v)$), (5) **id-NCA**, (6) **label-NCA**. In (5) above, it is assumed that identities containing $O(\log n)$ bits are assigned to the vertices by an adversary, and $id - NCA(u, v)$ is the identity of $NCA(u, v)$. In (6) above, it is assumed that each vertex can freely select its own identity (as long as all identities remain unique). In this case, the identities may also be referred to as labels.

Labeling schemes and c-query labeling schemes: Let f be a function defined on pairs of vertices. An f-*labeling scheme* $\pi = \langle \mathcal{M}, \mathcal{D} \rangle$ for a family of graphs $\mathcal{F}$ is composed of the following components:

1. A *marker* algorithm $\mathcal{M}$ that given a graph $G \in \mathcal{F}$, assigns a label $\mathcal{M}(v)$ to each vertex $v \in G$.
2. A (polynomial time) *decoder* algorithm $\mathcal{D}$ that given the labels $\mathcal{M}(u)$ and $\mathcal{M}(v)$ of two vertices u and v in some graph $G \in \mathcal{F}$, outputs $f(u, v)$.

The most common measure used to evaluate a labeling scheme $\pi = \langle \mathcal{M}, \mathcal{D} \rangle$, is the *label size*, i.e., the maximum number of bits used in a label $\mathcal{M}(v)$ over all vertices v in all graphs $G \in \mathcal{F}$.

Let c be some constant integer. Informally, in contrast to an f-labeling scheme, in a c-query f-labeling scheme, given the labels of two vertices u and v, the decoder may also consult the labels of c other vertices. More formally, a c-*query* f-*labeling scheme* $\varphi = \langle \mathcal{M}, Q, \mathcal{D} \rangle$ is composed of the following components:

1. A *marker* algorithm $\mathcal{M}$ that given a graph $G \in \mathcal{F}$, assigns a label $\mathcal{M}(v)$ to each vertex $v \in G$. This label is composed of two sublabels, namely, $\mathcal{M}^{index}(v)$ and $\mathcal{M}^{data}(v)$, where it is required that the index sublabels are unique, i.e., for every two vertices v and u, $\mathcal{M}^{index}(v) \neq \mathcal{M}^{index}(u)$. (In other words, the index sublabels can serve as identities.)
2. A (polynomial time) *query* algorithm Q that given the labels $\mathcal{M}(u)$ and $\mathcal{M}(v)$ of two vertices u and v in some graph $G \in \mathcal{F}$, outputs $Q(\mathcal{M}(u), \mathcal{M}(v))$ which is a set containing the indices (i.e., the first sublabels) of c vertices in G.
3. A (polynomial time) *decoder* algorithm $\mathcal{D}$ that given the labels $\mathcal{M}(u)$ and $\mathcal{M}(v)$ of two vertices u and v and the labels of the vertices in $Q(\mathcal{M}(u), \mathcal{M}(v))$, outputs $f(u, v)$.

As in the case of f-labeling schemes, we evaluate a c-query f-labeling scheme $\varphi = \langle \mathcal{M}, Q, \mathcal{D} \rangle$ by its *label size*, i.e., the maximum number of bits used in a label $\mathcal{M}(v)$ over all vertices v in all graphs $G \in \mathcal{F}$. We note that all the schemes in this paper use $c = 1$. Let us comment also that clearly, since the index sublabels must be disjoint, any c-query f-labeling scheme on any family of n-node graphs must have label size $\Omega(\log n)$. See Section 7 for alternative definition for query labeling schemes.

2.1   Routing Schemes and β-Approximation Routing Schemes

A *routing scheme* is composed of a *marker algorithm* $\mathcal{M}$ for assigning each vertex v of a graph G with a label $\mathcal{M}(v)$, coupled with a *router* algorithm $\mathcal{R}$ whose inputs are the header of a message, $\mathcal{M}(v)$ and the label $\mathcal{M}(y)$ of a destination vertex y. If a vertex x wishes to send a message to vertex y, it first prepares and attaches a header to the message. Then the router algorithm x outputs a port of x on which the message is delivered to the next vertex. This is repeated in every vertex until the message reaches the destination vertex y. Each intermediate vertex u on the route may replace the header of the message with a new header and may perform a local computation. The requirement is that the weighted length of resulting path connecting x and y is the same as the distance between x and y in G. For a constant β, a β-approximation routing scheme is the same as a routing scheme except that the requirement is that the length of resulting route connecting x and y is a β-approximation for the distance between x and y in G.

In addition to the label size, we also measure a routing scheme (and a β-approximation routing scheme) by the *header size*, i.e., the maximum number of bits used in a header of a message.

3   Labeling Schemes with One Query

In this section we demonstrate that the query model allows for significantly shorter labels. In particular, we describe simple 1-query labeling schemes with labels that beat the lower bounds in the following well studied cases: for the family of n-node trees, schemes supporting the routing (in the fixed-port model), distance, separation level, and the id-NCA functions; for the family of n-node general graphs, a scheme supporting the flow function. We note that all the schemes we present use asymptotically optimal labels.

Most of the 1-query labeling schemes obtained in this section use the label-NCA labeling scheme $\pi_{NCA} = \langle \mathcal{M}_{NCA}, \mathcal{D}_{NCA} \rangle$ described in [5]. Given an n-node, the marker algorithm $\mathcal{M}_{NCA}$ assigns each vertex v a distinct label $\mathcal{M}_{NCA}(v)$ using $O(\log n)$ bits. Given the labels $\mathcal{M}_{NCA}(v)$ and $\mathcal{M}_{NCA}(u)$ of two vertices v and u in the tree, the decoder $\mathcal{D}_{NCA}$ outputs the label $\mathcal{M}_{NCA}(w)$.

3.1   Id-NCA Function in Trees

We first describe a 1-query scheme $\varphi_{id-NCA} = \langle \mathcal{M}_{id-NCA}, \mathcal{Q}_{id-NCA}, \mathcal{D}_{id-NCA} \rangle$ that demonstrates how easy it is to support the id-NCA function on $\mathcal{T}(n)$ using one query and $O(\log n)$-bit labels. (Recall that the lower bound on schemes without queries is $\Omega(\log^2 n)$ [45].)

Informally, the idea behind φ_{id-NCA} is to have the labels of u and v (their first sublabels) be the labels given by the label-NCA labeling scheme $\pi_{NCA}(v)$. Hence, they are enough for the query algorithm to find the π_{NCA} label of their nearest common ancestor w. Then, the decoder algorithm finds w's identity simply in the second sublabel of w.

Let us now describe the 1-query labeling scheme φ_{id-NCA} more formally. Given a tree T, recall that it is assumed that each vertex v is assigned a unique identity $id(v)$ by an adversary and that each such identity is composed of $O(\log n)$ bits. The marker algorithm $\mathcal{M}_{id-NCA}$ labels each vertex v with the label $\mathcal{M}_{id-NCA}(v) = \langle \mathcal{M}^{index}_{id-NCA}(v), \mathcal{M}^{data}_{id-NCA}(v) \rangle = \langle \mathcal{M}_{NCA}(v), id(v) \rangle$. Given the labels $\mathcal{M}_{id-NCA}(v)$ and $\mathcal{M}_{id-NCA}(u)$ of two vertices v and u in the tree, the query algorithm $\mathcal{Q}_{id-NCA}$ uses the decoder $\mathcal{D}_{NCA}$ applied on the corresponding first sublabels to output the sublabel $\mathcal{M}^{index}_{id-NCA}(w) = \mathcal{M}_{NCA}(w)$, where w is the NCA of v and u. Given the labels $\mathcal{M}_{id-NCA}(v)$, $\mathcal{M}_{id-NCA}(u)$ and $\mathcal{M}_{id-NCA}(w)$ where w is the NCA of v and u, the decoder $\mathcal{D}_{id-NCA}$ simply outputs the second sublabel of w, i.e., $\mathcal{M}^{data}_{id-NCA}(w) = id(w)$. The fact that φ_{id-NCA} is a correct 1-query labeling scheme for the id-NCA function on $\mathcal{T}(n)$ follows from the correctness of the label-NCA labeling scheme π_{NCA}. Since the label size of $\pi_{NCA}(v)$ is $O(\log n)$ and since the identity of each vertex v is encoded using $O(\log n)$ bits, we obtain that the label size of φ_{id-NCA} is $O(\log n)$. As mentioned before, since the index sublabels must be disjoint, any query labeling scheme on $\mathcal{T}(n)$ must have label size $\Omega(\log n)$. The following lemma follows.

Lemma 1. *The label size of a 1-query id-NCA labeling scheme on $\mathcal{T}(n)$ is $\Theta(\log n)$.*

3.2   Distance and Separation Level in Trees

The above method can be applied for other functions. For example, let us now describe 1-query labeling schemes $\varphi_{sep-level}$ and φ_{dist} supporting the distance and separation level functions respectively on $\mathcal{T}(n, W)$. Both our scheme have label size $\Theta(\log n + \log W)$. Recall that any labeling scheme (without queries) supporting either the distance function or the separation level function on $\mathcal{T}(n, W)$ must have size $\Omega(\log^2 n + \log n \log W)$, [29,45]. The proof of the following lower bound claim is deferred to the full paper.

Claim. Let c be a constant. Any c-query labeling scheme supporting either the separation level function or the distance function on $\mathcal{T}(n, W)$ must have label size $\Omega(\log W + \log n)$.

The construction of our 1-query labeling schemes supporting the separation level and distance functions uses a similar method to the one described in Subsection

3.1. Both schemes are based on keeping the depth of a vertex in its data sub-label (instead of its identity). The correctness of the 1-query labeling scheme supporting the distance function is based on the following equation.

$$d(v, u) = depth(v) + depth(u) - 2 \cdot depth(NCA(v, u)). \qquad (1)$$

The description of these schemes as well as the proof of the following lemma is deferred to the full paper.

Lemma 2. *The label size of a 1-query labeling scheme supporting either the separation-level or the distance function on $T(n, W)$ is $\Theta(\log n + \log W)$.*

3.3   Routing in Trees Using One Query

As mentioned before, any 1-query routing labeling scheme on $T(n)$ must have label size $\Omega(\log n)$. In this subsection, we establish a 1-query routing labeling scheme φ_{fix} in the fixed-port model using $O(\log n)$-bit labels.

In [21], they give a routing scheme $\pi_{des} = \langle \mathcal{M}_{des}, \mathcal{D}_{des} \rangle$ for the designer port model in $T(n)$. Given a tree $T \in T(n)$, for every vertex $v \in T$, and every neighbor u of v, let $port_{des}(v, u)$ denote the port number (assigned by the designer of the routing scheme π_{des}) leading from v to u. In particular, the port number leading from each non-root vertex v to its parent $p(v)$ is assigned the number 1, i.e., $port_{des}(v, p(v)) = 1$. Given the labels $\mathcal{M}_{des}(v)$ and $\mathcal{M}_{des}(w)$ of two vertices v and w in T, the decoder $\mathcal{D}_{des}$ outputs the port number $port_{des}(v, u)$ at v leading from v to the next vertex u on the shortest path connecting v and w.

Let T be an n-node tree. We refer to a port number assigned by the designer of the routing scheme π_{des} as a *designer port number* and to a port number assigned by the adversary as an *fixed-port number*. Let *port* be some port of a vertex in the fixed-port model. Besides having a fixed-port number assigned by the adversary, we may also consider *port* as having a designer port number, the number that would have been assigned to it had we been in the designer port model. For a port leading from vertex v to vertex u, let $port_{fix}(v, u)$ denote its fixed-port number and let $port_{des}(v, u)$ denote its designer port number.

We now describe our 1-query routing scheme $\varphi_{fix} = \langle \mathcal{M}_{fix}, \mathcal{Q}_{fix}, \mathcal{D}_{fix} \rangle$ which operates in the fixed-port model. Given a a tree $T \in T(n)$ and a vertex $v \in T$, the index sublabel of v is composed of two fields, namely, $\mathcal{M}^{index}(v) = \langle \mathcal{M}_1^{index}(v), \mathcal{M}_2^{index}(v) \rangle$ and the data sublabel of v is composed of three fields, namely, $\mathcal{M}^{data}(v) = \langle \mathcal{M}_1^{data}(v), \mathcal{M}_2^{data}(v), \mathcal{M}_3^{data}(v) \rangle$. If v is not the root then the index and data sublabels of v are $\mathcal{M}^{index}(v) = \langle \mathcal{M}_{des}(p(v)), port_{des}(p(v), v) \rangle$ and $\mathcal{M}^{data}(v) = \langle \mathcal{M}_{des}(v), port_{fix}(p(v), v), port_{fix}(v, p(v)) \rangle$. Note that we use the designer port number as a part of the label in the fixed-port model. More-over, the designer port number at the parent is used to label the child in the fixed-port model. Also note that the index sublabel is unique, since $\mathcal{M}_{des}(x)$ must be unique for π_{des} to be a correct routing scheme.

The index sublabel of the root r of T is $\langle 0, 0 \rangle$ and the data sublabel of r is $\mathcal{M}^{data}(r) = \langle \mathcal{M}_{des}(r), 0, 0 \rangle$. Note that since the labels given by the marker algorithm $\mathcal{M}_{des}$ are unique, the index sublabels of the vertices are unique.

Given the labels $\mathcal{M}(v)$ and $\mathcal{M}(w)$ of two vertices v and w, the decoder $\mathcal{D}$ first checks whether $\mathcal{D}_{des}(\mathcal{M}_1^{data}(v), \mathcal{M}_1^{data}(w)) = 1$, i.e., whether the next vertex on the shortest path leading from v to w is v's parent. In this case, the query algorithm is ignored and the decoder $\mathcal{D}_{fix}$ simply outputs $\mathcal{M}_3^{data}(v)$ which is the (fixed) port number at v leading to its parent. Otherwise, the query algorithm Q_{fix} outputs $\langle \mathcal{M}_1^{data}(v), \mathcal{D}_{des}(\mathcal{M}_1^{data}(v), \mathcal{M}_1^{data}(w)) \rangle = \langle \mathcal{M}_1^{data}(v), \mathcal{D}_{des}$ $(\mathcal{M}_{des}(v), \mathcal{M}_{des}(w)) \rangle$ which is precisely the index sublabel of u, the next vertex on the shortest path leading from v to w (and a child of v), i.e., $\langle \mathcal{M}_1^{data}(v),$ $port_{des}(v, u) \rangle$. Therefore, given labels $\mathcal{M}_{fix}(v), \mathcal{M}_{fix}(w)$ and label $\mathcal{M}_{fix}(u)$, the decoder $\mathcal{D}_{fix}$ outputs $\mathcal{M}_2^{data}(u)$ which is the desired port number $port_{fix}(v, u)$. Since the label size of π_{des} is $O(\log n)$ and since each port number is encoded using $O(\log n)$ bits, we obtain the following lemma.

Lemma 3. *In the fixed-port model, the label size of a 1-query routing scheme on $\mathcal{T}(n)$ is $\Theta(\log n)$.*

3.4   Flow in General Graphs

We now consider the family $\mathcal{G}(n, W)$ of connected n-node weighted graphs with maximum edge capacities W, and present a 1-query flow labeling scheme φ_{flow} for this family using $O(\log n + \log W)$-bit labels. Recall that any labeling scheme (without queries) supporting the flow function on $\mathcal{G}(n, W)$ must have size $\Omega(\log^2 n + \log n \log W)$ [36]. The proof of the following lemma is deferred to the full paper.

Lemma 4. *The label size of a 1-query flow labeling scheme on $\mathcal{G}(n, W)$ is $\Theta(\log n + \log W)$.*

4   A 3-Approximation Routing Scheme in the Fixed-Port Model

We construct a 3-approximation routing scheme (without queries) on $\mathcal{T}(n)$ by applying the method described in Subsection 3.3 to the traditional model. Our 3-approximation routing labeling scheme π_{approx} operates in the fixed-port model and has label size and header size $O(\log n)$. Recall that any (precise) routing scheme on $\mathcal{T}(n)$ must have label size $\Omega(\log^2 n / \log \log n)$ [22]. We note that our ideas for translating routing schemes from the designer port model to the fixed-port model implicitly appear in [1], however, a 3-approximation routing scheme (without queries) on $\mathcal{T}(n)$ is not explicitly constructed there. The description of the 3-approximation routing labeling scheme π_{approx} as well as the proof of the following lemma is deferred to the full parer.

Lemma 5. *π_{approx} is a correct 3-approximation routing scheme on $\mathcal{T}(n)$ operating in the fixed port model. Moreover, its label size and header size are $\Theta(\log n)$.*

5 Adapting the 1-Query Schemes on Trees to the Dynamic Setting

In this section we show how to translate our 1-query labeling schemes on trees to the dynamic settings, i.e, to the leaf-increasing and leaf-dynamic tree models, [35,38] (see also "Related work" in Section 1). In a dynamic scheme, the marker protocol updates the labels after every topological change. We show that the reduction in the label sizes obtained by introducing a single query in the dynamic scenario is similar to the reduction in the static case.

To describe the adaptation fully, we need to give many details about the methods of [38,35,21]. Unfortunately, this is not possible in this extended abstract.

The initial idea is to apply the methods introduced in [38,35] to convert labeling schemes for static networks to work on dynamic networks too. Unfortunately, we cannot do this directly, since these methods were designed for traditional labeling schemes and not for 1-query labeling schemes.

The next idea is to perform the conversion indirectly. That is, recall (Section 3) that our 1-query labeling schemes utilize components that are schemes in the traditional model (with no queries). That is, some utilize π_{NCA}, the label-NCA labeling scheme of [5] and some utilize π_{des}, the routing scheme of [21]. Hence, one can first convert these components to the dynamic setting. Second, one can attempt to use the resulted dynamic components in a similar way that we used the static components in Section 3. This turns out to be simple in the cases of the distance, separation level and id-NCA functions, but more involved in the case of the routing function. The necessary modifications of the schemes, as well as the proof of the following theorem, appear in the full paper.

Theorem 1. *Consider the fixed-port model and let $k(x)$ be any function satisfying that $k(x)$, $\log_{k(x)} x$ and $\frac{k(x)}{\log k(x)}$ are nondecreasing functions and that $k(\Theta(x)) = \Theta(k(x))$.[1] There exist dynamic 1-query labeling schemes supporting the distance, separation level, id-NCA and routing functions on trees with the following complexities.*

1. *In the leaf-increasing tree model, with label size $O(\log_{k(n)} n \cdot \log n)$ and amortized message complexity $O(k(n) \cdot \log_{k(n)} n)$.*
2. *In the leaf-increasing tree model, if an upper bound n_f on the number vertices in the dynamically growing tree is known in advance, with label size $O(\frac{\log^3 n_f}{\log \log n_f})$ and amortized message complexity $O(\frac{\log n_f}{\log \log n_f})$.*
3. *In the leaf-dynamic tree model, with label size $O(\log_{k(n)} n \cdot \log n)$ and amortized message complexity $O\left(\sum_i k(n_i) \cdot \log_{k(n_i)} n \cdot \frac{MC(\pi, n_i)}{n_i}\right) + O(\sum_i \log^2 n_i)$.*

6 Applications in a Non-distributed Environments

Distance queries in trees: Harel and Tarjan [33] describe a linear time algorithm to preprocess a tree and build a data structure allowing NCA queries to be

---

[1] The above requirements are satisfied by most natural sublinear functions such as $\alpha x^\epsilon \log^\beta x$, $\alpha \log^\beta \log x$ etc.

answered in constant time on a RAM. Subsequently, simpler algorithms with better constant factors have been proposed in [47,13,25,51,14]. On a pointer machine, [33] show a lower bound of $\Omega(\log \log n)$ on the query time, which matches the upper bound of [53]. In the leaf-increasing tree model, [20,10] show how to make updates in amortized constant time while keeping the constant worst-case query time on a RAM, or the $O(\log \log n)$ worst-case query time on a pointer machine. In [16] they show how to maintain the above mentioned results on a RAM in the leaf-dynamic tree model with worst case constant update. See [5] for a survey.

Simply by adding a pointer from each vertex to its depth and using Equation 1, we obtain the following lemma.

Lemma 6. *The results of [33,47,13,25,51,14,33,53,20,10,16] can be translated to support either distance queries or separation level queries (instead of NCA queries).*

We note that other types of dynamic models were studied in the non-distributed environment regarding NCA queries (e.g. [16,10]). However, in these types of topological changes, our transformation to distance queries is not efficient since any such changes may effect the depth of too many vertices.

Flow queries in general graphs: Let $G \in \mathcal{G}(n, W)$ and let $u_1, u_2, \cdots, u_n$ be the set of vertices of G. Recall that in in [36], they show how to construct a weighted tree $\tilde{T}_G \in \mathcal{T}(O(n), W \cdot n)$ with n leaves $v_1, v_2, \cdots, v_n$ such that $flow_G(u_i, u_j) = sep - level_{T_G}(v_i, v_j)$. Using Lemma 6 applied on the results of [33], we can preprocess $\tilde{T}_G$ with $O(n \cdot \max\{1, \frac{\log W}{\log n}\})$ space such that separation level queries can be answered in $O(\max\{1, \frac{\log W}{\log n}\})$ time. The exact model needed to prove the following lemma formally is deferred to the full paper.

Lemma 7. *Any graph $G \in \mathcal{G}(n, W)$ can be preprocessed using $O(n \cdot \max\{1, \frac{\log W}{\log n}\})$ space such that flow queries can be answered in $O(\max\{1, \frac{\log W}{\log n}\})$ time.*

7   Conclusion and Open Problems

In this paper we demonstrate that considering the labels of three vertices, instead of two, can lead to a significant reduction in the sizes of the labels. Inspecting two labels, and inspecting three, are approaches that lie on one end of a spectrum. On the other end of the spectrum would be a representation for which the decoder inspects the labels of all the nodes before answering (n-query labeling schemes). It is not hard to show that for any graph family $\mathcal{F}$ on n node graphs, and for many functions (for example, distance or adjacency), one can construct an n-query labeling scheme on $\mathcal{F}$ using asymptotically optimal labels, i.e, $\log |\mathcal{F}|/n + \Theta(\log n)$-bit labels (though the decoder may not be polynomial). The idea behind such a scheme is to enumerate the graphs in $\mathcal{F}$ arbitrarily. Then, given some $G \in \mathcal{F}$ whose number in this enumeration is i, distribute the binary representation of i among the vertices of G. In this way, given the labels of all nodes, the decoder can reconstruct the graph and answer the desired query. Therefore, a natural question is to examine other points in this spectrum, i.e, examine c-query labeling schemes for $1 < c < n$.

There are other dimensions to the above question. For example, by our definition, given the labels of two vertices, the c vertices that are chosen by the query algorithm Q are chosen simultaneously. Alternatively, one may define a possibly stronger model in which these c vertices are chosen one by one, i.e., the next vertex is determined using the knowledge obtained from the labels of previous vertices.

References

1. Abraham, I., Gavoille, C., Malkhi, D.: Routing with improved communication-space trade-off. In: Proc. 18th Int. Symp. on Distributed Computing (October 2004)
2. Angluin, D., Aspnes, J., Chen, J., Wu, Y., Yin, Y.: Fast construction of overlay networks. In: Proc. SPAA 2005, pp. 145–154 (2005)
3. Afek, Y., Awerbuch, B., Plotkin, S.A., Saks, M.: Local management of a global resource in a communication. J. of the ACM 43, 1–19 (1996)
4. Alstrup, S., Bille, P., Rauhe, T.: Labeling schemes for small distances in trees. In: Proc. 14th ACM-SIAM Symp. on Discrete Algorithms (January 2003)
5. Alstrup, S., Gavoille, C., Kaplan, H., Rauhe, T.: Nearest Common Ancestors: A Survey and a new Distributed Algorithm. Theory of Computing Systems 37, 441–456 (2004)
6. Alstrup, S., Holm, J., Thorup, M.: Maintaining Center and Median in Dynamic Trees. In: Proc. 7th Scandinavian Workshop on Algorithm Theory (July 2000)
7. Abiteboul, S., Kaplan, H., Milo, T.: Compact labeling schemes for ancestor queries. In: Proc. 12th ACM-SIAM Symp. on Discrete Algorithms (January 2001)
8. Alstrup, S., Rauhe, T.: Improved Labeling Scheme for Ancestor Queries. In: Proc. 19th ACM-SIAM Symp. on Discrete Algorithms (January 2002)
9. Alstrup, S., Rauhe, T.: Small induced-universal graphs and compact implicit graph representations. In: Proc. 43rd annual IEEE Symp. on Foundations of Computer Science (November 2002)
10. Alstrup, S., Thorup, M.: Optimal pointer algorithms for finding nearest common ancestors in dynamic trees. J. of Algorithms 35(2), 169–188 (2000)
11. Breuer, M.A.: Coding the vertexes of a graph. IEEE Trans. on Information Theory IT-12, 148–153 (1966)
12. Breuer, M.A., Folkman, J.: An unexpected result on coding the vertices of a graph. J. of Mathematical Analysis and Applications 20, 583–600 (1967)
13. Bender, M.A., Farach-Colton, M.: The LCA problem revised. In: 4th LATIN, pp. 88–94 (2000)
14. Berkman, O., Vishkin, U.: Recursive star-tree parallel data structure. SIAM J. on Computing 22(2), 221–242 (1993)
15. Cidon, I., Gopal, I.S., Kutten, S.: New models and algorithms for future networks. IEEE Transactions on Information Theory 41(3), 769–780 (1995)
16. Cole, R., Hariharan, R.: Dynamic LCA Queries on Trees. SIAM J. on Computing 34(4), 894–923 (2005)
17. Cohen, E., Halperin, E., Kaplan, H., Zwick, U.: Reachability and Distance Queries via 2-hop Labels. In: Proc. 13th ACM-SIAM Symp. on Discrete Algorithms (January 2002)

18. Cohen, E., Kaplan, H., Milo, T.: Labeling dynamic XML trees. In: Proc. 21st ACM Symp. on Principles of Database Systems (June 2002)
19. Eppstein, D., Galil, Z., Italiano, G.F.: Dynamic Graph Algorithms. In: Atallah, M.J. (ed.) Algorithms and Theoretical Computing Handbook, CRC Press, Boca Raton, FL (1999)
20. Gabow, H.N.: Data Structure for Weighted Matching and Nearest Common Ancestor with Linking. In: Proc.1st Annual ACM Symp. on Discrete Algorithms, pp. 434–443 (January 1990)
21. Fraigniaud, P., Gavoille, C.: Routing in trees. In: Orejas, F., Spirakis, P.G., van Leeuwen, J. (eds.) ICALP 2001. LNCS, vol. 2076, pp. 757–772. Springer, Heidelberg (2001)
22. Fraigniaud, P., Gavoille, C.: A space lower bound for routing in trees. In: Proc. 19th Int. Symp. on Theoretical Aspects of Computer Science, pp. 65–75 (March 2002)
23. Fraigniaud, P., Gauron, P.: D2B: A de Bruijn based content-addressable network. Theor. Comput. Sci. 355(1), 65–79 (2006)
24. Feigenbaum, J., Kannan, S.: Dynamic Graph Algorithms. In: Handbook of Discrete and Combinatorial Mathematics, CRC Press, Boca Raton, FL (2000)
25. Gabow, H.N., Bentley, J.L., Tarjan, R.E.: Scaling and related techniques for geometry problems. In: Proc. 16th Annual ACM Symp. on Theory of Computing (May 1984)
26. Gavoille, C., Paul, C.: Split decomposition and distance labeling: an optimal scheme for distance hereditary graphs. In: Proc. European Conf. on Combinatorics, Graph Theory and Applications (September 2001)
27. Gavoille, C., Peleg, D.: Compact and Localized Distributed Data Structures. J. of Distributed Computing 16, 111–120 (2003)
28. Gavoille, C., Katz, M., Katz, N.A., Paul, C., Peleg, D.: Approximate Distance Labeling Schemes. In: Meyer auf der Heide, F. (ed.) ESA 2001. SV-LNCS, vol. 2161, pp. 476–488. Springer, Heidelberg (2001)
29. Gavoille, C., Peleg, D., Pérennes, S., Raz, R.: Distance labeling in graphs. J. of Algorithms 53(1), 85–112 (2004)
30. Harel, H.: A linear time algorithm for lowest common ancestors problem. In: 21st Annual IEEE Symp. on Foundation of Computer Science (November 1980)
31. Harchol-Balter, M., Leighton, F.T., Lewin, D.: Resource Discovery in Distributed Networks. In: Proc. PODC 1999, pp. 229–237 (1999)
32. Holm, J., Lichtenberg, K., Thorup, M.: Poly-logarithmic deterministic fully-dynamic algorithms for connectivity, minimum spanning tree, 2-edge, and biconnectivity. J. of the ACM 48(4), 723–760 (2001)
33. Harel, H., Tarjan, R.E.: Fast algorithms for finding nearest common ancestors. SIAM J. on Computing 13(2), 338–355 (1984)
34. Jamrozik, H.A., Feeley, M.J., Voelker, G.M., Evans, J., Karlin, A.R., Levy, H.M., Vernon, M.K.: Reducing network latency using subpages in a global memory environment. In: Proc. the 7th ACM Conference on Architectural Support for Programming Languages and Operating Systems (1996)
35. Korman, A.: General Compact Labeling Schemes for Dynamic Trees. In: Proc. 19th Int. Symp. on Distributed Computing (September 2005)
36. Katz, M., Katz, N.A., Korman, A., Peleg, D.: Labeling schemes for flow and connectivity. SIAM Journal on Computing 34, 23–40 (2004)
37. Korman, A., Peleg, D.: Labeling Schemes for Weighted Dynamic Trees. In: Proc. 30th Int. Colloq. on Automata, Languages & Prog., SV LNCS (July 2003)

38. Korman, A., Peleg, D., Rodeh, Y.: Labeling schemes for dynamic tree networks. Theory of Computing Systems 37, 49–75 (2004)
39. Korman, A., Peleg, D., Rodeh, Y.: Constructing Labeling schemes through Universal Matrices. In: Proc. 17th Int. Symp. on Algorithms and Computation (December 2006)
40. Kannan, S., Naor, M., Rudich, S.: Implicit representation of graphs. In SIAM J. on Discrete Math 5, 596–603 (1992)
41. Markatos, E.P., Dramitinos, G.: Remote Memory to Avoid Disk Thrashing: A Simulation Study. In: Proc. MASCOTS 1996, pp. 69–73 (1996)
42. Malkhi, D., Naor, M., Ratajczak, D.: Viceroy: A scalable and dynamic emulation of the butterfly. In: Proc. 21st annual ACM symposium on Principles of distributed computing (2002)
43. Naor, M., Wieder, U.: Novel architectures for P2P applications: the continuous-discrete approach. In: Proc. SPAA 2003, pp. 50–59 (2003)
44. Peleg, D.: Proximity-preserving labeling schemes and their applications. In: Proc. 25th Int. Workshop on Graph-Theoretic Concepts in Computer Science (June 1999)
45. Peleg, D.: Informative labeling schemes for graphs. In: Nielsen, M., Rovan, B. (eds.) MFCS 2000. SV-LNCS, vol. 1893, pp. 579–588. Springer, Heidelberg (2000)
46. Peleg, D.: Distributed Computing: A Locality-Sensitive Approach. SIAM (2000)
47. Powell, P.: A further improved LCA algorithm. Technical Report TR90-01, University of Minneapolis (1990)
48. Ratnasamy, S., Francis, P., Handley, M., Karp, R., Shenker, S.: A scalable content-addressable network. In: Proc. ACM SIGCOMM 2001, pp. 161–172 (August 2001)
49. Stocia, I., Morris, R., Karger, D., Kaashoek, F., Balakrishnan, H.: Chord: A Scalable Peer-to-peer Lookup Service for Internet Applications. In: Proc. ACM SIGCOMM 2001, San Diego, CA (August 2001)
50. Sleator, D.D., Tarjan, R.E.: A data structure for dynamic trees. Journal of Computer and System Sciences 26(1), 362–391 (1983)
51. Schieber, B., Vishkin, U.: On finding lowest common ancestors: Simplification and parallelization. SIAM J. on Computing 17, 1253–1262 (1988)
52. Thorup, M.: Compact oracles for reachability and approximate distances in planar digraphs. J. of the ACM 51, 993–1024 (2004)
53. Tsakalides, A.K., van Leeuwen, J.: An optimal pointer machine algorithm for finding nearest common ancestors. Technical Report RUU-CS-88-17, Department of CS, University of Utrecht (1988)
54. Thorup, M., Zwick, U.: Compact routing schemes. In: Proc. 13th ACM Symp. on Parallel Algorithms and Architecture, pp. 1–10 (July 2001)
55. Zhao, B., Kubiatowicz, J., Joseph, A.: Tapestry: An Infrastructure for Fault-tolerant Wide-area Location and Routing. Tech. rep, Univ. of California, Berkeley (2001)

A Simple Optimistic Skiplist Algorithm

Maurice Herlihy[1,3], Yossi Lev[1,3], Victor Luchangco[3], and Nir Shavit[2,3]

[1] Brown University, Box 1910, Computer Science Department, Providence, RI 02912
[2] Tel Aviv University, School of Computer Science, Ramat Aviv, Israel 69978
[3] Sun Microsystems Laboratories, 1 Network Drive, Burlington, MA 01803-0903

Abstract. Because of their highly distributed nature and the lack of global rebalancing, skiplists are becoming an increasingly important logarithmic search structure for concurrent applications. Unfortunately, none of the concurrent skiplist implementations in the literature, whether lock-based or lock-free, have been proven correct. Moreover, the complex structure of these algorithms, most likely the reason for a lack of a proof, is a barrier to software designers that wish to extend and modify the algorithms or base new structures on them.

This paper proposes a simple new lock-based concurrent skiplist algorithm. Unlike other concurrent skiplist algorithms, this algorithm preserves the skiplist properties at all times, which facilitates reasoning about its correctness. Though it is lock-based, the algorithm is highly scalable due to a novel use of optimistic synchronization: it searches without acquiring locks, requiring only a short lock-based validation before adding or removing nodes. Experimental evidence shows that this simpler algorithm performs as well as the best previously known lock-free algorithm under the most common search patterns.

1 Introduction

Skiplists [11] are an increasingly important data structure for storing and retrieving ordered in-memory data. In this paper, we propose a new lock-based concurrent skiplist algorithm that appears to perform as well as the best existing concurrent skiplist implementation under most common usage conditions. The principal advantage of our implementation is that it is much simpler, and thus much easier to reason about.

The original lock-based concurrent skiplist implementation by Pugh [10] is rather complex due to its use of pointer-reversal; to the best of our knowledge, it has never been proven correct. The ConcurrentSkipListMap, written by Doug Lea [8] based on work by Fraser and Harris [2] and released as part of the Java[TM] SE 6 platform, is the most effective concurrent skiplist implementation that we are aware of. This algorithm is lock-free, and performs well in practice. Its principal limitation is that it too is complicated: certain interleavings can cause the usual skiplist invariants to be violated, sometimes transiently, and sometimes indefinitely. These violations do not seem to affect performance or correctness, but they make it difficult to reason about the algorithm. By contrast, the algorithm presented here is lock-based and preserves the skiplist invariants

G. Prencipe and S. Zaks (Eds.): SIROCCO 2007, LNCS 4474, pp. 124–138, 2007.
© Springer-Verlag Berlin Heidelberg 2007

at all times. The algorithm is simple enough that we are able to provide a straightforward proof of correctness.

The key to our novel lock-based algorithm is the combination of two complementary techniques. First, it is *optimistic*: the methods traverse the list without acquiring locks. Moreover, they are able to ignore locks acquired by other threads while the list is being traversed. Only when a method finds the items it is seeking, does it lock the item and its predecessors, and then validates that these nodes are unchanged. Second, our algorithm is *lazy*: removing an item involves *logically* deleting it by marking it before it is *physically* removed (unlinked) from the list.

Lea [7] observes that in the most common search structure usage patterns search operations significantly dominate inserts, and inserts dominate deletes. A typical pattern is 90% search operations, 9% inserts, and only 1% deletes (see also [3]). Preliminary experimental tests conducted on a Sun Fire^{TM} T2000 multi-core and a Sun Enterprise^{TM} 6500 show that despite its simplicity, our new optimistic lock-based algorithm performs as well as the Lea's `ConcurrentSkipListMap` algorithm under this common usage pattern. In fact, its performance is slightly inferior to the `ConcurrentSkipListMap` algorithm only under extreme contention in multiprogrammed environments. This is because our raw experimental implementation did not have any added contention control.

We therefore believe the algorithm proposed here provides a viable alternative to the `ConcurrentSkipListMap` algorithm, especially in applications where programmers need to understand and possibly modify the basic skiplist structure.

2   Background

A skiplist [11] is a collection of sorted linked lists, each at a given "level", that mimics the behavior of a search tree. The list at each level, other than the bottom level, is a sublist of the list at the level beneath it. Each node is assigned a random level, up to some maximum, and participates in the lists up to that level. Figure 1 shows a skiplist with integer keys. The number of nodes in each list

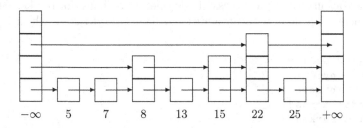

Fig. 1. A skiplist with a maximum level of 4. The number below each node (i.e., array of next pointers) is the key of that node, with $-\infty$ and $+\infty$ as the keys for the left and right sentinel nodes respectively.

decreases exponentially with the level, implying that we can find a key quickly by searching first at higher levels, skipping over large numbers of shorter nodes, and progressively working downward until a node with the desired key is found, or the bottom level is reached. Thus, the expected time complexity of skiplist operations is logarithmic in the length of the list.

It is convenient to have *left sentinel* and *right sentinel* nodes, at the beginning and end of the lists respectively. These nodes have the maximum allowed level, and initially, when the skiplist is empty, the right sentinel is the successor of the left sentinel at every level. The left sentinel's key is smaller, and the right sentinel's key is greater, than any key that may be added to the set. Searching the skiplist thus always begins at the left sentinel.

3   The New Algorithm

We present our concurrent skiplist algorithm in the context of an implementation of a set object supporting three methods, add, remove and contains: add(v) adds v to the set and returns true iff v was not already in the set; remove(v) removes v from the set and returns true iff v was in the set; and contains(v) returns true iff v is in the set. We show that our implementation is *linearizable* [5]; that is, every operation appears to take place atomically at some point (the *linearization point*) between its invocation and response. We also show that the implementation is deadlock-free, and that the contains operation is *wait-free*; that is, a thread is guaranteed to complete a contains operation within a finite number of steps regardless of the activity of other threads.

Our algorithm builds on the lazy-list algorithm of Heller et al. [4], a simple concurrent linked-list algorithm with an optimistic fine-grained locking scheme for the add and remove operations, and a wait-free contains operation: we use lazy-lists at each level of the skiplist. As in the lazy-list algorithm, the key of each node is strictly greater than the key of its predecessor, and each node has a marked flag, which is used to make remove operations appear atomic. However, here we may have to link the node in at several levels, and thus might not be able to insert a node using a single atomic instruction, which could serve as the linearization point of a successful add operation. Thus, for the lazy skiplist, we augment each node with an additional flag, fullyLinked, which is set to

```
4  class Node {
5    int key;
6    int topLevel;
7    Node* nexts [];
8    bool marked;
9    bool fullyLinked;
10   Lock lock;
11 };
```

Fig. 2. A node

true after a node has been linked in at all its levels; setting this flag is the linearization point of a successful add operation in our skiplist implementation. Figure 2 shows the fields of a node.

A key is in the abstract set if and only if there is an unmarked, fully linked node with that key in the list (i.e., reachable from the left sentinel).

To maintain the skiplist invariant—that is, that each list is a sublist of the list at lower levels—changes are made to the list structure (i.e., the nexts pointers) only when locks are acquired for all nodes that need to be modified. (There is one exception to this rule involving the add operation, discussed below.)

In the following detailed description of the algorithm, we assume the existence of a garbage collector to reclaim nodes that are removed from the skiplist, so nodes that are removed from the list are not recycled while any thread might still access them. In the proof (Section 4), we reason as though nodes are never recycled. In a programming environment without garbage collection, we can use solutions to the repeat offenders problem [6] or hazard pointers [9] to achieve the same effect. We also assume that keys are integers from MinInt+1 to MaxInt-1. We use MinInt and MaxInt as the keys for LSentinel and RSentinel, which are the left and right sentinel nodes respectively.

Searching in the skiplist is accomplished by the findNode helper function (see Figure 3), which takes a key v and two maximal-level arrays preds and succs of node pointers, and searches exactly as in a sequential skiplist, starting at the highest level and proceeding to the next lower level each time it encounters a node whose key is greater than or equal to v. The thread records in the preds array the last node with a key less than v that it encountered at each level,

```
33 int findNode(int v,
34              Node* preds[],
35              Node* succs[]) {
36   int lFound = -1;
37   Node* pred = &LSentinel;
38   for (int level = MaxHeight-1;
39        level >= 0;
40        level--) {
41     Node* curr = pred->nexts[level];
42     while (v > curr->key) {
43       pred = curr; curr = pred->nexts[level];
44     }
45     if (lFound == -1 && v == curr->key) {
46       lFound = level;
47     }
48     preds[level] = pred;
49     succs[level] = curr;
50   }
51   return lFound;
52 }
```

Fig. 3. The findNode helper function

and that node's successor (which must have a key greater than or equal to v) in thesuccs array. If it finds a node with the sought-after key, findNode returns the index of the first level at which such a node was found; otherwise, it returns −1. For simplicity of presentation, we have findNode continue to the bottom level even if it finds a node with the sought-after key at a higher level, so all the entries in both preds and succs arrays are filled in after findNode terminates (see Section 3 for optimizations used in the real implementation). Note that findNode does not acquire any locks, nor does it retry in case of conflicting access with some other thread. We now consider each of the operations in turn.

The add operation (Figure 4), calls findNode to determine whether a node with the key is already in the list. If so (lines 59–66), and the node is not marked, then the add operation returns false, indicating that the key is already in the set. However, if that node is not yet fully linked, then the thread waits until it is (because the key is not in the abstract set until the node is fully linked). If the node is marked, then some other thread is in the process of deleting that node, so the thread doing the add operation simply retries.

If no node was found with the appropriate key, then the thread locks and *validates* all the predecessors returned by findNode up to the level of the new node (lines 69–84). This level, denoted by topLevel, is determined at the very beginning of the add operation using the randomLevel function.[1] Validation (lines 81–82) checks that for each level $i \leq$ topLevel, preds[i] and succs[i] are still adjacent at level i, and that neither is marked. If validation fails, the thread encountered a conflicting operation, so it releases the locks it acquired (in the finally block at line 97) and retries.

If the thread successfully locks and validates the results of findNode up to the level of the new node, then the add operation is guaranteed to succeed because the thread holds all the locks until it fully links its new node. In this case, the thread allocates a new node with the appropriate key and level, links it in, sets the fullyLinked flag of the new node (this is the linearization point of the add operation), and then returns true after releasing all its locks (lines 86–97). The thread writing newNode->nexts[i] is the one case in which a thread modifies the nexts field for a node it has not locked. It is safe because newNode will not be linked into the list at level i until the thread sets preds[i]->nexts[i] to newNode, *after* it writes newNode->nexts[i].

The remove operation (Figure 5), likewise calls findNode to determine whether a node with the appropriate key is in the list. If so, the thread checks whether the node is "okay to delete" (Figure 6), which means it is fully linked, not marked, and it was found at its top level.[2] If the node meets these requirements, the

---

[1] This function is taken from Lea's algorithm to ensure a fair comparison in the experiments presented in Section 5. It returns 0 with probability $\frac{3}{4}$, i with probability $2^{-(i+2)}$ for $i \in [1, 30]$, and 31 with probability 2^{-32}.

[2] A node found not in its top level was either not yet fully linked, or marked and partially unlinked, at some point when the thread traversed the list at that level. We could have continued with the remove operation, but the subsequent validation would fail.

```
54 bool add(int v) {
55   int topLevel = randomLevel(MaxHeight);
56   Node* preds[MaxHeight], succs[MaxHeight];
57   while (true) {
58     int lFound = findNode(v, preds, succs);
59     if (lFound ≠ −1) {
60       Node* nodeFound = succs[lFound];
61       if (!nodeFound->marked) {
62         while (!nodeFound->fullyLinked) {}
63         return false;
64       }
65       continue;
66     }
67     int highestLocked = −1;
68     try {
69       Node *pred, *succ, *prevPred = null;
70       bool valid = true;
71       for (int level = 0;
72            valid && (level ≤ topLevel);
73            level++) {
74         pred = preds[level];
75         succ = succs[level];
76         if (pred ≠ prevPred) {
77           pred->lock.lock();
78           highestLocked = level;
79           prevPred = pred;
80         }
81         valid = !pred->marked && !succ->marked &&
82                 pred->nexts[level]==succ;
83       }
84       if (!valid) continue;

86       Node* newNode = new Node(v, topLevel);
87       for (int level = 0;
88            level ≤ topLevel;
89            level++) {
90         newNode->nexts[level] = succs[level];
91         preds[level]->nexts[level] = newNode;
92       }

94       newNode->fullyLinked = true;
95       return true;
96     }
97     finally { unlock(preds, highestLocked); }
98   }
```

Fig. 4. The add function

```
101 bool remove(int v) {
102   Node* nodeToDelete = null;
103   bool isMarked = false;
104   int topLevel = -1;
105   Node* preds[MaxHeight], succs[MaxHeight];
106   while (true) {
107     int lFound = findNode(v, preds, succs);
108     if (isMarked ||
109         (lFound != -1 && okToDelete(succs[lFound],lFound))){

111         if (!isMarked) {
112           nodeToDelete = succs[lFound];
113           topLevel = nodeToDelete->topLevel;
114           nodeToDelete->lock.lock();
115           if (nodeToDelete->marked) {
116             nodeToDelete->lock.unlock();
117             return false;
118           }
119           nodeToDelete->marked = true;
120           isMarked = true;
121         }
122         int highestLocked = -1;
123         try {
124           Node *pred, *succ, *prevPred = null;
125           bool valid = true;
126           for (int level = 0;
127                valid && (level <= topLevel);
128                level++) {
129             pred = preds[level];
130             succ = succs[level];
131             if (pred != prevPred) {
132               pred->lock.lock();
133               highestLocked = level;
134               prevPred = pred;
135             }
136             valid = !pred->marked && pred->nexts[level]==succ;
137           }
138           if (!valid) continue;

140           for (int level = topLevel; level >= 0; level--) {
141             preds[level]->nexts[level] = nodeToDelete->nexts[level];
142           }
143           nodeToDelete->lock.unlock();
144           return true;
145         }
146         finally { unlock(preds,highestLocked); }
147       }
148       else return false;
149   }
150 }
```

Fig. 5. The remove function

```
152 bool okToDelete(Node* candidate, int lFound) {
153    return (candidate->fullyLinked
154            && candidate->topLevel==lFound
155            && !candidate->marked);
156 }
```

Fig. 6. The okToDelete method

thread locks the node and verifies that it is still not marked. If so, the thread marks the node, which logically deletes it (lines 111–121); that is, the marking of the node is the linearization point of the remove operation.

The rest of the procedure accomplishes the "physical" deletion, removing the node from the list by first locking its predecessors at all levels up to the level of the deleted node (lines 124–137), and splicing the node out one level at a time (lines 140–142). To maintain the skiplist structure, the node is spliced out of higher levels before being spliced out of lower ones (though, to ensure freedom from deadlock, as discussed in Section 4, the locks are acquired in the opposite order, from lower levels up). As in the add operation, before changing any of the deleted node's predecessors, the thread validates that those nodes are indeed still the deleted node's predecessors. If the validation fails, then the thread releases the locks on the old predecessors (but not the deleted node) and tries to find the new predecessors of the deleted node by calling findNode again. However, at this point it has already set the local isMarked flag so that it will not try to mark another node. After successfully removing the deleted node from the list, the thread releases all its locks and returns true.

If no node was found, or the node found was not "okay to delete" (i.e., was marked, not fully linked, or not found at its top level), then the operation simply returns false (line 148). It is easy to see that this is correct if the node is not marked because for any key, there is at most one node with that key in the skiplist (i.e., reachable from the left sentinel) at any time, and once a node is put in the list (which it must have been to be found by findNode), it is not removed until it is marked. However, the argument is trickier if the node is marked, because at the time the node is found, it might not be in the list, and some unmarked node with the same key may be in the list. However, as we argue in Section 4, in that case, there must have been some time during the execution of the remove operation at which the key was not in the abstract set.

Finally, we consider the contains operation (Figure 7), which just calls findNode and returns true if and only if it finds a unmarked, fully linked node with the appropriate key. If it finds such a node, then it is immediate from the definition that the key is in the abstract set. However, as mentioned above, if the node is marked, it is not so easy to see that it is safe to return false. We argue this in Section 4.

We implemented the algorithm in the Java™ programming language, in order to compare it with Doug Lea's nonblocking skiplist implementation in the java.util.concurrent package. The array stack variables in the pseudocode are replaced by thread-local variables, and we used a straightforward

```
158 bool contains(int v) {
159    Node* preds[MaxHeight], succs[MaxHeight];
160    int lFound = findNode(v, preds, succs);
161    return (lFound ≠ -1
162            && succs[lFound]->fullyLinked
163            && !succs[lFound]->marked);
164 }
```

Fig. 7. The contains method

lock implementation (we could not use the built-in object locks because our acquire and release pattern could not always be expressed using synchronized blocks).

The pseudocode presented was optimized for simplicity, not efficiency, and there are numerous obvious ways in which it can be improved, many of which we applied to our implementation. For example, if a node with an appropriate key is found, the add and contains operations need not look further; they only need to ascertain whether that node is fully linked and unmarked. If so, the contains operation can return true and the add operation can return false. If not, then the contains operation can return false, and the add operation either waits before returning false (if the node is not fully linked) or else must retry. The remove operation does need to search to the bottom level to find all the predecessors of the node to be deleted, however, once it finds and marks the node at some level, it can search for that exact node at lower levels rather than comparing keys.[3] This is correct because once a thread marks a node, no other thread can unlink it.

Also, in the pseudocode, findNode always starts searching from the highest possible level, though we expect most of the time that the highest levels will be empty (i.e., have only the two sentinel nodes). It is easy to maintain a variable that tracks the highest nonempty level because whenever that changes, the thread that causes the change must have the left sentinel locked. This ease is in contrast to the nonblocking version, in which a race between concurrent remove and add operations may result in the recorded level of the skiplist being less than the actual level of its highest node.

4   Correctness

In this section, we sketch a proof for our skiplist algorithm. There are four properties we want to show: that the algorithm implements a linearizable set, that it is deadlock-free, that the contains operation is wait-free, and that the underlying data structure maintains a correct skiplist structure, which we define more precisely below.

---

[3] Comparing keys is expensive because, to maintain compatibility with Lea's implementation, comparison invokes the compareTo method of the Comparable interface.

4.1   Linearizability

For the proof, we make the following simplifying assumption about initialization: Nodes are initialized with their key and height, their `nexts` arrays are initialized to all `null`, and their `fullyLinked` and `marked` fields are initialized to `false`. Furthermore, we assume for the purposes of reasoning that nodes are never reclaimed, and there is an inexhaustible supply of new nodes (otherwise, we would need to augment the algorithm to handle running out of nodes).

We first make the following observations: The key of a node never changes (i.e., `key` = k is stable), and the `marked` and `fullyLinked` fields of a node are never set to `false` (i.e., `marked` and `fullyLinked` are stable). Though initially `null`, `nexts[i]` is never written to `null` (i.e., `nexts[i]` $\neq$ `null` is stable). Also, a thread writes a node's `marked` or `nexts` fields only if it holds the node's lock (with the one exception of an `add` operation writing `nexts[i]` of a node before linking it in at layer i).

From these observations, and by inspection of the code, it is easy to see that in any operation, after calling `findNode`, we have `preds[i]->key` < v and `succs[i]->key` $\geq$ v for all i, and `succs[i]->key` > v for i > 1Found (the value returned by `findNode`). Also, for a thread in `remove`, `nodeToDelete` is only set once, and that unless that node was marked by some other thread, this thread will mark the node, and thereafter, until it completes the operation, the thread's `isMarked` variable will be `true`. We also know by `okToDelete` that the node is fully linked (and indeed that only fully linked nodes can be marked).

Furthermore, the requirement to lock nodes before writing them ensures that after successful validation, the properties checked by the validation (which are slightly different for `add` and `remove`) remain true until the locks are released.

We can use these properties to derive the following fundamental lemma:

Lemma 1. *For a node n and $0 \leq i \leq n$->topLayer:*

$$n\text{->}nexts[i] \neq null \implies n\text{->}key < n\text{->}nexts[i]\text{->}key$$

We define the relation $\rightarrow_i$ so that $m \rightarrow_i n$ (read "m leads to n at layer i") if `m->nexts[i]` $= n$ or there exists m' such that $m \rightarrow_i m'$ and `m'->nexts[i]` $= n$; that is, $\rightarrow_i$ is the transitive closure of the relation that relates nodes to their immediate successors at layer i. Because a node has (at most) one immediate successor at any layer, the $\rightarrow_i$ relation "follows" a linked list at layer i, and in particular, the layer-i list of the skip list consists of those nodes n such that `LSentinel` $\rightarrow_i n$ (plus `LSentinel` itself). Also, with Lemma 1, if $m \rightarrow_i n$ and $m \rightarrow_i n'$ and `n->key` < `n'->key` then $n \rightarrow_i n'$.

Using these observations, we can show that if $m \rightarrow_i n$ in any reachable state of the algorithm, then $m \rightarrow_i n$ in any subsequent state unless there is an action that splices n out of the layer-i list, that is, an execution of line 141. This claim is proved formally for the lazy-list algorithm [1], and that proof can be adapted to this algorithm. Thus, the only action that adds a key to the abstract set is the setting of the `fullyLinked` flag of a node with that key. `remove` Because n must already be marked before being spliced out of the list, and because the fullyLinked flag is never set to false (after its initialization), this claim implies that a key can be removed from the abstract set only by marking its node, which we argued earlier is the linearization point of a successful remove operation.

Similarly, we can see that if LSentinel $\rightarrow_i n$ does *not* hold in some reachable state of the algorithm, then it does not hold in any subsequent state unless there is some execution of line 91 with $n = $ newNode (as discussed earlier, the previous line doesn't change the list at layer-i because newNode is not yet linked in then). However, the execution of that line occurs while newNode is being inserted and before newNode is fully linked. Thus, the only action that adds a node to a list at any level is the setting of the node's fullyLinked flag.

Finally, we argue that if a thread finds a marked node then the key of that node must have been absent from the list at some point during the execution of the thread's operation. There are two cases: If the node was marked when the thread invoked the operation, the node must have been in the skip list at that time because marked nodes cannot be added to the skip list (only a newly allocated node can be added to the skip list), and because no two nodes in the skip list can have the same key, no unmarked node in the skip list has that key. Thus, at the invocation of the operation, the key is not in the skip list. On the other hand, if the node was not marked when the thread invoked the operation, then it must have been marked by some other thread before the first thread found it. In this case, the key is not in the abstract set immediately after the other thread marked the node. This claim is also proved formally for the simple lazy-list [1], and that proof can be adapted to this algorithm.

4.2 Maintaining the Skiplist Invariant

Our algorithm guarantees that the skiplist invariant are preserved at all times. By "skiplist invariant", we mean that the list at each layer is a sublist of the lists at lower layers.

To see that the algorithm preserves the skiplist structure, note that linking new nodes into the skip list always proceeds from bottom to top, and while holding the locks on all the soon-to-be predecessors of the node being inserted. On the other hand, when a node is being removed from the list, the higher layers are unlinked before the lower layers, and again, while holding locks on all the immediate predecessors of the node being removed.

This property is not guaranteed by the lock-free algorithm. In that algorithm, after linking a node in the bottom layer, an add operation links the node in the rest of the layers from top to bottom. This may result in a state of a node that is linked only in its top and bottom layers, so that the list at the top layer is *not* a sublist of the list at the layer immediately beneath it, for example. Moreover, attempts to link in a node at any layer other than the bottom are not retried, and hence this state of nonconformity to the skiplist structure may persist indefinitely.

4.3   Deadlock Freedom and Wait-Freedom

The algorithm is deadlock-free because a thread always acquires locks on nodes with larger keys first. More precisely, if a thread holds a lock on a node with key v then it will not attempt to acquire a lock on a node with key greater than or equal to v. The `contains` operation is wait-free because it does not acquire any locks and never retries; it searches the list only once.

5   Performance

We evaluated our skiplist algorithm by implementing it in the Java programming language, as described earlier. We compared our implementation against Doug Lea's nonblocking skiplist implementation in the `ConcurrentSkipListMap` class [8] of the `java.util.concurrent` package, which is part of the Java[TM] SE 6 platform; to our knowledge, this is the best widely available concurrent skiplist implementation. We also implemented a straightforward sequential skip -list, in which methods were `synchronized` to ensure thread safety, for use as a baseline in these experiments. We describe some of the results we obtained from these experiments in this section.

We present results from experiments on two multiprocessor systems with quite different architectures. The first system is a Sun Fire[TM] T2000 server, which is based on a single UltraSPARC® T1 processor containing eight computing cores, each with four hardware strands, clocked at 1200 MHz. Each four-strand core has a single 8-KByte level-1 data cache and a single 16-KByte instruction cache. All eight cores share a single 3-MByte level-2 unified (instruction and data) cache, and a four-way interleaved 32-GByte main memory. Data access latency ratios are approximately 1:8:50 for L1:L2:Memory accesses. The other system is an older Sun Enterprise[TM] 6500 server, which contains 15 system boards, each with two UltraSPARC® II processors clocked at 400 MHz and 2 Gbytes of RAM for a total of 30 processors and 60 Gbytes of RAM. Each processor has a 16-KByte data level-1 cache and a 16-Kbyte instruction cache on chip, and a 8-MByte external cache. The system clock frequency is 80 MHz.

We present results from experiments in which, starting from an empty skiplist, each thread executes one million (1,000,000) randomly chosen operations. We varied the number of threads, the relative proportion of `add`, `remove` and `contains` operations, and the range from which the keys were selected. The key for each operation was selected uniformly at random from the specified range.

In the graphs that follow, we compare the throughput in operations per millisecond, and the results shown are the average over six runs for each set of parameters. Figure 8 presents the results of experiments in which 9% of the operations were `add` operations, 1% were `remove` operations, and the remaining 90% were `contains` operations, where the range of the keys was either two hundred thousand or two million. The different ranges give different levels of contention, with significantly higher contention with the 200,000 range, compared with the 2,000,000 range. As we can see from these experiments, both our algorithms and Lea's scale well (the sequential algorithm, as expected, does not).

Sun Fire™ T2000   Sun Enterprise™ 6500

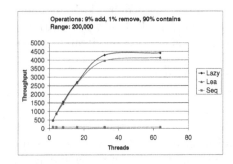

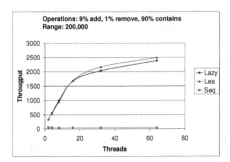

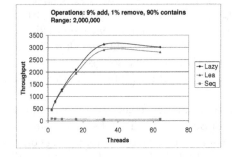

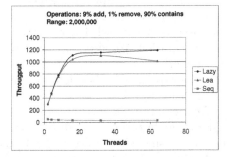

Fig. 8. Throughput in operations per millisecond of 1,000,000 operations, with 9% add, 1% remove, and 90% contains operations, and a range of either 200,000 or 2,000,000

In all but one case (with 200,000 range on the older system), our implementation has a slight advantage.

In the next set of experiments, we ran with higher percentages of add and remove operations, 20% and 10% respectively (leaving 70% contains operations). The results are shown in Figure 9. As can be seen, on the T2000 system, the two implementations have similar performance, with a slight advantage to Lea in a multiprogrammed environment when the range is smaller (higher contention). The situation is reversed with the larger range. This phenomenon is more noticeable on the older system: there we see a 13% advantage to Lea's implementation on the smaller range with 64 threads, and 20% advantage to our algorithm with the same number of threads when the range is larger.

To explore this phenomenon, we conducted an experiment with a significantly higher level of contention: half add operations and half remove operations with a range of 200,000. The results are presented in Figure 10. As can be clearly seen, under this level of contention, our implementation's throughput degrades rapidly when approaching the multiprogramming zone, especially on the T2000 system. This degradation is not surprising: In our current implementation, when an add or remove operation fails validation, or fails to acquire a lock immediately, it simply calls yield; there is no proper mechanism for managing contention.

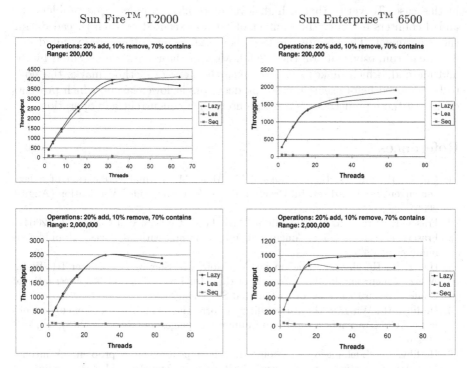

Fig. 9. Throughput in operations per millisecond of 1,000,000 operations with 20% **add**, 10% **remove**, and 70% **contains** operations, and range of either 200,000 or 2,000,000

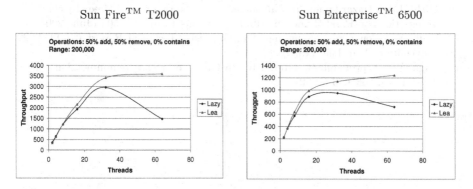

Fig. 10. Throughput in operations per millisecond of 1,000,000 operations, with 50% **add** and 50% **remove** operations, and a range of 200,000

Since the **add** and **remove** operations require that the predecessors seen during the search phase be unchanged until they are locked, we expect that under high contention, they will repeatedly fail. Thus, we expect that a back-off mechanism, or some other means of contention control, would greatly improve performance in this case. To verify that a high level of conflict is indeed the problem, we added counters to count the number of retries executed by each thread during the experiment. The counters indeed show that many retries are executed in a 64-thread run, especially on the T2000. Most of the retries are executed by the add method, which makes sense because the remove method marks the node to be removed before searching its predecessors in lower layers, which prevents change of these predecessor's next pointers by a concurrent add operation.

References

1. Colvin, R., Groves, L., Luchangco, V., Moir, M.: Formal verification of a lazy concurrent list-based set. In: Proceedings of Computer-Aided Verification (August 2006)
2. Fraser, K.: Practical Lock-Freedom. PhD thesis, University of Cambridge (2004)
3. Fraser, K., Harris, T.: Concurrent programming without locks. Unpublished manuscript (2004)
4. Heller, S., Herlihy, M., Luchangco, V., Moir, M., Shavit, N., Scherer III, W.N.: A lazy concurrent list-based set algorithm. In: Proceedings of 9th International Conference on Principles of Distributed Systems (2005)
5. Herlihy, M., Wing, J.: Linearizability: A correctness condition for concurrent objects. ACM Transactions on Programming Languages and Systems 12(3), 463–492 (1990)
6. Herlihy, M., Luchangco, V., Moir, M.: The repeat offender problem: A mechanism for supporting dynamic-sized, lock-free data structures. In: Proceedings of Distributed Computing: 16th International Conference (2002)
7. Lea, D.: Personal communication (2005)
8. Lea, D.: Concurrent Skip List Map. In java.util.concurrent
9. Michael, M.: Hazard pointers: Safe memory reclamation for lock-free objects. IEEE Transactions on Parallel and Distributed Systems 15(6), 491–504 (June 2004)
10. Pugh, W.: Concurrent maintenance of skip lists. Technical Report CS-TR-2222, University of Maryland (1990)
11. Pugh, W.: A probabilistic alternative to balanced trees. Communications of the ACM 33(6), 668–676 (1990)

Data Aggregation in Sensor Networks: Balancing Communication and Delay Costs

Peter Korteweg[1,*], Alberto Marchetti-Spaccamela[2], Leen Stougie[1,3], and Andrea Vitaletti[2]

[1] TU Eindhoven
Fax: + 31 40 246 5995
p.korteweg@tue.nl
[2] University of Rome "La Sapienza"
[3] CWI Amsterdam

Abstract. In a sensor network the sensors, or nodes, obtain data and have to communicate these data to a central node. Because sensors are battery powered they are highly energy constrained. Data aggregation can be used to combine data of several sensors into a single message, thus reducing sensor communication costs at the expense of message delays. Thus, the main problem of data aggregation is to balance the communication and delay costs.

In this paper we study the data aggregation problem as a bicriteria optimization problem; the objectives we consider are to minimize maximum energy consumption of a sensor and a function of the maximum latency costs of a message. We consider distributed algorithms under an asynchronous time model, and under an almost synchronous time model, where sensor clocks are synchronized up to a small drift. We use competitive analysis to assess the quality of the algorithms.

Keywords: distributed algorithms, sensor networks, data aggregation, bicriteria optimization.

1 Introduction

A wireless sensor network (WSN) consists of sensor nodes and one or more central nodes or *sinks*. Sensor nodes are able to monitor events, to process the sensed information and to communicate the sensed data. Sinks are powerful base stations which gather data sensed in the network; sinks either process this data or act as gateways to other networks. Sensors send data to the sink through multi-hop communication.

A particular feature of sensor nodes is that they are battery powered, making sensor networks highly energy constrained. Replacing batteries on hundreds of nodes, often deployed in inaccessible environments, is infeasible or too costly and, therefore, the key challenge in a sensor network is the reduction of energy

---

* Corresponding author: TU Eindhoven, PO Box 513, 5600 MB Eindhoven, The Netherlands.

G. Prencipe and S. Zaks (Eds.): SIROCCO 2007, LNCS 4474, pp. 139–150, 2007.

consumption. Energy consumption can be divided into three domains: sensing, communication and data processing [1]. Communication is most expensive because a sensor node spends most of its energy in data transmission and reception [7]. This motivates the study of techniques to reduce overall data communication, possibly exploiting processing capabilities available at each node. Data aggregation is one such technique. It consists of aggregating redundant or correlated data in order to reduce the overall size of sent data, thus decreasing the network traffic and energy consumption. In this paper we comply with most of the literature on sensor networks concentrating on *total aggregation*, i.e. data packets are assumed to have the same size and aggregation of two or more incoming packets at a node results in a single outgoing packet. Total aggregation is possible if data are completely correlated, or can be described by a single value, e.g. when the required data is maximum temperature. Observe that even if total aggregation might be considered a simplistic assumption in other cases, it allows us to provide an upper bound on the expected benefits of data aggregation in terms of power consumption.

WSN deal with real world environments. In many cases, sensor data must be delivered within time constraints so that appropriate observations can be made or actions taken [11]. We assume that the routing network is a tree; this is a common assumption in data aggregation network problems [1,4].

In [3] we studied the Data Aggregation Sensor Problem as a unicriterion problem where we minimized the maximum communication costs subject to a budget on the latency costs. Here the budget constraint was a hard constraint.

The dynamics governing the monitored phenomena are often not well understood and/or defined at the beginning of the monitoring process. For this reason a strict constraint on latency could be unappropriate.

A common assumption in literature on data aggregation is that value of information degrades over time. E.g. Broder and Mitzenmacher [5] describe a data aggregation model where there is a reward function on the data collected by a server; the function increases with the quantity of data collected and decreases over time. A similar tradeoff holds for data aggregation in sensor networks: delaying data decreases the information value of the data, but increases network lifetime.

Both the above discussed tradeoffs and the partial knowledge of the monitored process at the beginning, suggest to use a bicriteria objective function to asses the quality of the algorithms instead of hard constraints.

The bicriteria data aggregation sensor problem
The Data Aggregation Sensor Problem (DASP) is to send all messages to the sink such as to minimize the communication costs, and to minimize the latency costs. For the first objective we have chosen to minimize the maximum communication costs per node. This is a natural objective in sensor networks because of limited and unreplenishable energy at nodes. The objective maximizes the network lifetime, i.e. the time that all sensors can communicate. For the second objective we have chosen to minimize the maximum latency cost.

The two objectives conflict with each other. We can easily find algorithms with low communication costs by delaying messages and aggregating them into packets. As communication costs are independent of the size of packets sent, but linear in the number of packets sent, aggregation reduces the communication costs, at the expense of increased latency costs. Similarly we can find algorithms with low latency costs at the expense of high communication costs. The objective is to find algorithms where both costs are relatively good.

We formulate the problem as a bicriteria optimization problem: minimizing one of the objectives under a budget restriction on the other objective. We call a bicriteria optimization problem an (B, A)-bicriteria problem if we minimize objective A under a budget on objective B. In this paper we study the (B, A)-sensor problem where objective A is maximum communication costs and objective B is maximum latency costs. Quality of algorithms is assessed through the concept of (β, α)-approximation: allowing an excess of multiplicative factor β on the budget of objective B, the value produced is worst-case within ratio α from optimal with respect to objective A. For network design problems this was formalized in [9,10]. The concept is general in the sense that results hold regardless which of the two objectives is minimized, and which is budgeted.

Sensor nodes are equipped with a clock that can be used to measure the latency of messages. We distinguish three distributed on-line models, which are common in literature on distributed algorithms, see [12]. In the *synchronous* model all nodes are equipped with a *common clock*, i.e. the times indicated at all clocks are identical. A common clock may facilitate synchronization of actions in various nodes. In the *asynchronous* model there is no such common clock. In the *almost synchronous* model, all nodes are equipped with a clock and the clocks are almost synchronous, i.e. there is a relatively small drift between any two clocks. In practice, these clocks can easily drift seconds per day, accumulating significant errors over time [12].

Results

In this paper we present distributed on-line algorithms for sensor networks with a routing intree. The first main contribution is that we study for the first time sensor network problems in a bicriteria optimization framework. In Section 2 we formalize the model.

In Section 3, for the asynchronous model we present an algorithm which balances communication and latency costs. If δ is tree depth, and U is the ratio between maximum and minimum allowed delay, then the algorithm is $(2\delta^\lambda, 2\delta^{1-\lambda} \log U)$-competitive, for any λ, $0 < \lambda \leq 1$. The algorithm is member of a class of memoryless algorithms for which we show that no better competitiveness than $(\delta^\lambda, \delta^{1-\lambda})$ exists.

In Section 4 we present the second main contribution, which is the analysis of algorithms for sensor networks in which clocks in various nodes show small drifts. For this so-called almost synchronous model we present an algorithm which for sensors with a clock drift of at most Δ between any two nodes and latency budget L is $(1 + \Delta\delta/L, \log^2 \delta)$-competitive. For small drift, i.e. $\Delta\delta/L$ small, the competitiveness comes close to the best possible competitiveness in the synchronous model. We notice that no previous results are known for this model, which is in fact the more realistic one.

Related work

In [3] we studied the Data Aggregation Sensor Problem as a unicriterion problem where we minimized the maximum communication costs subject to a budget on the latency costs. Here the budget constraint was a hard constraint. Interpreted in the bicriteria setting the results imply $(1, O(\log U))$ for synchronous, and $(1, \delta \log U)$ for the asynchronous models. No results were given for the almost synchronous model.

In the past, many bicriteria optimization problems were formulated as a unicriterion optimization problem with as single objective a weighted sum of the two objectives. For aggregation problems with objectives to minimize communication costs and latency costs such a formulation as a unicriterion optimization problem can be found in [2,4,6,8].

Both Khanna et al. [8] and Brito et al. [4] consider the Multicast Aggregation Problem (MAP), or TCP Acknowledgment problem, on a tree. The Multicast Aggregation Problem is equivalent to the Data Aggregation Sensor Problem in the sense that messages, which arrive over time, have to be sent to a sink in the graph. The main difference with our problem is in the objectives. First, the objective of MAP is to minimize the sum of communication costs; this is a natural objective if nodes have permanent access to energy. This is not true for sensor networks, for which minimizing energy cost per node is more suitable. The other objective of MAP is to minimize the sum of latency costs, and latency costs do not depend on communication time to the sink. Second, the authors analyze the problem using a single objective which is a weighted sum of communication costs and latency costs.

A main drawback of formulating the problem using a single objective is that the choice of the weights influences the outcome. Especially if the objectives are measured in different units, e.g. energy and time, then the choice of weights is highly arbitrary. Thus, we believe that a bicriteria setting is more appropriate in this case.

2   Preliminaries

We study sensor networks $G = (V, A)$, which are *intrees* rooted at a *sink node* $s \in V$. Nodes represent sensors and arcs represent the possibility of communication between two sensors. Over time, n messages, $N := \{1, \ldots, n\}$, arrive at nodes and have to be sent to the sink. Message j arrives at its *release node* v_j at its *release date* r_j; message j arrives at the sink via the unique $v_j - s$-path. Thus, each message is completely defined by the pair (v_j, r_j).

A *packet* is a set of messages which are sent simultaneously along an arc. Each initial message is a packet and two packets j and j' can be aggregated at a node v into a single packet. The resulting packet can be recursively aggregated with other packets.

Communication of a message along an arc takes time and energy cost. In this paper we assume that the communication time $\tau : A \rightarrow \mathbb{R}_{\geqslant 0}$ and communication cost $c : A \rightarrow \mathbb{R}_{\geqslant 0}$ are independent of packet size. We often refer to the *communication cost* of a node as the communication cost of its unique outgoing arc. This models the situation in which all messages have more or less the same size and where *total aggregation* is possible, as discussed in the introduction. For the sake of simplicity we also assume that all communication times $\tau(a)$ are equal, namely we set $\tau(a) = 1 \ \forall a \in A$.

For $v \in V$, τ_v is the total communication time of the path from v to s. We define $r'_j := r_j + \tau_{v_j}$ as *earliest possible arrival time* of j at s. We assume that each node v knows its total communication time τ_v to the sink. Finally, we define $\delta := \max_v \tau_v$ as the depth of the network in terms of the communication time. We assume $\delta \geq 2$, avoiding the trivial case of $\delta = 1$.

The value of information degrades over time. To model this we define the quality degradation cost of a message. Let d_j be the arrival time of message j. We assume that the quality degradation of a message j depends on the latency of a message $l_j := d_j - r_j$. In this paper we choose the latency as our quality degradation function, i.e. our function increases linearly over time. We also refer to these costs as *latency costs* and we say that a solution is L-bounded if $l_j \leq L$ for all i. Since $\delta = \max_v \tau_v$ a L-bounded feasible solution must satisfy $L \geq \delta$, as otherwise it is impossible to send all messages j to the sink such that their latency costs are within budget L.

The budget on the latency imposes an arrival time interval $I_j := [r_j + \tau_{v_j}, r_j + L]$ of any L-bounded solution. It also imposes a transit interval for each node u on the $v_j - s$ path: $I_j(u) := [r_j + \tau_{v_j} - \tau_u, r_j + L - \tau_u]$. I.e. in each L-bounded solution message j should transit at u in interval $I_j(u)$. Finally, we define $U = \frac{\max_j |I_j|}{\max\{1, \min_j |I_j|\}}$. Since $L \geq \delta$ we have $U \leq \delta$.

Given a solution S the communication cost of node v_i is the total energy cost spent by v_i and it is given by the total number of messages sent by v_i times the communication cost of v_i. We are interested minimizing maximum communication cost over all nodes.

Given a bound L on the latency and β, $\beta \geq 1$, we study the communication cost of algorithms that provide βL-bounded feasible solution: a βL-bounded feasible solution is (β, α)-approximate if its communication cost is at most α times the communication cost of the optimal L-bounded solution. An interesting special case is to find a minimum γ such that there exists a (γ, γ)-approximate algorithm [9].

In this paper we consider *distributed on-line* algorithms, in which nodes communicate independently of each other and messages are released over time. Therefore, at any time t the input of each node's algorithm is given by packets that have been released at or forwarded from that node in the period $[0, t]$. An algorithm is (β, α)-competitive if it is an (β, α)-approximation and the algorithm is an online algorithm.

2.1   The Synchronous Model

For the synchronous model we presented an algorithm for the latency constrained sensor aggregation problem in [3]. In the following we restate the algorithm in a bicriteria setting, because we use the algorithm as a subroutine in our algorithm for the almost synchronous model. The algorithm is based on the following lemma.

Lemma 1. *[3] Given any interval $[a, b]$, such that $b - a \geq 1$. Let $i^* = \max\{i \in \mathbb{N} \mid \exists k \in \mathbb{N} : k2^i \in [a, b]\}$, then k^* for which $k^* 2^{i^*} \in [a, b]$ is odd and unique.*

We use notation $t(I)$ to represent the unique point in the interval $I = [a, b]$ which equals $k^* 2^{i^*}$ with i^* and k^* as defined in Lemma 1. The algorithm sends messages j to the sink at time $t(I)$ where interval I depends on message j and budget L on the latency costs. We choose as interval the interval of an L-bounded solution, i.e. I_j.

Algorithm:CommonClock (CC): Message j is sent from v_j at time $t(I_j) - \tau_{v_j}$ to arrive at s at time $t(I_j)$ unless some other packet passes v_j in the interval $[r_j, t(I_j) - \tau_{v_j}]$, in which case j is aggregated and the packet is forwarded directly.

The analysis of the competitive ratio of CC is based on the following lemma that will be used in the sequel. The lemma bounds the competitive ratio for instances in which the arrival intervals I_j differ by a factor at most 2 in length.

Lemma 2. *[3] CC is $(1, 3)$-competitive if there exists an $i \in \mathbb{N}$ such that $2^{i-1} < |I_j| \leq 2^i \; \forall j$.*

This result immediately implies the following theorem.

Theorem 1. *[3] CC is $(1, O(\log U))$-competitive[1].*

In the the CC algorithm no message incurs a delay cost which exceeds its budget. A simple modification of the CC algorithm which balances the communication and delay costs can be obtained by replacing $t(I_j)$ by $t(I_j^*)$ as follows. Let $\mu := \max\{1, \min_j(L_j - \tau_{v_j})\}$, and let $N_m = \{j \in N \mid (\frac{\log U}{\log \log U})^{m-1} \mu \leq |I_j| < (\frac{\log U}{\log \log U})^m \mu\}$ for $m \in \mathbb{N}$. The algorithm sends messages $j \in N_m$ to the sink at time $t(I_j^*)$ where $I_j^* = [r_j + \tau_{v_j}, r_j + \tau_{v_j} + (\frac{\log U}{\log \log U})^m \mu]$.

The proof of the following theorem is omitted.

Theorem 2. *There exists an algorithm that is $(\frac{\log U}{\log \log U}, \frac{\log U}{\log \log U})$-competitive.*

3   The Asynchronous Model

For the asynchronous model we present a modification of the algorithm Spread Latency (SL), as proposed in [3]. The algorithm assigns to message j a total waiting time of $2(\tau_{v_j})^\lambda$ times the allowed latency minus communication time, for some λ, $0 < \lambda \leq 1$. SL equally divides this waiting time over the nodes:

---

[1] All logarithms in this paper are base 2.

at each node of the $v_j - s$ path message j is assigned a waiting time of $2(L - \tau_{v_j})/(\tau_{v_j})^{1-\lambda}$ time units. When messages are simultaneously at the same node they get aggregated into a packet, which is sent over the outgoing arc as soon as the waiting time of at least one of these messages has passed.

Theorem 3. *Algorithm* SL *is* $(2\delta^\lambda, 2\delta^{1-\lambda} \log U)$*-competitive for* λ, $0 < \lambda \leq 1$.

Proof. Consider algorithm SL for fixed λ, $0 < \lambda \leq 1$. First note that because no message is delayed due to aggregation the latency of each message j is at most

$$\tau_{v_j} 2(L - \tau_{v_j})/\tau_{v_j}^{1-\lambda} + \tau_{v_j} \leq 2\delta^\lambda L.$$

We prove that for all $a \in A$ the number of packets SL sends through a is at most $2\delta^{1-\lambda} \log U$ times that number in an optimal L-bounded solution. This proves the theorem.

Let $\mu := \max\{1, \min_j (L - \tau_{v_j})\}$. Consider a packet P of messages sent by an optimal L-bounded solution through (u, v) at t. To bound the number of packets sent by SL that contain at least one message from P, define $P_i := \{j \in P \mid 2^{i-1}\mu \leq L - \tau_{v_j} < 2^i\mu\}$, for $i = 1, \ldots, \lceil \log U \rceil$. We charge any sent packet to the message that caused the packet to be sent due to its waiting time being over. It suffices to prove that the number of packets charged to messages in P_i is $2\delta^{1-\lambda}$. Since the waiting time of messages $j \in P_i$ at node u is at least $2 \cdot 2^{i-1}\mu/\delta^{1-\lambda}$, the delay between any two packets that are charged to messages in P_i is at least $2^i\mu/\delta^{1-\lambda}$. Since the optimal solution sends packet P at t through (u, v), we get $t \in I_j(u) \; \forall j \in P$ and thus $I_j(u) \subseteq [t - 2^i\mu, t + 2^i\mu] \; \forall j \in P_i$. Thus, the number of packets charged to messages in P_i is at most $2 \cdot 2^i\mu/(2^i\mu/\delta^{1-\lambda}) = 2\delta^{1-\lambda}$. $\square$

SL determines the waiting time of each message at the nodes it traverses independently of all other messages. We call such an algorithm a *memoryless* algorithm. To be precise, in a memoryless algorithm node v determines the waiting time of message j based only on the message characteristics (v_j, r_j), budget L, communication time to the sink τ_{v_j} and clock time. The following lower bound shows that the competitive ratio of SL cannot be beaten by more than a factor $\log U$ by any other memoryless algorithm. In the derivation of the lower bound we restrict to memoryless algorithms that employ the same algorithm in all nodes with the same communication time to s. This is not a severe restriction, given that communication time to s is the only information about the network that a node has.

Theorem 4. *No deterministic asynchronous memoryless algorithm is better than* $(\delta^\lambda, \delta^{1-\lambda})$*-competitive, for fixed* λ, $0 \leq \lambda \leq 1$.

Proof. Consider any deterministic asynchronous memoryless algorithm with latency costs at most δ^λ times the budget on the latency costs for fixed λ, $0 \leq \lambda \leq 1$. An adversary chooses a binary tree with root s and all leaves at distance δ from s. The adversary releases message 1 with latency L at time r_1 in a leaf v_1. There must be a node u where message 1 waits at most $\delta^\lambda(L - \tau_{v_1})/\delta$. The adversary releases message $j, j = 2, \ldots, \delta^{1-\lambda}$ at time $r_1 + j(L - \tau_{v_1})/\delta^{1-\lambda}$ such that all messages j are sent over node u, and no two messages can be aggregated before reaching v. Because $\tau_{v_j} = \tau_{v_1}$ $\forall j$ and we assumed that any memoryless algorithm applies the same algorithm in nodes at equal distance, all messages are sent non-aggregated to and from u, whereas they are aggregated as early as possible in an optimal solution, in particular at u.   □

Theorems 3 and 4 immediately imply the following corollary.

Corollary 1. *There exists a deterministic asynchronous algorithm that is $(\sqrt{\delta}, \sqrt{\delta}\log U)$-competitive and no deterministic asynchronous memoryless algorithm is better than $(\sqrt{\delta}, \sqrt{\delta})$-competitive.*

If we assume that $L \geq 2\delta$, which in practice is not a severe restriction at all, essentially the same analysis as in the proof of Theorem 3 gives $(2\delta^\lambda, 2\delta^{1-\lambda})$-competitiveness. Thus, in this case SL is a best possible on-line algorithm up to a constant multiplicative factor.

4   The Almost Synchronous Model

Typically in sensor networks clocks have a small drift. The CC-algorithm is not robust in the sense that its competitive ratio may be much worse if we assume existence of such clock drifts. However, the idea underlying the CC-algorithm gives rise to algorithms which have good competitive ratio even in the almost synchronous model. In this section we present such an algorithm. We assume that the difference between the time indicated at any two clocks is at most Δ. We assume all communication times to be equal and of unit length, i.e. $\tau(a) = 1$ $\forall a$. We also divide nodes into classes; a node v is of class p if p is the maximal integer such that $\tau_v = h2^p + 1$ for some integer h, and v is of class 0 if $\tau_v = 1$. Note that $p \in \{0, \ldots, \lceil \log \delta \rceil\}$. The algorithm is the following:

> **Algorithm:AlmostSynchronousClock** (ASC) Message j incurs 3 kinds of delay:
> 1. a delay of $t(I_j) - \tau_{v_j} - r_j$ at its release node v_j;
> 2. a delay of Δ at each node it traverses;
> 3. a delay which sums to $2^{p+1}\Delta$ at the first node of class $p, p > 0$, it traverses.

The waiting time of message j at a node v is the sum of the delays. A message is sent from a node v once its waiting time is over, unless some other message (packet) is sent from v earlier in which case j is aggregated with this packet.

Note that if $\Delta = 0$ the algorithm is identical to the CC-algorithm. To illustrate delay of the third kind we give an example: if a message traverses nodes of classes 1-4 in order 1,2,3,4 then its delay of the third kind of these nodes is respectively $4\Delta, 4\Delta, 8\Delta, 16\Delta$. If the order is 4,1,2,3 then its delay of the third kind is 32Δ at the node of class 4 and 0 elsewhere.

Now we analyze the competitive ratio of ASC. Let $V_k := \{v | 2^{k-1} < \tau_v \leq 2^k\}$ for some $k \in \mathbb{N}$, for $k = 1, \ldots, \lceil \log \delta \rceil$. First, we analyze the behavior of the algorithm for instances in which the release nodes of all messages is in V_k for some $k \in \mathbb{N}$.

Lemma 3. *If the* CC-*solution sends a packet from* v, *the* ASC-*solution sends at most* $(k+1)$ *packets from* v *which contain a message of the* CC-*packet, if* $\forall j$ $v_j \in V_k$ *for some* $k \in \mathbb{N}$.

Proof. Each packet, either CC or ASC, contains at least one message whose waiting time is completely over when the packet is forwarded. Hence without loss of generality we only consider messages whose waiting time is completely over when counting packets.

Consider a packet P_{CC} sent by the CC-solution from some node v at time t. In the remainder of the proof we only consider the messages in this packet. We analyze the number of ASC-packets which contain a message of P_{CC}. The delays of messages in P_{CC} are chosen such that all messages in this packet which traverse v, i.e. v is not the release node, arrive at this node at time t. As the delay of the first kind in the ASC-algorithm is identical to the delay incurred by the CC-algorithm we focus on the deviation from this time to analyze the number of packets ASC sends. This deviation may be caused either by delay of kind 2 and 3, or by the clock drift.

If $k = 0$ the lemma trivially holds, because all messages which are sent over some node $v \in V_0$ have this node v as release node. Hence, if they are sent in a single packet by the CC-solution they are also sent in a single packet in the ASC-solution.

For $k \geq 1$ we introduce the following notation: $V_{p,k} = \{v \in V_k | v$ is of class p, $\forall v' \in V_k$ of class $p, \tau_v \leq \tau_{v'}\}$ for $p \in \{0, \ldots, k-1\}$. $V_{p,k}$ is the set of nodes in V_k of class p with minimal communication time to the sink. Define $\tau(V_{p,k}) := \tau_v$ for some $v \in V_{p,k}$. The nodes of V_k are partitioned into layers $U_{p,k}$ for $p \in \{0, \ldots, k-1\}$ as follows:

$$U_{p,k} := \{v \in V_k | \tau(V_{p,k}) \leq \tau_v < \tau(V_{p+1,k})\} \text{ for } p \in \{1, \ldots, k-3\},$$
$$U_{k-2,k} := \{v \in V_k | \tau(V_{k-2,k}) \leq \tau_v\},$$
$$U_{k-1,k} := V_{k-1,k}.$$

Note that $V_{p,k} \subseteq U_{p,k}$ for all p. Further, each message j with $v_j \in U_{p,k}$ traverses some node in $V_{p,k}$. See Figure 1 for a sketch of the layer structure.

We characterize a set of nodes S by its depth, which is $\max_{v \in S} \tau_v - \min_{v \in S} \tau_v$ and the *class string*. The class string is an ordered string representing the class of nodes in S by increasing communication time to the sink. I.e. V_3 has depth 4 and class string $\{2010\}$. In general, set V_k has depth 2^{k-1}. Node sets S and S' are *equivalent* if they have the same depth and class string.

We observe that all messages j with $v_j \in V_k$ are sent to a node in V_{k-1} from some node in $V_{k-1,k}$, i.e. a node of class $k-1$. Also, there are no nodes of higher class in V_k and this is the only node of class $k-1$ a node traverses in V_k. From these observations we may derive that all messages j with $v_j \in V_k$ which are sent over the same node $v \in V_{k-1,k}$ are sent from this node in a single packet. This can be seen as follows. The total accumulated delay of kind 2 and 3 that any message has incurred when sent from v is at least $2^k \Delta + \Delta$ because v is of class $k-1$. The total accumulated delay of kind 2 and 3 that any message has incurred

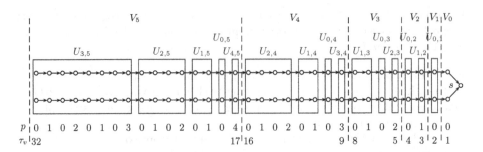

Fig. 1. Node set V_k and layers $U_{0,k}, \ldots, U_{k-1,k}$ for $k = 1, \ldots, 5$

when it arrives at v is at most $2^{k-1}\Delta + 2^{k-1}\Delta$, because the maximum class of any other node in V_k is $k-2$ and each message has traversed at most 2^{k-1} nodes. As the clock drift is bounded by Δ and the difference between the minimum and maximum delay of any two messages is at most $(2^k\Delta + \Delta) - (2^{k-1}\Delta + 2^{k-1}\Delta) = \Delta$ all messages j with $v_j \in V_k$ which are sent over v in P_{CC} must be sent from this node in a single ASC-packet.

Now we are in position to prove our lemma using induction on k. Suppose the lemma holds for $V_0, \ldots, V_k$. Consider set V_{k+1}; this set is partitioned into layers $U_{0,k+1}, \ldots, U_{k,k+1}$. For $\ell = 0, \ldots, k-1$ layer $U_{\ell,k+1}$ is *equivalent* to set $V_{\ell+1}$, hence all messages j with $v_j \in U_{\ell,k+1}$ which are sent from the same node in $U_{\ell,k+1}$ are sent in a single packet. Thus there are at most k packets which arrive at any node $v \in U_{k,k+1}$. As $U_{k,k+1}$ has depth 1, all messages which have v as their release node, are sent from this node in a single packet. Hence, the total number of packets sent from any node in V_{k+1} is bounded by $k + 1$. This proves the lemma. □

Theorem 5. ASC *is* $(1 + 4\Delta\delta/L, \log^2 \delta)$-*competitive.*

Proof. Consider a packet P sent by the optimal solution. Let $\mathcal{P}_{ASC}$ be the set of packets sent by the ASC-algorithm which contain at least one message from P. Let $N_{i,k} = \{j \in N_i | \tau_{v_j} \in V_k\}$, for $i, k \in \mathbb{N}$, $1 \leq i \leq \lceil \log U \rceil$, $1 \leq k \leq \lceil \log \delta \rceil$. Observe that for any choice of budget on the latency L, there are at most $2 \log \delta$ nonempty sets $N_{i,k}$. Using this, it follows from Lemma 2 and Lemma 3 that $|\mathcal{P}_{ASC}| = O(\log^2 \delta)$. Hence, the communication costs of the ASC-solution are at most $O(\log^2 \delta)$ times the cost of an optimal L-bounded solution.

The latency of any message j is at most $L + \Delta\tau_{v_j} + 2\Delta\tau_{v_j} + \Delta$, where the sum consists of the delay of kind 1,2,3 and the clock drift. Thus, the latency of message j is at most $(1 + 4\Delta\delta/L)$ times the budget on the latency. □

If the drift is very small, competitiveness of ASC approaches the lower bound of $(1, \log \delta)$ of the synchronous case, which we proved in [3]. If the drift is of the same order as the latency, i.e. $\Delta = O(L)$, then the SL algorithm, with $\lambda = 1$, has strictly better (β, α)-competitive ratio, than the ASC algorithm. In case of such drifts, it is not plausible anymore to consider the clocks to be synchronized in any sense.

5   Conclusions and Open Problems

We presented on-line distributed algorithms for data aggregation in sensor networks. We considered algorithms under two different models for sensor clocks. For the almost synchronous time model we presented an algorithm which minimizes communication costs under a small excess of the latency budget. These are the first analyses of algorithms for this model, which models actual sensor networks closer than the known ones. We emphasize that the results depend linearly on the drift, and that if the drift is very small our algorithms approach best possible competitive ratios.

For the asynchronous time model we presented an algorithm which balances the communication and latency costs up to a factor $\log U$, where U is the ratio between maximum and minimum allowed delay. We showed that no memoryless algorithm can have a competitive ratio which is more than a factor $\log U$ better than ours, and in case the latency budget is not too small our algorithm is best possible within the class of memoryless algorithms.

The competitive ratio of our asynchronous algorithm is almost balanced; it would be interesting to find an algorithm with balanced ratios, equal to the lower bounds we presented in this paper. Another path for future research is to make a more careful analysis of the almost synchronous time model, in order to determine the maximum clock drift for which almost synchronous algorithms have better competitive ratio than asynchronous algorithms.

References

1. Akyildiz, I., Su, W., Sanakarasubramaniam, Y., Cayirci, E.: Wireless sensor networks: A survey. Computer Networks Journal 38(4), 393–422 (2002)
2. Albers, S., Bals, H.: Dynamic TCP acknowledgment: Penalizing long delays. SIAM Journal Discrete Mathematics 19(4), 938–951 (2005)
3. Becchetti, L., Korteweg, P., Marchetti-Spaccamela, A., Skutella, M., Stougie, L., Vitaletti, A.: Latency constrained aggregation in sensor networks. In: Azar, Y., Erlebach, T. (eds.) ESA 2006. LNCS, vol. 4168, pp. 88–99. Springer, Heidelberg (2006)
4. Brito, C., Koutsoupias, E., Vaya, S.: Competitive analysis of organization networks or multicast acknowledgement: how much to wait. In: Proceedings of the 15th Annual ACM-SIAM Symposium on Discrete Algorithms (SODA), pp. 627–635 (2004)
5. Broder, A., Mitzenmacher, M.: Optimal plans for aggregation. In: Proceedings of the 21st Annual Symposium on Principles of Distributed Computing (PODC), pp. 144–152 (2002)
6. Dooly, D.R., Goldman, S.A., Scott, S.D.: On-line analysis of the TCP acknowledgment delay problem. Journal of the ACM 48(2), 243–273 (2001)
7. Heinzelman, W., Chandrakasan, A., Balakrishnan, H.: Energy efficient communication protocols for wireless microsensor networks. In: Proceedings of Hawaiian International Conference on Systems Science, pp. 3005–3014 (2000)

8. Khanna, S., Naor, J., Raz, D.: Control message aggregation in group communication protocols. In: Widmayer, P., Triguero, F., Morales, R., Hennessy, M., Eidenbenz, S., Conejo, R. (eds.) ICALP 2002. LNCS, vol. 2380, pp. 135–146. Springer, Heidelberg (2002)
9. Marathe, M.V., Ravi, R., Sundaram, R., Ravi, S.S., Rosenkrantz, D.J., Hunt III, H.B.: Bicriteria network design problems. Journal of Algorithms 28(1), 142–171 (1998)
10. Ravi, R., Marathe, M.V., Ravi, S.S., Rosenkrantz, D.J., Hunt III, H.B.: Many birds with one stone: Multi-objective approximation algorithms (extended abstract). In: Proceedings of the 25th Annual ACM Symposium on Theory of Computing (STOC), pp. 438–447 (1993)
11. Stankovic, J.A.: Research challenges for wireless sensor networks. SIGBED Rev. 1(2), 9–12 (2004)
12. Sundararaman, B., Buy, U., Kshemkalyani, A.D.: Clock synchronization for wireless sensor networks: a survey. Ad. Hoc. Networks 3(3), 281–323 (2005)

Optimal Moves for Gossiping
Among Mobile Agents

Tomoko Suzuki[1], Taisuke Izumi[2], Fukuhito Ooshita[1], Hirotsugu Kakugawa[1],
and Toshimitsu Masuzawa[1]

[1] Graduate School of Information Science and Technology, Osaka University
1-3 Machikaneyama, Toyonaka, 560-8531, Japan
{t-suzuki, f-oosita, kakugawa, masuzawa}@ist.osaka-u.ac.jp
[2] Graduate School of Engineering, Nagoya Institute of Technology,
Gokiso-tyo, Syowa-ku, Nagoya, 466-8555, Japan
t-izumi@nitech.ac.jp

Abstract. Mobile-agent-based distributed systems are attracting
widespread attention as the adaptive and flexible systems: mobile agents
traverse the distributed system and carry out a task at each node. In
such mobile-agent-based systems, *gossip* is the most fundamental scheme
supporting cooperation among mobile agents. It requires to accomplish
all-to-all information exchange over all agents so that each agent can obtain the all information each agent initially has. Rendezvous algorithms,
which require that all the agents rendezvous on a node at a time, can
achieve this requirement, however it takes excessive cost for our objective. In this paper, we newly introduce the mobile agent gossip problem.
In this problem, an agent can obtain the information of another agent
by meeting the agent itself or the agent that has already got the information. The gossip scheme is expected to accomplish the all-to-all
information exchange with a smaller number of agents' moves than the
rendezvous algorithms. We propose mobile agent gossip algorithms on
several network topologies, and prove that all proposed algorithms are
asymptotically optimal in term of the number of moves.

1 Introduction

1.1 Background

A *mobile agent system* is one of the most promising frameworks to implement
distributed applications. Mobile agents are autonomous programs that can mi-
grate from one node to another on the network, and traverse the distributed
system to carry out a task at each node. Since the adaptability and flexibility
of mobile agents simplify the design of distributed systems, several mobile agent
systems have been proposed and developed. In typical mobile agent systems,
multiple mobile agents are used to improve system performance: for example,
each agent traverses the network to collect load information of nodes and links
in network management systems. In such mobile-agent-based systems, *gossip* is
one of the most fundamental schemes for cooperation among mobile agents. It

G. Prencipe and S. Zaks (Eds.): SIROCCO 2007, LNCS 4474, pp. 151–165, 2007.
© Springer-Verlag Berlin Heidelberg 2007

requires to accomplish all-to-all information exchange over all mobile agents such that each agent can obtain the all information each agent initially has. By using the gossip scheme, a negotiation with other agents and information collection of the whole network are easily realized in distributed systems.

One naive approach to implement the gossip is to use *rendezvous* algorithms [2,11,12,14], which require that all the agents on a network rendezvous on a node at a time: all agents exchange their own information at the rendezvous point. However, in some cases, the use of rendezvous algorithms takes excessive cost to implement the gossip. For example, suppose the gossip over k agents on a line network of N nodes. Then, to achieve the gossip, the following scenario is allowable: Let agent p be the leftmost agent. The agent p migrates to the right end of the line to collect information of all agents, and then, returns to the left end to deliver the information to all agents. As the result, each agent can obtain the information of all agents. While the rendezvous problem has the trivial $\Omega(kN)$ lower bound on the total number of agent moves, the above scenario takes only $2N$ moves. That is, in this case, rendezvous algorithms are quite costly for our objective.

1.2   Related Works

Rendezvous algorithm is one of approaches to implement the gossip and have been studied by many researchers [2,11,12,14]. Kranakis et.al. have summarized the recent study results about rendezvous in [11]. Barriere et.al. have indicated in [2] that the computabilities of the rendezvous problem and the election problem among agents (i.e., a single agent is elected among agents) are equivalent. It is obvious that the gossip can be solved by a rendezvous algorithm and that the election problem can be solved by a gossip algorithm. Thus, the computabilities of the gossip, the rendezvous problem and the election problem are also equivalent. The election problem have also been studied in [1,5,4]. In [4,12], it is indicated that rendezvous and election problem can not be solved if agents are anonymous and know neither the network size N nor the number of agents k. Therefore, most of these studies assume that each agent knows the network size N. However, it is unrealistic to assume that each node initially knows the global information N in distributed systems. Thus, in this paper, we assume that agents are identifiable. While most of studies about rendezvous focus on time complexity and memory complexity, we focus on move complexity. Some studies also focus on move complexity, but they assume that each agent initially knows the network size N [4,14]. The gossip problem among nodes has been extensively investigated [3,6]. However, the gossip among agents has not been considered before, we propose the mobile agent gossip problem in this paper.

1.3   Our Contribution

Motivated by the above observation, we newly formulate the *mobile agent gossip problem* (MAGP), and investigate its solutions. The goal of this problem is that each agent collects all information other agents initially have with the smallest

Table 1. Our Contribution

Graph	System model	Sense of direction	Total number of agents' moves	
			Upper Bound	Lower Bound
Ring	asynchronous	without	$O(N \log k + N)$	$\Omega(N \log k + N)$
	synchronous		$O(N)$	$\Omega(N)$
Tree	asynchronous	without	$O(N)$	$\Omega(N)$
Complete	asynchronous	without	$O(N \log k + N)$	$\Omega(N \log k + N)$
		with	$O(N)$	$\Omega(N)$
Arbitrary	asynchronous	without	$O(N \log k + |E|)$	$\Omega(N \log k + |E|)$

N, $|E|$ and k are the numbers of nodes, links and agents respectively.

number of moves. Different from the rendezvous, the gossip problem allows relay of information: an agent p_i can obtain the information of other agent p_j directly from p_j or via other agents. Therefore, we can expect that the gossip algorithm inherently takes a smaller number of total moves than rendezvous.

We propose algorithms for MAGP on several network topologies; rings, trees, complete networks and arbitrary networks. Table 1 summarizes our contribution about MAGP in this paper. The property of sense of direction implies that every link is locally labeled at its connecting nodes in a globally consistent way. For all system models, we also prove the lower bounds on the total number of moves. These results indicate that all the proposed algorithms are asymptotically optimal in term of the total number of moves. Interestingly, all the algorithms have the move complexities sublinear in k. Especially, the complexity of the algorithms for synchronous rings, asynchronous trees and complete networks with sense of direction is independent of k. Since the trivial lower bound for the rendezvous problem on trees and rings is $\Omega(kN)$, our results imply that MAGP inherently has lower complexity for the total number of moves than the rendezvous problem.

Some of the above results are based on the investigation of the relation between MAGP and the process leader election. More precisely, in several models, we show that some of the upper/lower bounds for MAGP can be obtained from those for the process leader election. Actually, all our proposed algorithms consist of two phases. At the first phase, exactly one agent is elected as a leader from k agents. On this phase, in some of our algorithms, agents elect their leader by simulating the algorithm for the process leader election. On the other hand, the lower bounds for MAGP are proved by the reduction of the process leader election to MAGP.

The rest of this paper is organized as follows. In Section 2, we present the model of mobile agent systems, and define the mobile agent gossip problem(MAGP). In Section 3, we investigate the relation between MAGP and the process leader election. Section 4 presents the algorithms and the lower bounds on the total number of moves for MAGP on several networks. Section 5 concludes the paper.

2 Preliminaries

2.1 System Models

The network is modeled as an undirected graph $G = (V, E)$, where V and E are respectively the node set and the link set in the network G. A link in E connects two distinct nodes in V. The link between nodes u and v is denoted by e_{uv} or e_{vu}. The number of nodes is denoted by N (i.e., $N = |V|$), and the number of mobile agents in the network is denoted by k. An agent is an autonomous program that can migrate from one node to another on the network. We assume that each agent and each node has distinct identifier[1]. Each agent does not initially know identifiers of other agents or nodes. We also assume that each agent has prior knowledge of neither the number of nodes N nor the number of agents k. There is no assumption about the initial location of each agent: more than one agent may be initially located on a same node. The node on which an agent is initially located is called the *home node* of the agent. Agents on node $u \in V$ can migrate to node $v \in V$ only when link e_{uv} is contained in E. For each node, links connecting to it are locally labeled, so that an agent on the node can distinguish the links. Let $\lambda_u(e_{uv})$ be the label of link e_{uv} on node u. Each node v is provided with a *whiteboard*, i.e., a local storage where agents on v can write, read and erase information. Each agent can perform the following operations on the whiteboard at each node.

- *write(v, infos)* : an agent writes information *infos* on node v's whiteboard.
- *read(v)* : an agent reads information written on node v's whiteboard
- *delete(v)* : an agent deletes information written on node v's whiteboard.

Access to a whiteboard is done exclusively: when multiple agents on a node execute their operations, the operations are sequentially executed in an arbitrary order. Agents are said to be asynchronous if migration time and local processing time of agents are unpredictable but finite. In contrast, agents are synchronous if its execution is partitioned into rounds; in each round, every agent arrives at a node, executes local computation on the node, and stays on the node or starts migration to one of the neighboring nodes.

A state of an agent is represented by the set of variables the agent has and the set of information the agent collects, and a state of a node is represented by the state of its own whiteboard. A system configuration C is represented by the states of all nodes, the states of all agents, and the locations of all agents. A system configuration is changed by events of agents (e.g., migration to a neighboring node or access to a whiteboard on a node). Let C_0 be the initial configuration of a system and Ev_i be the set of events that occurs simultaneously at the configuration C_i. An execution of a distributed system is an alternative sequence of configurations and sets of events $EX = C_0, Ev_0, C_1, Ev_1, C_2, \cdots$, such that occurrence of events Ev_{i-1} changes the configuration from C_{i-1} to C_i.

---

[1] In our algorithm for MAGP, node IDs are not required if agents have unique IDs. Notice that it is not essential whether nodes have unique IDs or not because the naming is possible by agents' traverse.

2.2   Mobile Agent Gossip Problem

In this paper, we consider the *mobile agent gossip problem*(MAGP). At an initial configuration, each agent p_j has only its own information I_j. The goal of this problem is that every agent collects the information of all agents. The MAGP is defined as follows.

Let $S_j(C_i)$ be the set of information an agent p_j has at a configuration C_i. At an initial configuration C_0, the set of information $S_j(C_0)$ each agent p_j has includes only its own information I_j:

$$S_j(C_0) = \{I_j\}. \tag{1}$$

The mobile agent gossip problem is solved when all k agents terminate after the following condition is satisfied at a configuration C_i:

$$\forall j (0 \le j < k)\ S_j(C_i) = \bigcup_{0 \le l < k} \{I_l\} \tag{2}$$

We make no assumption on the size of information each agent initially has. To avoid introducing huge space for whiteboard, we disallow each agent p_j to leave the set of information $S_j(C_i)$ on a whiteboard. Agents can write only the control information on a whiteboard, e.g., some number of identifiers and counter values. Instead, we allow agents on a same node to exchange the set of information with each other. When h agents $p_0, p_1, \cdots .p_{h-1}$ are located on a same node at a configuration C_i, then the following holds for the configurations C_{i+1}:

$$\forall j (0 \le j < h)\ S_j(C_{i+1}) = \bigcup_{0 \le l < h} S_l(C_i) \tag{3}$$

We define one move as a migration of an agent from one node to its neighbor. The complexity of the MAGP is measured by the total number of moves until all agents terminate in the worst case.

3   Reduction of the Leader Election Problem to MAGP

By reduction from the leader election problem of nodes, we obtain the lower bounds on the number of total moves for MAGP. In this section, we prove that the leader election problem of nodes (LEP) can be reduced to MAGP. Notice that we consider the leader election among "nodes" in message passing systems, instead of agents. Informally, the objective of LEP is that each node eventually decides a single common leader node. This paper considers LEP under the assumption that only k nodes spontaneously initiate an algorithm for LEP and each other nodes initiates an algorithm for LEP when it receives a message. More precisely, the leader election problem is specified as follows:

Definition 1 (Leader Election Problem of k Initiator Nodes (LEP))
Let $v_0, \ldots, v_{k-1}(1 \le k < N)$ be the initiator nodes on the network G. An algorithm is said to solve the leader election problem if it satisfies the following conditions:

- *Exactly one of the nodes is elected as the leader and all the other nodes existing in the network G know the identifier of the leader node.*
- *Once a node decides whether to be as the leader, the node never changes the decision.*

We can obtain the following relation between LEP and MAGP.

Theorem 1. *When the total moves for MAGP is m_g and the total moves required for an agent to travel the whole of the network is m_t, LEP can be solved with $m_g + m_t$ messages in an message passing system.*

Proof . *We prove this theorem by showing that LEP can be solved by using an MAGP algorithm in agent systems. In message passing systems, the behavior of agents can be simulated in the same number of messages as the agents' moves. Each initiator node v_i creates an agent p_i that has information including v_i's identifier. By applying an algorithm for MAGP, the k agents created by k initiator nodes can collect all identifiers of the initiator nodes. Each agent can elect exactly one leader agent from the k agents based on the initiator nodes' identifiers. The leader agent, say p_j, travels the whole network so that the node v_j becomes the leader and the other nodes know the identifier of v_j.* □

4   Algorithms for MAGP

In this section, we present the upper and the lower bounds for MAGP. All proposed algorithms for MAGP consist of two phases. At the first phase (called *election phase*), a leader agent is elected from k agents, and at the second phase (called *traverse phase*), the leader agent travels the whole network to collect and deliver information each agent has while all the other agents. When more than one agent is initially located on a same node, one agent is elected among the agents based on their identifiers and the elected agent executes the election phase. Therefore, in what follows, we consider the initial configuration where at most one agent is located on a node.

4.1   Non-rooted Tree Networks

In this subsection, we present an algorithm for MAGP on a non-rooted tree.

At the election phase, each agent joins a tournament to become the leader. For easy understanding, suppose an initial configuration where each leaf node (i.e., a node with a single incident link) has a single agent. We ignore all the agents on internal nodes, that is, they make no action. Then, the leader agent is elected as follows: the agent on a leaf node migrates to the neighboring node u. On the node u, the agent waits until agents come from all neighboring nodes except only one node, and then, one of the agents on u wins this stage of the tournament and migrates to the connecting node no agent has come from.

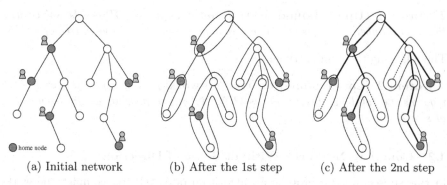

| (a) Initial network | (b) After the 1st step | (c) After the 2nd step |

Fig. 1. Example of agents' behavior on a non-rooted tree

By repeating the above actions, only one agent wins the tournament[2]. The tournament needs $O(N)$ moves since each link is used exactly once[3].

In MAGP, however, there is no assumption of initial locations of agents: we cannot assume that each leaf node has an agent at the initial configuration. Therefore, at the start of the election phase, agents create a configuration where one agent stays on each leaf node.

In what follows, we explain the agent's behavior at the election phase. On the first step at the election phase, each agent on a leaf node or an internal node travels on the tree by the depth-first-search (DFS) to construct its territory so that territories of different agents have no common node and those of all agents cover the whole network: the first agent that arrives at a node v becomes the owner of v (Fig. 1 (b)). When the agent can construct its territory including all nodes in the tree, that is, there is only one agent in the tree, the agent becomes the leader. Otherwise, each agent proceeds the second step. On the second step, each agent travels again in its territory (a subtree) by the DFS. Each agent considers its home node as a root node of the subtree. Then, on each link e_{uv} (node v is one of children nodes of node u), the agent cuts the link e_{uv} when the node v has no descendant node connecting with the territory of another agent. At the configuration after the cutting links on all nodes, each leaf node has a single agent (Fig.1 (c)). And then, each agent on leaf nodes competes at the tournament including only the non-cutted links as mentioned above. Agents on internal nodes after the second step do not join the tournament and wait until the leader agent collects and delivers the information.

At the traverse phase, the leader agent travels the whole network by DFS with $O(N)$ moves.

On the first and the second step, the total number of agents' moves is $O(N)$ since each agent travels in only its territory by the DFS. The leader election with the tournament needs $O(N)$ moves. Thus, the following theorem holds.

---

[2] At the final stage, both of two remaining agents may be elected if they stay two endpoints of the last unused link l and concurrently move along l. However, We can easily break such symmetry case by writing the winner agent IDs on the whiteboards.

[3] Exceptionally, the last unused link may be used twice.

Theorem 1 (Upper bound on tree networks). *MAGP can be solved with the total number of moves $O(N)$ in any asynchronous non-rooted tree of size N.*

The following theorem is trivial.

Theorem 2 (Lower bound on tree networks). *Any algorithm for MAGP requires the total number of moves $\Omega(N)$ in any asynchronous non-rooted tree of size N.*

4.2 Complete Networks Without Sense of Direction

In this subsection, we propose an algorithm of MAGP on complete networks without sense of direction. That is, links connecting to a node have arbitrary local labels.

At the election phase, a leader agent is elected by simulating the algorithm for LEP on complete networks without sense of direction proposed in [10]. In the algorithm, each node repeatedly captures other nodes, and finally, the node that has captured more than half nodes on the network becomes the leader.

In what follows, we briefly describe the election phase in our algorithm simulating the algorithm proposed in [10]. We define three states of a node:

- a *candidate* node is a home node of an agent that tries to capture other nodes and that has a chance of becoming the leader agent.
- A *captured* node is the node captured by an agent.
- A *passive* node is one that is neither a *candidate* node nor a *captured* node.

Let i be the identifier of agent p_i. Each agent p_i writes its identifier i and the number of captured nodes on the whiteboard at its home node v (the number of captured nodes is initially set to zero). Let n_i be the number of captured nodes by an agent p_i. Each agent p_i repeats the following actions until p_i captures more than half nodes of the network or its home node v becomes *passive* or *captured* : p_i on its home node v migrates one of its neighbors u to try to capture it. Let n_u and id_u be the number of nodes and the identifier written on u's whiteboard respectively. The next action of the agent p_i depends on the state of u:

- (**Case1**) The node u is *passive* node: the agent p_i captures u.
- (**Case2**) The node u is *candidate* node: if n_i is larger than n_u, or if n_i is the same number as n_u and i is larger than id_u, p_i captures u. Otherwise, the agent p_i returns to its home node v and makes v *passive*.
- (**Case3**) The node u is *captured* node: the agent p_i attacks the node w that is the home node of the agent capturing u. The agent p_i migrates to the node w, and then, p_i captures u and makes w *passive* if one of the following conditions is satisfied: (1) the node w is a *passive* or *captured* node, (2) n_i is larger than n_w, or (3) n_i is the same number as n_w and i is larger than id_w. Otherwise, the agent p_i returns to its home node v and makes v *passive*.

When the agent p_i captures a node u, p_i marks the link connecting to its home node so that other agents visiting u can move to the home node of p_i.

The agent p_i increments the number of n_v on the whiteboard at its home node v every time p_i captures a new node.

At the traverse phase, the leader agent travels the whole network with $O(N)$ moves.

The message complexity of the algorithm in [10] has been proved to be $O(N \log N)$. In our algorithm, agents simulate one-message transmission by one migration among nodes. The number of agents that can capture $N/2 + 1$ nodes is one, and the number of agents that can capture $N/2$ nodes is at most two. Similarly, the number of agents that can capture N/h nodes is at most h. Therefore, the captures of nodes are done at most $(N/2 + 1) + \sum_{h=2}^{k} 1/h \cdot N \le N \log k + N/2 + 1$ times in total. The capture of one node by an agent needs at most 4 moves. Thus, the following theorem holds.

Theorem 3 (Upper bound on complete networks without sense of direction). *MAGP can be solved with the total number of moves $O(N \log k + N)$ in any asynchronous complete network without sense of direction of size N, where k is the number of agents.*

It is proved in [10] that the message complexity of any LEP algorithm on complete networks without sense of direction is $\Omega(N \log N)$. In the case of k initiator nodes, we can prove the $\Omega(N \log k + N)$ lower bound on the message complexity for LEP by similar proof as one in [10]. From Theorem 1 and the above fact, we can show the following lower bound on the number of agents' moves for MAGP on complete networks without sense of direction.

Theorem 4 (Lower bound on complete networks without sense of direction). *Any algorithm for MAGP requires the total number of moves $\Omega(N \log k + N)$ in any asynchronous complete network without sense of direction of size N, where k is the number of agents.*

4.3   Complete Networks with Sense of Direction

In this subsection, we propose an algorithm for MAGP on complete networks with sense of direction. The sense of direction is given at each node as follows: nodes are denoted by $v_0, v_1, \cdots, v_{N-1}$, numbered clockwise in the ring, and for every $i, j (0 \le i, j \le N - 1, i \ne j)$, the link $e_{v_i v_j}$ is labeled by $(j - i) \bmod N$ at v_i and $(i - j) \bmod N$ at v_j. Figure 2(a) shows an complete network with the sense of direction of six nodes.

At the election phase, the behavior of each agent is similar to the algorithm for LEP proposed in [13]. In this algorithm, each node sends messages including its identifier along the *ring* in both directions. On the first stage, the ring consists of the links labeled 1 or $N - 1$. When a node receives the messages, the node decides whether it survives or not, based on their identifiers. To reduce the message cost, the winner node uses the chordal links between the winner and the looser , and the looser and the looser nodes on the next stage. That is, the ring is shrunk into a smaller one. On the next stage, the winner nodes repeat the above actions

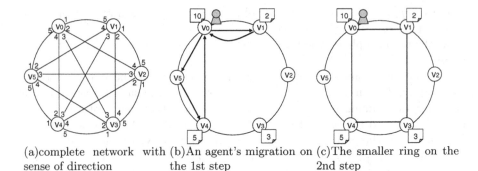

(a)complete network with (b)An agent's migration on (c)The smaller ring on the
sense of direction   the 1st step   2nd step

Fig. 2. An example of agents' behavior on a complete network with sense of direction

on the smaller ring. These actions are continued until only one node wins over.
The final winner becomes the leader.

In our algorithm for MAGP, each agent simulates the algorithm proposed in
[13] to elect a leader agent. Each agent first writes its identifiers on the whiteboard at its home node. Figure 2(b) shows an example of an agent's migrations
on the first stage. On the first stage, each agent migrates along the links labeled
1 with counting the number of passed nodes. When the agent finds the identifier
of another agent on the whiteboard at visited node v, it checks the identifier
and directly returns to its home node through the chordal link between v and
its home node, which can be found using the number of passed nodes. Similarly,
the agent repeats the above actions in the reverse direction. The agent that wins
both of its neighboring agents based on their identifiers can survive into the
next stage. The looser agents stop their migration and wait on their home node.
On the next stage, the winner repeats the above actions on the smaller ring,
which consists of the chordal links connecting the home nodes of the winner and
the looser, or the looser and the looser agents on the previous stage (Fig. 2(c)).
When only one winner exists on the ring, the election phase terminates and the
agent that wins all other agents becomes the leader.

At the traverse phase, by migrating along the links labeled 1, the leader agent
travels the whole network with $O(N)$ moves.

On each stage, each link included in a ring is used at most four times for
agents' migrations. The ring size on the first stage is N, and on the second stage
is k since the ring on the second stage is constructed by only agents' home nodes.
On the h-th stage, the ring size is at most $k/2^{h-2}$, and the number of winner
agents is at most $k/2^{h-1}$. Therefore, by the $\log k$-th stage, the leader agent is
elected. Thus, the following theorem holds.

**Theorem 5 (Upper bound on complete networks with sense of direc-
tion).** *MAGP can be solved with the total number of moves $O(N)$ in any asynchronous complete network with sense of direction of size N.*

The lower bound on complete networks is trivial.

Theorem 6 (Lower bound on complete networks with sense of direction). *Any algorithm for MAGP requires the total number of moves $\Omega(N)$ in any asynchronous complete network with sense of direction of size N.*

4.4   Arbitrary Networks

In this subsection, we propose an algorithm for MAGP on arbitrary networks, where no sense of direction is assumed.

The outline of the election phase is as follows: each agent first constructs its territory that has no common node with territories of other agents by the DFS: each agent captures nodes in its territory on a first-come-first-capture basis. Let T_i be the territory of agent p_i. And then, each agent merges its territory with other agents' territories. When the territories of agents p_i and p_j are merged, either p_i or p_j becomes the owner of the merged territory. Finally, the agent that captures all nodes in its territory becomes the leader.

To merge territories, each agent behaves similar actions with Gallager's algorithm for constructing minimum spanning tree (MST) proposed in [8]. In Gallager's algorithm, the MST is constructed by merging subtrees based on weights of links. To simulate Gallager's algorithm, in our algorithm, each link is assigned a value as its weight. More precisely, the label $(\min\{id(u), id(v)\}, \max\{id(u), id(v)\})$ is assigned to each link e_{uv} as its weight, where $id(u)$ is the identifier of node u. Two weight labels (u_1, v_1) and (u_2, v_2) are compared with lexicographic order, that is, $(u_1, v_1) < (u_2, v_2) \leftrightarrow u_1 < u_2 \vee (u_1 = u_2 \wedge v_1 < v_2)$. Each territory can be considered as a sub-tree constructed by Gallager's algorithm. Therefore, by simulating Gallager's algorithm for constructing MST, only one agent can construct territory including all nodes.

In what follows, we explain the outline of Gallager's algorithm. A *minimum outgoing link* of a territory T_i is the link with the lowest weight connecting a node in T_i and a node in another territory. Let MOL_i be the minimum outgoing link for territory T_i. Gallager's algorithm repeats two procedures in each territory: one is to search MOL_i in the territory T_i, and another is to merge the territory T_i with the other territory connecting via MOL_i. We define the level L_i of territory T_i as the number of times to search MOL_i in T_i (the initial level of each territory is set to zero). Note that the level L_i is not the number of times that territory T_i merges other territories, and is the number of times to find MOL_i by searching all of outgoing links in T_i. To reduce the message cost, Gallager's algorithm executes the search of MOL_i and the merger of territories according to levels of each territory.

1. The territory T_i merges the other territory $T_j (i > j)$ only if one of the following conditions is satisfied.
   (a) $L_i > L_j$ and MOL_j connects a node in the territory T_i: the territory T_i merges the territory T_j. The level of T_i is not changed. That is, the search of MOL_i is not restarted in T_i.
   (b) $L_i = L_j$ and $MOL_i = MOL_j$: the territories T_i merges the territory T_j. The level of T_i is set to $L_i + 1$. That is, the search of MOL_i is executed in the merged territory.

2. If the territory T_i has an outgoing link connecting with the territory T_j that has the level L_j smaller than the level L_i ($L_j < L_i$), T_i might merge T_j on the link. Hence, if the territory T_i finds such outgoing link while searching MOL_i, T_i suspends searching MOL_i until T_i knows whether or not T_i merges T_j.

Gallager had proved in [8] the level of each territory is at most $\log N$ by applying the above rules, where N is the number of nodes that executes the algorithm.

At the election phase in our algorithm, each agent simulates Gallager's algorithm. Each agent p_i writes its identifier and the level of its territory on the whiteboards at nodes in its territory. The search of MOL_i is done by the DFS traversing whole of the territory T_i, and needs $O(|T_i|)$ moves. The state of each connecting link whether the link is outgoing link or not and the weight of outgoing link are written on whiteboard at each node. Let link e_{vu} be the MOL_i, where nodes v and u are included in the territories T_i and $T_j (i \neq j)$ respectively. When the agent p_i finds the link e_{vu} is MOL_i, p_i migrates to the node u and reads the level l_u written on u's whiteboard. Then, the agent p_i behaves in accordance with the above rules of merger for the levels L_i and l_u. When the territory T_i is merged by the territory T_j (i.e., $L_i < l_u$), the agent p_i travels whole of the territory T_i by the DFS writing the identifier j and the level l_u on each whiteboard. When the territories T_i and T_j are merged (i.e., $L_i = l_u$ and $MOL_i = MOL_j$), either agent p_i or p_j is elected (depending on their identifiers) as the owner of the merged territory, and the another is not the owner and waits on the staying node. Let p_i be the owner of the new territory. The agent p_i travels whole of the territories T_i and T_j by DFS writing the identifier i and the new level $L_i + 1$. The agent that captures all nodes in its territory becomes the leader.

At the traverse phase, by the DFS on the MST, the leader agent travels the whole network with $O(N)$ moves. The checks of all links to construct first territory of each agent are required $O(|E|)$ moves, where E is the set of links. Every link is checked at most once whether the link is outgoing link or not, so the check of all link is required $O(|E|)$ moves in total. The searches of minimum outgoing link on each level is required $O(N)$ moves in total on each level. Since the number of territories at the start of constructing MST is k, the level of any territory is at most $\log k$. Therefore, the following theorem holds.

Theorem 7 (Upper bound on arbitrary networks). *MAGP can be solved with the total number of moves $O(N \log k + |E|)$ in any asynchronous network of size N, where k is the number of agents and $|E|$ is the number of links.*

The lower bound on message complexity for LEP has proved to be $\Omega(N \log k + |E|)$ in [8]. We can get the following lower bound on moves for MAGP from the trivial lower bound $\Omega(|E|)$, and the traverse cost $O(|E|)$ and Theorem 1.

Theorem 8 (Lower bound on arbitrary networks). *Any algorithm for MAGP requires the total number of moves $\Omega(N \log k + |E|)$ in any asynchronous network of size N, where k is the number of agents and $|E|$ is the number of links.*

4.5 Asynchronous Ring Networks

In this subsection, we propose an algorithm for MAGP on asynchronous ring networks. On ring networks, the network has no sense of direction. That is, links connecting to a node have arbitrary local labels.

From Theorem 7, the following theorem holds.

Theorem 9 (Upper bound on asynchronous ring networks). *MAGP can be solved with the total number of moves $O(N \log k + N)$ in any asynchronous ring network of size N, where k is the number of agents.*

As in the case of complete networks without sense of direction, we can get the following lower bound on moves for MAGP from Theorem 1 and the lower bound $\Omega(N \log k)$ on message complexity for LEP proved in [9].

Theorem 10 (Lower bound on asynchronous ring networks). *Any algorithm for MAGP requires the total number of moves $\Omega(N \log k + N)$ in any asynchronous ring network of size N, where k is the number of agents.*

By using the ring structure, we can simplify the election phase of the algorithm for MAGP on arbitrary networks as stated in the following. At the election phase, the agent simulates the algorithm for LEP proposed in [9]. Each agent writes its identifier on the whiteboard at its home node and acts the following actions repeatedly: the agent migrates in a direction along the ring and checks the identifier of its neighboring agent, and then, does the same actions in the reverse direction. If the agent has the larger identifier than its two neighboring agents, the agent merges its territory with the two neighboring territories and becomes the owner of the merger territory. Otherwise, the agent is not the owner and waits on the staying node. If the agent can return to its home node without meeting other owners, the agent becomes the leader. As the results, the maximum level of territory is at most $\log k$.

4.6 Synchronous Ring Networks

In this subsection, we present an algorithm for MAGP on synchronous rings. On synchronous ring, the number of required moves can be reduced by using synchronous clock each agent has.

At the election phase, agents simulate the algorithms for LEP proposed in [7]. On synchronous rings, each agent waits on a node for a time depending on its identifier to reduce the number of moves: each agent p_i waits for $2^i - 1$ rounds every time p_i arrives at each node. That is, the agent p_i migrates to one of neighbors at most once every 2^i rounds.

The outline of the election phase is as follows: each agent first writes its identifier on the whiteboard at its home node, and migrates in a direction along the ring. When the agent finds an identifier of other agent, the agent decides, according to the identifier, whether it should continue its migration or not. If the agent has the larger identifier than the written identifier, the agent stops its migration and waits on the node. The only one agent can return to its home node and then it becomes the leader.

At the traverse phase, the leader agent travel the whole network by migrating in a direction with $O(N)$ moves.

The message complexity of the algorithm for LEP proposed in [7] has been proved to be $O(N)$. At the election phase in our algorithm, the agents simulate one-message transmission by one migration among nodes: a migration event of an agent from a node v to u corresponds to a send event of a message from the node v to u in an execution of the algorithm proposed in [7]. Thus, the following theorem can holds.

Theorem 11 (Upper bound on synchronous ring networks). *MAGP can be solved with the total number of moves $O(N)$ in any synchronous ring network size N.*

The lower bound on synchronous rings is trivial.

Theorem 12 (Lower bound on synchronous ring networks). *Any algorithm for MAGP requires the total number of moves $\Omega(N)$ in any synchronous ring network of size N.*

5   Conclusions

In this paper, we have considered the mobile agent gossip problem (MAGP), which has been newly proposed. The gossip is the most fundamental scheme supporting cooperation among mobile agents. The goal of MAGP is that all agents obtain all information each agent initially has with the smallest number of moves. MAGP inherently has lower complexity for the total number of moves than the rendezvous problem.

We have proposed algorithms for MAGP on several network topologies: rings, trees, complete networks and arbitrary networks. For all network topologies, we can prove that all proposed algorithms are asymptotically optimal in term of the total number of moves. All the proposed algorithms have the move complexities sublinear in k. Especially, the complexity of the algorithms for synchronous rings, trees and complete networks with sense of direction is independent of k. Some of our results are based on the investigation of the relation between MAGP and the process leader election problem (LEP): some of the upper/lower bounds for MAGP have been obtained from those for LEP.

Acknowledgement. This work is supported in part by a Grant-in-Aid for Scientific Research((B)(2)15300017) of JSPS, Grand-in-Aid for Young Scientists ((B)18700059), Grant-in-Aid for Scientific Research((B)17300020) of JSPS, Grant-in-Aid for Scientific Research on Priority Areas(16092215) of MEXT, Grant-in-Aid for JSPS Fellows (2005, 50673), "The 21st Century Center of Excellence Program" of MEXT, and the Ookawa Foundation Research Grant.

References

1. Barriere, L., Flocchini, P., Fraigniaud, P., Santoro, N.: Can we elect if we cannot compare. In: Proceedings of the 15th ACM Symposium on Parallelism in Algorithms and Architectures (SPAA 2003), pp. 324–332 (June 2003)
2. Barriere, L., Flocchini, P., Fraigniaud, P., Santoro, N.: Rendezvous and election of mobile agents: Impact of sense of direction. Theory of Computing Systems 40(2), 143–162 (2007)
3. Bruck, J., Ho, C.-T., Kipnis, S., Upfal, E., Weathersby, D.: Efficient algorithms for all-to-all communications in multiport message-passing systems. IEEE Transactions on Parallel and Distributed Systems 8(11), 1143–1156 (1997)
4. Das, S., Flocchini, P., Nayak, A., Santoro, N.: Effective elections for anonymous mobile agents. In: Asano, T. (ed.) ISAAC 2006. LNCS, vol. 4288, pp. 732–743. Springer, Heidelberg (2006)
5. Deugo, D.: Mobile agents for electing a leader. In: Proceedings of the 4th International Symposium on Autonomous Decentralized Systems (ISADS 1999), pp. 324–327 (March 1999)
6. Flammini, M., Perennes, S.: On the optimality of general lower bounds for broadcasting and gossiping. SIAM Journal on Discrete Mathematics 14(2), 267–282 (2001)
7. Frederickson, G.N., Lynch, N.: Electing a leader in a synchronous ring. Journal of the ACM 31(1), 98–115 (1987)
8. Gallager, R.G., Humblet, P.A., Spira, P.M.: A distributed algorithm for minimum-weight spanning tree. ACM Transactions on Programming Languages and Systems 5(1), 66–77 (1983)
9. Hirschberg, D.S., Sinclair, J.B.: Decentralized extrema-finding in circular configurations of processors. Communications of the ACM 23(11), 627–628 (1980)
10. Korach, E., Moran, S., Zaks, S.: Optimal lower bounds for some distributed algorithms for a complete network of processors. Theoretical Computer Science 64(1), 125–132 (1989)
11. Kranakis, E., Krizanc, D., Rajsbaum, S.: Mobile agent rendezvous: A survey. In: Proceedings of the 13rd Colloquium on Structural Information and Communication Complexity (SIROCCO 2007), pp. 1–9 (July 2007)
12. Kranakis, E., Santoro, N., Sawchuk, C., Krizanc, D.: Mobile agent rendezvous in a ring. In: Proceedings of the 23rd International Conference on Distributed Computing Systems (ICDCS 2003), pp. 592–599 (May 2003)
13. Loui, M.C., Matsushita, T.A., West, D.B.: Election in complete networks with sense of direction. Information Processing Letters 22(4), 185–187 (1986)
14. Marco, G.D., Gargano, L., Kranakis, E., Krizanc, D., Pelc, A., Vaccaro, U.: Asynchronous deterministic rendezvous in graphs. Theoretical Computer Science 355(3), 315–326 (2006)

Swing Words to Make Circle Formation Quiescent

Yoann Dieudonné and Franck Petit

LaRIA CNRS, Université de Picardie Jules Verne
Amiens, France

Abstract. In this paper, we first introduce the *swing words*. Based on intrinsic properties of these words, we present a new approach to solve the Circle Formation Problem in the semi-synchronous model (SSM)—no two robots are supposed to be at the same position in the initial configuration. The proposed protocol is *quiescent*— all the robots are eventually *motionless* in the desired configuration, which remains true thereafter. In SSM, the improvement of the latest recent work for this problem is twofold: (1) the protocol works for any number n of weak robots, except if $n = 4$, and (2) no robot is required to reach its computed destination in one step.

Finally, starting from a biangular configuration, our protocol also solves CFP in the fully asynchronous model (CORDA). To our best knowledge, it is the first CFP protocol for SSM which is compatible with CORDA.

Keywords: Distributed Coordination, (Uniform) Circle Formation, Mobile Robot Networks, Self-Deployment.

1 Introduction

Consider a distributed system where the computing units are *mobile weak robots* (*sensors* or *agents*), i.e., devices equipped with sensors and designed to move in a two-dimensional plane. By weak, we mean that the robots are *anonymous*, *autonomous*, *disoriented*, and *oblivious*, i.e., devoid of (1) any local parameter (such that an identity) allowing to differentiate any of them, (2) any central coordination mechanism or scheduler, (3) any common coordinate mechanism or common sense of direction, and (4) any way to remember any previous observation nor computation performed in any previous step. Furthermore, all the robots follow the same program (*uniform* or *homogeneous*), and there is no kind of explicit communication medium. The robots implicitly "communicate" by observing the position of the others robots in the plane, and by executing a part of their program accordingly.

In such a weak and unrealistic model, there has been considerable interest in the design of deterministic coordination protocols (or algorithms). One of the common features of these works is the study of the minimal level of ability the robots are required to have to achieve the desired task. So far, the studied tasks are geometric problems, so that *pattern formation*, *line formation*, *gathering*, and *circle*

G. Prencipe and S. Zaks (Eds.): SIROCCO 2007, LNCS 4474, pp. 166–179, 2007.

formation—refer to [18,11,14,15] for this problems. Basically, every protocol for these problems aim to be *quiescent*, i.e., in every execution, all the robots are eventually *motionless* in the desired configuration, which remains true thereafter.

The *Circle Formation Problem* (CFP) belongs to the class of pattern formations. It consists in the design of a protocol insuring that starting from an initial arbitrary configuration, all the robots eventually form a circle with equal spacing between any two adjacent robots. (This problem is sometime referred as the *uniform circle formation Problem* [15].) In other words, the system is expected to converge to a configuration where the robots stop by forming a *regular n-gon*.

Related Works. An informal CFP algorithm was first given in [4]. Several CFP protocols were subsequently proposed. An heuristics based algorithm was proposed in [16]. A protocol for non-oblivious robots is given in [18]. Deterministic CFP algorithms were first presented in [5,1]. All the above solutions guaranteed only asymptotical convergence toward a configuration in which the robots are uniformly distributed on the boundary of a circle. In other words, these solutions are not quiescent, i.e., the robots move infinitely often and never reach the desired final configuration.

The above solutions work in the semi-synchronous model (SSM) [17] in which the cycles of all the robots are synchronized and their actions are atomic. In each cycle, every robot is either active or inactive (at least one of them is active), and only active robots perform their cycle. Every execution in SSM is assumed to be *fair*, i.e., every robot becomes active infinitely often.

The solution in [12] works in a fully asynchronous model, called CORDA [13]. In CORDA, each robot infinitely and independently from the other robots, asynchronously cycles through a WAIT state, an OBSERVE state (the robot observes its environment), a COMPUTE state (the robot computes its destination point based on the current locations of the other robots), and a MOVE state (the robot moves toward the computed destination of an unpredictable amount of space assumed to be neither infinite, nor infinitesimally small. Each state and the distance traveled while in the MOVE state are assumed to be finite. Since CORDA is a weaker model than SSM, solutions designed in CORDA also work in SSM [14]—the reverse is not true.

On top of working in CORDA, the solution in [12] is quiescent, i.e., the robots eventually stop in a final configuration where they are regularly spread out along the border of a circle. However, if n is even, the robots may only achieve a biangular circle—the distance between two adjacent robots is alternatively either α or β. In [7], a deterministic quiescent CFP protocol is proposed—the exact n-gon is eventually built—using the useful properties of Lyndon words. However, the solution in [7] works in SSM and only for a prime number of robots.

A common strategy in order to solve a non trivial problem as CFP is to combine subproblems which are easier to solve. In general, CFP is separated into two distinct parts. The first subproblem consists in placing the robots along the boundary of a circle C, without considering their relative positions. The second subproblem, called *uniform transformation problem* (UTP), consists in starting from there, and arranging robots, without them leaving the circle C, evenly along the boundary of C. In [5], the authors conjecture that there is no deterministic solution solving UTP in finite time in the semi-synchronous model in SSM—the robots being uniform, anonymous, oblivious, and none of them sharing any kind of coordinate system or common sense of direction. In a recent paper [10], the validity of the conjecture is proven. The solution in [6] tackles this latter problem by providing a deterministic algorithm for any number of robots except 4, 6 and 8. The solution in [6] combines the solution in [12] and a non-trivial method based on a concentric circles to eventually achieve a regular n-gon. However, the solution in [6] works assume that no robot can stop before reaching its destination. Note that all the solutions so far assume that in the initial configuration, no two robots are located at the same position. This partially contradicts the specification of CFP since the initial configuration is assumed to be arbitrary. As already noticed in [5], this implies that none of them is self-stabilizing [9]. In a recent paper [8], we provide a randomized self-stabilizing protocol to scatter the robots at distinct positions.

Contribution. We first present a new approach to solve CFP, based on *Swing Words*. Informally, a finite non-empty word w over an ordered alphabet A ($|A| \geq 2$) is a *swing* word if and only if the five following conditions are true:

1. The length of w is even,
2. the word w contains at least two different letters,
3. each odd letter of w is greater than or equal to (respectively, lower than or equal to) its following letter,
4. each even letter of w but the last one is lower than or equal to (respectively, greater than or equal to) its following letter,
5. the last letter of w is lower than or equal to (respectively, greater than or equal to) the first letter of w.

For instance, over $A = \{a, b, c\}$, $a < b < c$, the words *abacbc*, *bcbc*, and *aaaaaaab* are swing words, whereas *abacba*, *aaaab*, and *baab* are not—more examples and a formal definition of swing words are given in Section 3.

Using intrinsic properties of Swing Words, we provide an original deterministic quiescent protocol for CFP in SSM (assuming that no two robots are initially at the same position). As in [6], the proposed protocol is not based on UTP. It is based on an original technique using the *convex hull* formed by the robots. It works for any number n of weak robots except if $n = 4$. So, it improves our previous result which was not working for $n = 6$ and $n = 8$. Moreover, in contrast with the solution proposed in [6], our protocol assume that no robot is required to reach its computed destination in one step. Finally, starting from a biangular configuration, our protocol solves CFP in CORDA. To our best knowledge, it is the first CFP protocol for SSM which is compatible with CORDA.

Outline of the Paper. In the next section (Section 2), we describe the distributed system and the problem we consider in this paper. In the same section, we present the Swing words and some properties. The deterministic algorithm is proposed in Section 3. Finally, we conclude this paper in Section 4.

2   Preliminaries

In this section, we define the distributed system, basic definitions and the considered problem.

Distributed Model. We adopt the model introduced in [17], below referred to as *SSM*. The *distributed system* considered in this paper consists of n robots $r_1, r_2, \cdots, r_n$—the subscripts $1, \ldots, n$ are used for notational purpose only. Each robot r_i, viewed as a point in the Euclidean plane, move on this two-dimensional space unbounded and devoid of any landmark. When no ambiguity arises, r_i also denotes the point in the plane occupied by that robot. It is assumed that the robots never collide and that two or more robots may simultaneously occupy the same physical location. Any robot can observe, compute and move with infinite decimal precision. The robots are equipped with sensors enabling to detect the instantaneous position of the other robots in the plane. Each robot has its own local coordinate system and unit measure. The robots do not agree on the orientation of the axes of their local coordinate system, nor on the unit measure. They are *uniform* and *anonymous*, i.e, they all have the same program using no local parameter (such that an identity) allowing to differentiate any of them. They communicate only by observing the position of the others and they are *oblivious*, i.e., none of them can remember any previous observation nor computation performed in any previous step.

Time is represented as an infinite sequence of time instants $t_0, t_1, \ldots, t_j, \ldots$ Let $P(t_j)$ be the multiset of the positions in the plane occupied by the n robots at time t_j ($j \geq 0$). For every t_j, $P(t_j)$ is called the *configuration* of the distributed system in t_j. $P(t_j)$ expressed in the local coordinate system of any robot r_i is called a *view*, denoted $v_i(t_j)$. At each time instant t_j ($j \geq 0$), each robot r_i is either *active* or *inactive*. The former means that, during the computation step (t_j, t_{j+1}), using a given algorithm, r_i computes in its local coordinate system a position $p_i(t_{j+1})$ depending only on the system configuration at t_j, and moves towards $p_i(t_{j+1})$—$p_i(t_{j+1})$ can be equal to $p_i(t_j)$, making the location of r_i unchanged. In the latter case, r_i does not perform any local computation and remains at the same position. In every single activation, the distance traveled by any robot r is bounded by σ_r. So, if the destination point computed by r is farther than σ_r, then r moves toward a point of at most σ_r. This distance may be different between two robots.

The concurrent activation of robots is modeled by the interleaving model in which the robot activations are driven by a *fair scheduler*. At each instant t_j ($j \geq 0$), the scheduler arbitrarily activates a (non empty) set of robots. Fairness means that every robot is infinitely often activated by the scheduler.

The Circle Formation Problem. In this paper, the term *"circle"* refers to a circle having a radius strictly greater than zero. Consider a configuration at time t_k ($k \geq 0$) in which the positions of the n robots are located at distinct positions on the circumference of a circle C. At time t_k, the *successor* r_j, $j \in 1 \ldots n$, of any robot r_i, $i \in 1 \ldots n$ and $i \neq j$, is the single robot such that no robot exists

between r_i and r_j on C in the clockwise direction. Given a robot r_i and its successor r_j on C centered in O:
1. r_i is said to be the *predecessor* of r_j,
2. r_i and r_j are said to be *adjacent*,
3 $\widehat{r_iOr_j}$ denotes the angle centered in O and with sides the half-lines $[O, r_i)$ and $[O, r_j)$ such that no robots (other than r_i and r_j) is on C inside $\widehat{r_iOr_j}$.

Definition 1 (regular n-gon). *A cohort of n robots $(n \geq 2)$ forms (or is arranged in) a regular n-gon if the robots take place on the circumference of a circle C centered in O such that for every pair r_i, r_j of robots, if r_j is the successor of r_i on C, then $\widehat{r_iOr_j} = \delta$, where $\delta = \frac{2\pi}{n}$. The angle δ is called the characteristic angle of the n-gon.*

The problem considered in this paper, called CFP (*Circle Formation Problem*) consists in the design of a distributed protocol which arranges a group of n ($n > 2$) mobile robots with initial distinct positions into a *regular n-gon* in finite time. In this paper, we ignore the trivial cases $n \leq 2$ because in that cases, they always form a regular n-gon.

Swing-Word. Let an ordered non-empty alphabet A be a finite set of letters. Denote $\prec$ an order on A. A non empty *word* w over A is a finite sequence of letters $a_0, \ldots, a_i, \ldots, a_{l-1}, l > 0$. The *concatenation* of two words u and v, denoted $u \circ v$ or simply uv, is equal to the word $a_0, \ldots, a_i, \ldots, a_{k-1}, b_0, \ldots, b_j, \ldots, b_{l-1}$ such that $u = a_0, \ldots, a_i, \ldots, a_{k-1}$ and $v = b_1, \ldots, b_j, \ldots, b_{l-1}$. Let ϵ be the *empty word* such that for every word w, $w\epsilon = \epsilon w = w$. The *length* of a word w, denoted by $|w|$, is equal to the number of letters of w—$|\epsilon| = 0$. The *mirror word* of a word $w = a_0, \ldots, a_{l-1}$, denoted by $\overline{w}$, is equal to the word $a_{l-1}, \ldots, a_0$—$\overline{\epsilon} = \epsilon$.

A word u is *lexicographically* smaller than or equal to a word v, denoted $u \preceq v$, iff there exists either a word w such that $v = uw$ or three words r, s, t and two letters a, b such that $u = ras$, $v = rbt$, and $a \prec b$.

Let k and j be two positive integers. The kth *power* of a word w is the word denoted s^k such that $s^0 = \epsilon$, and $s^k = s^{k-1}s$. The jth *rotation* of a word $w = a_0, \ldots, a_{|w|-1}$, notation $R_j(w)$, is defined by:

$$R_j(w) \overset{\text{def}}{=} \begin{cases} \epsilon & \text{if } w = \epsilon \\ a_j, \ldots, a_{|w|-1}, a_0, \ldots, a_{j-1} & \text{otherwise } (0 \leq j < |w|) \end{cases}$$

Let $u = a_0 a_1 \ldots a_{l-1}$ ($l \geq 2$) be a finite word over A. Denote Λ_u the subset of words $v = b_0 b_1 \ldots b_{l'-1}$ over $\{0,1\}$ such that (1) $l' = l$, and (2) for every $i \in 0 \ldots l' - 1$:

$$b_i = \begin{cases} 0 \text{ if } a_{i \bmod l} \leq a_{(i+1) \bmod l} \\ 1 \text{ if } a_{i \bmod l} \geq a_{(i+1) \bmod l} \end{cases}$$

Remark that the words in Λ_u are built on the cyclic representation of u. Let us consider the two following examples:

Example 1. Assume that $A = \{1, 2\}$ ($1 < 2$). Then, $\Lambda_{11} = \Lambda_{22} = \{00, 01, 10, 11\}$, $\Lambda_{12} = \{01\}$, $\Lambda_{112} = \{001, 101\}$, $\Lambda_{1112} = \{0001, 1001, 0101, 1101\}$, and $\Lambda_{1221} =$

$\{0110, 0111, 0010, 0011\}$. For instance, since both $1 \geq 1$ and $1 \leq 1$ are true, $\Lambda_{112} = \{001, 101\}$ because both $a_0 \leq a_1 \leq a_2 \geq a_0$ and $a_0 \geq a_1 \leq a_2 \geq a_0$ are true.

Example 2. Assume that $A = \{1, 2, 3\}$ $(1 < 2 < 3)$.
Then, $\Lambda_{11231} = \{00010, 00011, 10010, 10011\}$, and
$\Lambda_{311122} = \{101000, 100000, 110000, 111000, 101010, 100010, 110010, 111010\}$.

Definition 2 (Swing-Word). *A finite non-empty word* $w = a_0 a_1 \ldots a_{l-1}$ *(l $\geq$ 1) made over* A *is a Swing-Word iff the following two conditions are true:* (1) $w \neq a_0^l$, *and* (2) *there exists* $u \in \Lambda_w$ *such that* $u \in \{(01)^{\frac{l}{2}}, (10)^{\frac{l}{2}}\}$—*u is called an* associate *Swing-word of* w.

For instance, in Example 1 (resp., Example 2) above, 1112 (resp. 311122) is a *Swing*-word—$0101 \in \Lambda_{1112}$ (resp. $101010 \in \Lambda_{311122}$). (Note that even if Λ_{11} and Λ_{22} contain 01, both 11 and 22 are not swing words because they are equal to 1 and 2, respectively.)
   The following lemma directly follows from Definition 2:

Lemma 1. *If a word* $w = a_0 a_1 \ldots a_{l-1}$ *(l $\geq$ 1) is a Swing-word, then:*
1. $l = 2p$ $(p \geq 1)$;
2. $\overline{w}$ *is a Swing-word;*
3. *For every* $j \in 0 \ldots l - 1$, $R_j(w)$ *is a Swing-word.*

Lemma 2. *If a word* $w = a_0 a_1 \ldots a_{l-1}$ *(l $\geq$ 1) is a Swing-word, then* w *have a unique associate Swing-word.*

Proof. Let Λ_w^S be the subset of Λ_w such that $u \in \{(01)^{\frac{l}{2}}, (10)^{\frac{l}{2}}\}$. Since the length of w is finite, Λ_w^S contains at most 2 words. From Definition 2 , Λ_w^S contains at least one word. Assume by contradiction that $\Lambda_w^S = \{u_1, u_2\}$. Without lost of generality, $u_1 = (01)^{\frac{l}{2}}$, and $u_2 = (10)^{\frac{l}{2}}$. So, from Definition 2, we have both:

1. $a_0 \leq a_1 \geq \ldots \leq a_{l-1} \geq a_0$, and
2. $a_0 \geq a_1 \leq \ldots \geq a_{l-1} \leq a_0$.

So, $a_0 = a_1 = \ldots = a_{l-1} = a_0$. This contradicts the first condition of Definition 2.

3   Circle Formation Protocol

In this section, we present the main result of this paper. We first provide some basic definitions, followed by particular configurations of the system which we use for simplifying the design and proofs of the protocol. Next, the protocol is presented with the correctness proof.

3.1   Definitions and Basics Properties

Definition 3 (Biangular circle). *A cohort of* n *robots* $(n \geq 2)$ *forms (or is arranged in) a biangular circle if the robots take place on the circumference of* *a circle C centered in O and there exist two non zero angles α, β such that for every pair r_i, r_j of robots, if r_j is the successor of r_i on C, then $\widehat{r_i O r_j} \in \{\alpha, \beta\}$ and α and β alternate in the clockwise direction.*

Obviously, if $\alpha = \beta$ then, for any n value, the n robots form a regular n-gon. If $\alpha \neq \beta$, then n must be even ($n = 2p$, $p > 1$). In that case, the biangular circle is called a *strict* biangular circle—refer to Figure 1.

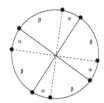

Fig. 1. An example showing a strict biangular circle ($\alpha \neq \beta$)

Definition 4 (Convex Hull). *: Given $n \geq 2$ points $p_1, p_2, \cdots, p_n$ on the plane, their Convex Hull $CH(p_1, p_2, \cdots, p_n)$ (CH for short, if no ambiguity arises) is the smallest polygon such that all the points are on its edges or inside it.*

Now, we introduce a definition of adjacent, predecessor and successor more general than the one presented in Section 2. Given a team R of n robots located at distinct positions on $CH(R)$, we said that two robots are CH-*adjacent* if and only if they are connected by an edge belonging to $CH(R)$. We say also that any robot r' is the CH-*successor* (resp. CH-*predecessor*) of r if and only if r' is the CH-adjacent robot in clockwise direction (resp. counterclockwise direction) on CH.

Observation 1. *We can associate an unique regular $2k$-gon to a regular k-gon ($k \geq 3$) centered in O, by applying the following construction (refer to Figure 2):*

1. *Consider one CH's edge $[p_1, p_2]$ of the regular k-gon, and place two points x_1, x_2 on this edge such that $\widehat{x_1 O x_2} = \frac{\pi}{k}$ and the distance between x_1 and p_1 is equal to the distance between x_2 and p_2.*
2. *Reiterate with the other CH's edges.*

These adding points form a regular $2k$-gon.

String of Edges. We use the subscript i in the notation of a robot r_i, $i \in 1 \ldots n$, to denote the order of the robots in an arbitrary clockwise direction on the Convex Hull CH. We proceed as follows: A robot is arbitrarily chosen as r_1 on CH. Next, for any $i \in 1 \ldots n - 1$, r_{i+1} denotes the CH-successor of r_i on CH (in the clockwise direction). Finally, the successor of r_n is r_1.

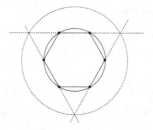

Fig. 2. An example showing a regular $2k$-gon associated to a regular k-gon—$k = 3$

Let the alphabet A be the set of k ($k \leq n$) strictly positive reals $x_1, x_2, \ldots, x_k$ such that $\forall i \in 1 \ldots n$, there exists $j \in 1 \ldots k$ such that $x_j =$ length of one of the CH's edges.

The order on A is the natural order ($<$) on the reals. So, the lexicographic order $\preceq$ on the words made over A is defined as follows:

$$u \preceq v \stackrel{\text{def}}{=} (\exists w | \ v = uw) \vee (\exists r, s, t, \ \exists a, b \in A | \ (u = ras) \wedge (v = rbt) \wedge (a < b))$$

For instance, if $A = \{1, 2\}$, then $1 \preceq 11 \preceq 12 \preceq 122 \preceq 2$.

For each robots r_i, let us define the word $SE(r_i)$ (respectively, $\overline{SE(r_i)}$) over A (SE stands for "string of edges") as follows:

$$SE(r_i) = a_i a_{i+1} \ldots a_{n-1} a_0 \ldots a_{i-1}$$
$$(\text{resp. } \overline{SE(r_i)} = a_{i-1} \ldots a_0 a_{n-1} a_{n-2} \ldots a_i)$$

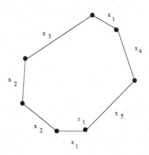

Fig. 3. $SE(r_1) = x_1 (x_2)^2 x_3 x_1 x_4 x_5$

An example showing a string of edges is drawn in Figure 3. Note that, if k robots are on CH, then for every robot r_i, $|SE(r_i)| = |\overline{SE(r_i)}| = k$. Moreover, if the configuration is a regular n-gon, then for every robot r_i, $SE(r_i) = \overline{SE(r_i)} = u^n$, where u is the common length of all the edges of CH.

Swing and Perfect Convex Hulls.

Definition 5 (Swing Convex Hull). *A Convex Hull CH is said to be a Swing convex hull (notation, Swing-CH) iff there exists any robot r_i on CH such that $SE(r_i)$ is a Swing-word.*

The following lemma directly follows from Lemmas 1 and 2:

Lemma 3. *If $SE(r_i)$ is a Swing-word, then for every $r_{i'}$ on CH, $SE(r_{i'})$ and $\overline{SE(r_{i'})}$ are swing words.*

Corollary 1. *If a Convex Hull CH is a Swing convex hull, then each robot r_i on CH can determine that CH is a Swing convex hull by locally computing its String of Edge, regardless the local clockwise direction of r_i.*

Let $SE(r_0) = s_0 s_1 \cdots s_{l-1}$ be a *Swing*-word on a *Swing-CH* and $u = u_0 u_1 \cdots u_{l-1}$ its associate *Swing*-word. We said that the edge $[r_i, r_{i+1}]$ is an *up-edge* (resp., a *down-edge*) iff $u_i = 1$ (resp. $u_i = 0$). From Lemmas 1 and 2 again, we can easily deduce that the up-edges and down-edges are the same for all the robots on a *Swing-CH*.

Given two robots robots r_i and $r_{i'}$ such that r_i and $r_{i'}$ are CH-adjacent on a *Swing-CH*, we say that they form a *couple* if and only if the CH's edge, linking r_i and $r_{i'}$ is a down-edge. In a *Swing-CH*, denote *SetLines* the set of lines (r_i, r_{i+1}) passing through both robots of the same couple. *IntersectionLines* (called shortly *IL*) is the set of intersection points between all lines $l_1 = (r_i, r_{i+1})$ and $l_2 = (r_{i+2}, r_{i+3})$ such that $l_1 \in SetCoupleLine, l_2 \in SetCoupleLine$ and r_{i+1} CH-adjacent to r_{i+2}.

Definition 6 (Perfect Convex Hull). *Let a team of $n = 2k$ robots on the convex hull CH. This latter is perfect if the four conditions holds:*

1. $n = 2k$ and $k \geq 3$.
2. CH is a *Swing-CH*.
3. *IL* is a regular k-gon.
4. *All the robots are on the edges of the regular $2k$-gon associated to IL.*

Note that Condition (1) in the above definition, excludes the case $k < 3$ because in this case *IL* does not exist.

We say that the cohort R is in Perfect Convex Hull if $CH(R)$ is perfect and all the robots are located at distinct position on it—refer to Figure 4 where non-perfect and perfect convex hull are shown.

Observation 2. *If all the robots are in strict biangular circle then they are in perfect convex hull.*

Definition 7 (Equivalence). *Two perfect convex hull are said to be equivalent if they share the same regular k-gon IL in a system configuration.*

The only possible difference between two equivalent perfect convex hulls is different positions of the robots on the regular n-gon associated to *IL*.

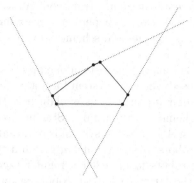

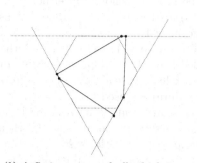

(a) A Swing convex hull which is not
perfect:
IL is not a regular 3-gon.

(b) A Swing convex hull which is not
perfect: There exists robots
which are not on the edges of the
regular 6-gon associated to IL.

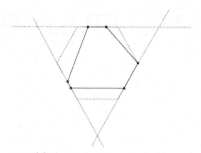

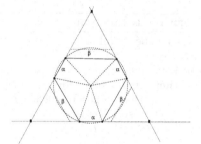

(c) A Perfect convex hull:
All the robots belong to the edges of the
6-gon.

(d) A strict biangular circle is
also a Perfect convex hull,
Illustrating Observation 2.

Fig. 4. Examples showing Swing convex hulls

3.2   The Protocol

Let us consider the overall scheme of our protocol presented in Algorithm 1. It is
mainly based on the perfect convex hull configuration presented in the previous
subsection. The proposed scheme is combined with the protocol presented in [12]
which leads a cohort of n robots from an arbitrary into a biangular configuration,
whether $n \geq 2$. In the remainder, we refer to the protocol in [12] as Procedure $<
A \rightsquigarrow B >$—from an Arbitrary configuration to a Biangular configuration.

Theorem 3 ([14]). *Any algorithm that correctly solves a problem P in Corda,
correctly solves P in SSM.*

The above result means that Procedure $< A \rightsquigarrow B >$ can be used in *SSM*.
Obviously, Procedure $< A \rightsquigarrow B >$ trivially solves the CFP if the number of
robots n is odd. So, to solve CFP for any number of robots, it remains to deal
with a system in a strict biangular configuration when n is even.

In the remainder, we consider that the system is in an arbitrary configuration if the robots do not form either (1) a regular n-gon, (2) a perfect convex hull, or (3) a strict biangular circle. Let us describe the general scheme provided by Algorithm 1.

Procedure $<A \leadsto B>$ excluded, the protocol mainly consists of one procedure called Procedure $<\text{PCH} \leadsto N\text{gon}>$ which is used when the system form a perfect convex hull. It leads the system into a regular n-gon. Let us explain how the procedures are used by giving the overall scheme of Algorithm 1. Starting from an arbitrary configuration, using Procedure $<A \leadsto B>$, the system is eventually in a biangular circle. If n is odd, then the robots form a regular n-gon and the system is done. Otherwise (n is even), the robots form either a regular n-gon or a strict biangular circle. Starting from the latter case, each robot executes Procedure $<\text{PCH} \leadsto N\text{gon}>$ (from Observation 2, we know that when the robots form a strict biangular circle, they form a perfect convex hull too). The resulting configuration of the execution of Procedure $<\text{PCH} \leadsto N\text{gon}>$ is a regular n-gon.

Algorithm 1. Procedure $<A \leadsto N\text{gon}>$ for any r_i in a cohort of $n \neq 4$ robots

---

n:= the number of robots;
if n is even
then if the robots do not form a regular n-gon
      then if the robots form a perfect convex hull
           then Execute $<\text{PCH} \leadsto N\text{gon}>$;
           else Execute $<A \leadsto B>$;
else   Execute $<A \leadsto B>$;

Theorem 4. *Assuming that initially no two robots are located at the same position, Procedure $<A \leadsto N\text{gon}>$ is a deterministic Circle Formation Protocol for any number $n \neq 4$ of robots in SSM.*

The above theorem follows from Procedure $<A \leadsto N\text{gon}>$, Algorithm 1, [12], and Lemmas 5. This last Lemma shows that, starting from a perfect convex hull, Procedure $<\text{PCH} \leadsto N\text{gon}>$ described below deterministically solves CFP.

Procedure $<\text{PCH} \leadsto N\text{gon}>$. Let us assume that in the initial configuration, the robots form a perfect convex hull. In such a configuration, every active robots r_i applies the following sheme—refer to Figure 5:

1. Robot r_i computes IL and the associated n-gon.
2. Robot r_i considers its neighbor r_i' such that $\{r_i, r_i'\}$ is a couple of the perfect convex hull. Then, r_i moves away from r_i' toward the vertex of the associated n-gon, on the line (r_i, r_i').

Lemma 4. *In SSM, using Procedure $<\text{PCH} \leadsto N\text{gon}>$, if all the robots are in a perfect convex hull at time t_j, then at time t_{j+1}, either the configuration is an equivalent perfect convex hull, or the n-gon is not formed.*

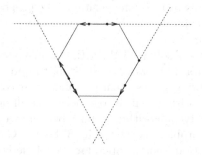

Fig. 5. An example describing the Procedure $<\text{PCH} \leadsto N\text{gon}>$

Proof. From Lemma 3 and Corollary 1, all the perfect convex hulls computed by the robots are all equivalent. Let $Unit$ be the distance between two adjacent vertices on the associated n-gon. Using Procedure $< \text{PCH} \leadsto N\text{gon} >$, at time t_{j+1}, the distance between two CH-adjacent robots forming a couple is lower or equal to $Unit$, whereas the distance between two CH-adjacent robots, not forming a couple, is greater or equal $Unit$. So, at time t_{j+1} the up-edge and the down-edge are the same ones than t_j, because $|down\ edge| \leq Unit \leq |up\ edge|$. Furthermore, each robot moves only on the edge of the associated regular n-gon without collision. So, each robot can recompute the same IL and the same associated regular n-gon.

The following lemma follows from Lemma 4 and fairness:

Lemma 5. *In SSM, Procedure $< \text{PCH} \leadsto N\text{gon} >$ is a deterministic algorithm transforming a perfect convex hull into a regular n-gon whether $n > 4$.*

Procedure $< A \leadsto N\text{gon} >$ in CORDA. Assume the fully asynchronous model CORDA, i.e., one or more robots can move while some others are either observing or computing—the case of waiting robots gets off the scope of this discussion. Let assume that the robots form a perfect convex hull. Lemma 4 shows that, while the regular n-gon is not formed, Procedure $< \text{PCH} \leadsto N\text{gon} >$ guarantees that the associated regular $\frac{n}{2}$-gon is preserved for every robot. So, even assuming CORDA, the perfect convex hull computed by the robots remains equivalent for every pair of robots while the regular n-gon is not formed. Thus, Lemma 4 also holds in CORDA. As a result, the following theorem holds:

Theorem 5. *If the initial configuration where the robots form a biangular circle, Procedure $< A \leadsto N\text{gon} >$ is a deterministic Circle Formation Protocol for any number $n \neq 4$ of robots in CORDA.*

4   Conclusion and Discussion

In this paper, we first introduced the Swing Words. We then provided a quiescent deterministic protocol for the Circle Formation Problem in SSM assuming that no two robots are initially at the same position. The proposed protocol works for any number $n \neq 4$ of weak robots, which are not assume to reach their computed destination in one step.

In CORDA, our protocol solves CFP starting from a biangular configuration— if $n \neq 4$. Starting from a different configuration, the protocol in [12] solves the CFP if n is odd, and brings the system in a biangular configuration otherwise. Hence, we have two algorithms which together cover all possible initial configurations. However, these two algorithms cannot be simply combined to solve CFP in CORDA. A similar problem is related in [3] for the Gathering Problem, and solved in [2]. This question remains open for the Circle Formation Problem.

References

1. Chatzigiannakis, I., Markou, M., Nikoletseas, S.: Distributed circle formation for anonymous oblivious robots. In: 3rd Workshop on Efficient and Experimental Algorithms, pp. 159–174 (2004)
2. Cieliebak, M., Flocchini, P., Prencipe, G., Santoro, N.: Solving the robots gathering problem. In: Baeten, J.C.M., Lenstra, J.K., Parrow, J., Woeginger, G.J. (eds.) ICALP 2003. LNCS, vol. 2719, pp. 1181–1196. Springer, Heidelberg (2003)
3. Cieliebak, M., Prencipe, G.: Gathering autonomous mobile robots. In: 9th International Colloquium on Structural Information and Communication Complexity (SIROCCO 9), pp. 57–72 (2002)
4. Debest, X A: Remark about self-stabilizing systems. Communications of the ACM 38(2), 115–117 (1995)
5. Defago, X., Konagaya, A.: Circle formation for oblivious anonymous mobile robots with no common sense of orientation. In: 2nd ACM International Annual Workshop on Principles of Mobile Computing (POMC 2002), pp. 97–104 (2002)
6. Dieudonné, Y., Labbani, O., Petit, F.: Circle formation of weak mobile robots. In: Datta, A.K., Gradinariu, M. (eds.) SSS 2006. LNCS, vol. 4280, pp. 262–275. Springer, Heidelberg (2006)
7. Dieudonné, Y., Petit, F.: Circle formation of weak robots and Lyndon words. Information Processing Letters 104(4), 156–162 (2007)
8. Dieudonné, Y., Petit, F.: Robots and Demons (The Code of the Origins). In: Fourth International Conference on Fun with Algorithms (FUN '07), To appear in Springer-Verlag Lecture Notes in Computer Science series (2007)
9. Dolev, S.: Self-Stabilization. The MIT Press, Cambridge, MA (2000)
10. Flocchini, P., Prencipe, G., Santoro, N.: Self-deployment algorithms for mobile sensors on a ring. In: Nikoletseas, S.E., Rolim, J.D.P. (eds.) ALGOSENSORS 2006. LNCS, vol. 4240, Springer, Heidelberg (2006)
11. Flocchini, P., Prencipe, G., Santoro, N., Widmayer, P.: Distributed coordination of a set of autonomous mobile robots. In: IEEE Intelligent Vehicule Symposium (IV 2000), pp. 480–485 (2000)
12. Katreniak, B.: Biangular circle formation by asynchronous mobile robots. In: Pelc, A., Raynal, M. (eds.) SIROCCO 2005. LNCS, vol. 3499, pp. 185–199. Springer, Heidelberg (2005)
13. Prencipe, G.: Corda: Distributed coordination of a set of autonomous mobile robots. In: Proceedings of the Fourth European Research Seminar on Advances in Distributed Systems (ERSADS 2001), pp. 185–190 (2001)

14. Prencipe, G.: Distributed Coordination of a Set of Autonomous Mobile Robots. PhD thesis, Dipartimento di Informatica, University of Pisa (2002)
15. Prencipe, G., Santoro, N.: Distributed algorithms for autonomous mobile robots. In: Sha, E., Han, S.-K., Xu, C.-Z., Kim, M.H., Yang, L.T., Xiao, B. (eds.) EUC 2006. LNCS, vol. 4096, Springer, Heidelberg (2006)
16. Sugihara, K., Suzuk, I.: Distributed algorithms for formation of geometric patterns with many mobile robots. Journal of Robotic Systems 3(13), 127–139 (1996)
17. Suzuki, I., Yamashita, M.: Agreement on a common x-y coordinate system by a group of mobile robots. Intelligent Robots: Sensing, Modeling and Planning, pp. 305–321 (1996)
18. Suzuki, I., Yamashita, M.: Distributed anonymous mobile robots - formation of geometric patterns. SIAM Journal of Computing 28(4), 1347–1363 (1999)

Distributed Algorithms for Partitioning a Swarm of Autonomous Mobile Robots

Asaf Efrima and David Peleg*

Department of Computer Science and Applied Mathematics,
The Weizmann Institute of Science, Rehovot, Israel 76100
david.peleg@weizmann.ac.il

Abstract. A number of recent studies address systems of mobile autonomous robots from a distributed computing point of view. Although such systems employ robots that are relatively weak and simple (i.e., dimensionless, oblivious and anonymous), they are nevertheless expected to have strong fault tolerance capabilities as a group. This paper studies the partitioning problem, where n robots must divide themselves into k size-balanced groups, and examines the impact of common orientation on the solvability of this problem. First, deterministic crash-fault tolerant algorithms are given for the problem in the asynchronous full-compass and semi-synchronous half-compass models, and a randomized algorithm is given for the semi-synchronous no-compass model. Next, the role of common orientation shared by the robots is examined. Necessary and sufficient conditions for the partitioning problem to be solvable are given in the different timing models. Finally, the problem is proved to be unsolvable in the no-compass asynchronous model.

1 Introduction

Background: Systems of multiple autonomous mobile robots (also known as *robot swarms*) are of interest for a variety of reasons, including decreased costs and a wide range of applications, such as military operations, search and rescue, fire fighting and space missions.

Most experimental and empirical studies of multiple robot systems rely on a central controller for managing the robots, and their coordination algorithms are based on heuristics. Recently, multiple robot systems have been studied from a *distributed computing* point of view [16, 10]. A number of distributed computation models were proposed in the literature, and a number of studies focused on characterizing the influence of the model on the ability of a robot swarm to perform its task. The common distributed models assume relatively weak and simple robots. In particular, these robots are assumed to be dimensionless, oblivious, anonymous and with no common coordinate system, orientation or scale, and no explicit communication. Each robot operates in simple "look-compute-move" cycles, basing its movements on viewing its surroundings and analyzing

---

* Supported in part by a grant from the Israel Science Foundation.

G. Prencipe and S. Zaks (Eds.): SIROCCO 2007, LNCS 4474, pp. 180–194, 2007.

the configuration of robot locations. A robot is capable of locating all robots within its visibility range on its private coordinate system, thereby calculating their position with respect to itself.

As the common models of multiple robot systems assume cheap, simple and relatively weak robots, the problem of possible failures becomes prominent, since in such systems one cannot possibly rely on assuming reliable hardware or software, especially when such robot systems are expected to operate in hazardous or harsh environments. At the same time, one of the main attractive features of multiple robot systems is their potential for enhanced fault tolerance; it seems plausible that the inherent redundancy of such systems may be exploited in order to enable them to perform their tasks even in the presence of faults.

Among the tasks studied so far are formation of geometric patterns (i.e., organizing the robots in a geometric form), gathering and convergence (i.e., collecting the robots to the same point), flocking (i.e., following a pre-designated leader), even distribution of robots within simple geometric patterns, searching a target within a bounded area, and more.

Another important task that has been studied to a lesser extent is that of *partitioning*. In this task, the robots must divide themselves into (size-balanced) groups. This task is closely related to that of converging. In this paper we examine the partitioning problem within various computation models.

Computation Models: The basic model studied in previous papers, e.g., [18, 19, 12, 6], can be summarized as follows. Each of the robots executes the same algorithm in cycles, with each cycle consisting of three steps:

1. "Look": Determine the current configuration by identifying the location of all visible robots and marking them on the robot's private coordinate system.
2. "Compute": Execute the algorithm, resulting in a goal point $\tilde{p}$.
3. "Move": Travel towards the point $\tilde{p}$. The robot might stop before reaching $\tilde{p}$, but is guaranteed to traverse at least a minimal distance unit s (unless reaching the goal first). The value of s is not known to the robots and they cannot use it in their computations.

In the common distributed model [16, 18, 17, 10, 4], the robots are assumed to be *dimensionless*, namely, treated as points that do not obstruct each other's visibility or movement, and *oblivious* or memoryless, namely, do not remember their previous actions or the previous positions of the other robots, and therefore cannot rely on information from previous cycles, or have alternating states. Also, the robots are indistinguishable and cannot identify each and every one of their peers. Moreover, the robots have no means of explicit communication. On the other hand, the robots are assumed to possess unlimited visibility, and sensors, computations and movements are assumed to be accurate.

The models considered here vary in two attributes. The first is timing models.

1. *Fully-synchronous (FSYNC) model*: the robots are driven by an identical clock, operate according to the same cycles and are active in every cycle.
2. *Semi-synchronous (SSYNC) model*: the robots operate according to the same cycles, but need not be active in every cycle. A fairness constraint guaran-

tees that each robot will eventually be active (infinitely many times) in any infinite execution.

3. *Asynchronous (ASYNC) model*: the robots operate on independent cycles of variable length. Formally, this can be modeled by each cycle starting with an additional "Wait" step.

The second attribute is orientation, referring to the local views of the robots in terms of their x-y coordinates. Elaborating on [12], the following sub-models of common orientation levels are considered.

1. *Full-compass*: Directions and orientations of both axes are common to all robots.
2. *Half-compass*: Directions of both axes are common to all robots, but the positive orientation of only one axis is common. (i.e., in the other axis, different robots may have different views of the positive orientation).
3. *Direction-only*: Directions of both axes are common to all robots, but the positive orientations of the axes are not common.
4. *Axes-only*: Directions of both axes are common to all robots, but the positive orientations of the axes are not common. In addition the robots do not agree on which of the two axes is the x axis and which is the y axis.
5. *No-compass*: There are no common axes.

In the no-compass and half-compass sub-models, the robots do not share the notion of "clockwise" or "right hand side". Note that the robots do not share a common unit distance or a common origin point even in the full-compass model.

Fault Tolerance: A major algorithmic aspect considered in this paper is fault tolerance. The algorithm may be required to operate in a model in which robots may fail. In such a setting, we may ask how well the algorithm can cope with one or more faulty robots.

The model discussed in this paper is the *crash-fault* model, in which a faulty robot simply stops moving. Since nonfaulty robots may also stay in place from time to time, there is no way for the active robots to identify a faulty robot. An f-fault-tolerant algorithm is one that operates correctly so long as there are no more than f faulty robots in the swarm. (The pattern and timing of failures can be thought of as controlled by an adversary, which may crash robots at any time in an adaptive manner, i.e., the faulty robots need not be picked in advance.) The exact requirements of the algorithm (i.e., the meaning of "operating correctly" in a faulty environment) is specific to the problem at hand.

For randomized algorithms we consider two possible adversary types. An adaptive adversary is allowed to make its decisions after learning the (possibly randomized) choices made by the algorithm. This means that in each cycle, first the robot computes its goal position, and then the adversary chooses the maximal distance the robot will reach in the direction of its goal point. In contrast, a non-adaptive adversary must make its decisions independently of the random choices of the algorithm. Namely, in each cycle, the adversary chooses the maximal distance the robot will reach before the robot computes its goal point (i.e., before knowing the direction in which the robot will move, which may be chosen randomly by the algorithm). Note that despite its name, there is some adaptiveness even in the non-adaptive adversary, since it still has control over the timing of the robots. We will normally assume the adaptive adversary, except when explicitly noted otherwise.

The Partitioning Problem: We consider the problem PARTITION(n, k), in which n robots, at initially distinct positions, must divide themselves into k size-balanced subsets. The robots in each subset must converge, so that some minimal distance is kept between robots of different subsets.

We use the following basic definitions. Let $dist(a, b)$ denote the Euclidean distance between points a and b. For sets of points X and Y, denote $dist(X, Y) = \min\{dist(x, y) \mid x \in X, y \in Y\}$. Denote the position of robot r_i at time t as $p_i[t] = (x_i[t], y_i[t])$. (We sometimes omit the parameter t when no confusion arises.) Denote the set of all robot positions at time t as $P[t]$.

Formally, the partitioning problem PARTITION(n, k) is defined as follows.

Input: A set of n robots $R = \{r_1, \ldots, r_n\}$, positioned in a 2-dimensional space, with initial positions $P_I = P[t_0] = \{p_1[t_0], \ldots, p_n[t_0]\}$, and an integer k. We assume that n is divisible by k and define $m = n/k$.
Goal: For some fixed $\eta > 0$, for every $\eta \geq \varepsilon > 0$, there is a time t_ε, such that for every time $t > t_\varepsilon$, R can be partitioned into k disjoint subsets $S_1, \ldots, S_k$ satisfying the following:

- **Partition:** $R = \bigcup_{i=1}^{k} S_i$ and $S_i \cap S_j = \emptyset$ for every $i \neq j$.
- **Size-balance:** The subsets are balanced, i.e., $|S_i| = m$ for every i.
- **Proximity:** Robots in the same subset are within ε of each other, i.e., $dist(r_w, r_l) < \varepsilon$ for every i and for every $r_w, r_l \in S_i$.
- **Separation:** Robots in different subsets are farther than 2η apart, i.e., $dist(S_i, S_j) > 2\eta$ for every $i \neq j$.

Although robots are dimensionless, we make the following assumption.
Non-overlap: No two robots have the same initial position, i.e., $p_i[t_0] \neq p_j[t_0]$ for every $i \neq j$.

In the general case, where n is not divisible by k, define $m = \lfloor n/k \rfloor$ and require that the subsets are *nearly-balanced*, i.e., $m \leq |S_i| \leq m + 1$ for every i.

Note that the choice of the separation distance as 2η is arbitrary, and any clear separation between the subsets will do. In practice, we may set $\eta = \frac{1}{2}d_{min}$, where d_{min} is the minimal distance between any two robots at time t_0. Note also that requiring the conditions to hold on every time $t > t_\varepsilon$ implies that the subsets S_i do not change after time t_η. (To see this, suppose they do change, so that two robots r_i, r_j that were in the same subset at time t_η are later separated. Then there must be a time $t > t_\eta$ whence $\eta < dist(r_i, r_j) < 2\eta$, in contradiction to both the proximity and separation requirements).

An algorithm for solving the PARTITION(n, k) problem is considered to be *f-fault-tolerant* if the following holds. In a non-faulty setting, the algorithm achieves a partitioning as defined above. In a faulty setting in which $\hat{f} \leq f$

robots crash, all the non-faulty robots are required to converge into k subsets. Each subset must be of size greater than $\max\{m - \hat{f}, 0\}$ and smaller than $m + 1$. Note that in case $f \geq m$, fewer than k actual subsets may be formed since some subsets may be empty.

Related Work: Most of the literature on distributed control algorithms for autonomous mobile robots has concentrated on the two basic tasks of gathering and convergence. Gathering requires the robots to occupy a single point within finite time, regardless of their initial configuration. Convergence is the closely related task in which the robots are required to converge to a single point, rather than reach it. More precisely, for every $\varepsilon > 0$ there must be a time t_ε by which all robots are within a distance of at most ε of each other. Hence the convergence problem may be considered as the special case PARTITION$(n, 1)$.

The problem of gathering autonomous mobile robots was studied extensively in two computational models. The first was the SSYNC model, introduced by Suzuki et al. [16, 19], and the second is the closely related CORDA model described by Prencipe et al. [12, 13], which is equivalent to our ASYNC model. In the SSYNC model, it was proven that it is impossible to gather *two* oblivious autonomous mobile robots that have no common sense of orientation under the SSYNC model [18, 19]. The algorithms presented therein for $n \geq 3$ robots rely on the assumption that a robot can identify a point p^* occupied by two or more robots (a.k.a. *multiplicity point*). This assumption was later proven to be essential for achieving gathering in all asynchronous and semi-synchronous models [14, 15]. Under this assumption, an algorithm is developed in [19] for gathering $n \geq 3$ robots in the SSYNC model. In the ASYNC model, an algorithm for gathering $n = 3, 4$ robots is presented in [14], and an algorithm for gathering $n \geq 5$ robots has been described in [3]. We use a similar assumption, stating that a robot can tell the number of robots in a multiplicity point. In [2] a gathering algorithm was given in a model in which the above assumption has been replaced by equipping the robots with an unlimited amount of memory.

Fault tolerance properties of the gathering problem are discussed in [1]. In the crash-fault model and the SSYNC model, an algorithm is given for gathering n robots with one crashed robot.

Some studies try to characterize the class of geometric patterns that the robots can form in various models. These results relate to the partitioning problem only in part. On the one hand, in partitioning the outcome is not restricted to one specific geometric shape. Instead, it is a collection of constraints that must be satisfied by the configuration. On the other hand, in a model where every geometric pattern is achievable, partitioning must be achievable as well (by forming a well-partitioned geometric pattern).

The effect of common orientation on the class of achievable geometric patterns (in the ASYNC model) is summarized in [12]. In the full-compass model, the robots can form an arbitrary given pattern. In the half-compass model the robots can form an arbitrary pattern only when n is odd (this is shown in [14] to hold also in a model in which the robots share axis directions only). In the no-compass model, with no common orientation, the robots cannot form an arbitrary given pattern. The class of patterns achievable by an even number of robots in the half-compass model is characterized in [11].

Non-oblivious robots in the SSYNC model are examined in [17,19]. The problem of agreement on a common x-y coordinate system is shown to be reducible to that of forming certain geometric patterns. The robots are always capable of agreeing on both the origin and a unit distance in this model, thus the difficulty lies in agreement on direction.

The convergence properties of Algorithm Go_to_COG are explored in [6,5]. In this simple algorithm a robot sets its goal point to be the center of gravity of all observed robot positions. Algorithm Go_to_COG is used extensively in the current paper. In [6] it is proven that the algorithm converges in the FSYNC and SSYNC models. In [5] it is proven to converge in the ASYNC model as well. In addition, the convergence rate is established in the FSYNC model. The number of cycles it takes to achieve gathering in the FSYNC model (in two dimensions) is $O(h/s)$, where h is the maximal width of the convex hull at time t_0, and s is the minimal movement distance unit. In the crash-fault model, it is shown that Algorithm Go_to_COG works for any number of faulty robots, and that in particular, the non-faulty robots will converge to the center of gravity of the crashed robots. Convergence and gathering with inaccurate sensors and movements are examined in [7]. Gathering is shown to be impossible for robots with inexact measurements, while a variant of Algorithm Go_to_COG is shown to converge for sufficiently small errors in measurements.

An algorithm for partitioning is shown in [16]. That algorithm uses a previous algorithm presented for flocking. It does not comply with the models presented above, mainly because it requires outside intervention (i.e., it requires an outside supervisor to move a few robots which the others will follow). Moreover, the robots are not indistinguishable, and the algorithm operates in two stages, thus requiring some memory.

Our Results: This paper discusses the partitioning problem on robot swarms, focusing on understanding the effects of common orientation on the solvability of the partitioning problem.

In Section 2 we present crash-fault tolerant partitioning algorithms for various levels of common orientation and different timing models. We start by presenting the basic Algorithm Part, which works for all timing models assuming a full compass. Variants of this algorithm are subsequently used throughout the paper. Next, an algorithm is given for the half-compass model in the FSYNC and SSYNC timing models. In the no-compass model we present a randomized algorithm that works in the SSYNC timing model against an adaptive adversary.

The role of common orientation levels shared by the robots, and its effect on their ability to achieve partitioning, is explored in Section 3. We examine a refined scale of common orientation levels (with respect to the directions and orientations of the axes). It is established that in the FSYNC and SSYNC timing models, having common axis directions is a necessary and sufficient condition for the feasibility of partitioning. In the ASYNC timing model, this is a necessary condition, and having also one common axis orientation is a sufficient condition.

In a companion paper [9], we examine the no-compass ASYNC model. As the partitioning problem is unsolvable in this model, we consider the effects of some simple modifications to the model on the solvability of the problem. We prove that if the initial configuration is not symmetric then partitioning is achievable. We then show that if the robots are identifiable, then the problem has an easy solution. In fact, the problem has a deterministic solution even in a setting where only one robot is identifiable. Finally, we prove that adding one bit of memory and communication makes the problem solvable by a randomized algorithm against a non-adaptive adversary.

2 Fault Tolerant Partitioning Algorithms

In this section we present deterministic algorithms that solve the PARTITION(n, k) problem in the full-compass ASYNC model and in the half-compass SSYNC model, and randomized algorithms for the no-compass FSYNC and SSYNC models. All of the algorithms presented are crash-fault tolerant.

2.1 The Full-Compass ASYNC Model

In this model the robots share a *full compass*, i.e., common x and y axes (directions and orientations). The algorithm described works in the asynchronous model.

The availability of a full compass permits a solution based on an ordering of the robots. Define the order relation $<_o$ by increasing coordinates on the x-axis and then on the y-axis, i.e., for positions p_i, p_j, we have that $p_i <_o p_j \iff (x_i < x_j) \vee (x_i = x_j \wedge y_i < y_j)$. Although the actual coordinates privately stored by each of the robots may differ, the common directions and orientations of the axes ensure that the order relations defined by the robots are the same. By the non-overlap assumption all initial positions are distinct, therefore at time t_0 we have a full ordering of the robots.

Lemma 1. *In the full-compass model, the robots can reach agreement on a total order relation $<_o$ of the robot locations at time t_0.*

Without loss of generality, denote the robots by $r_1, \ldots, r_n$ according to their order $<_o$ at t_0. Define the *order-based* partition of the robots, P_{OB}, by breaking the robots into k blocks of equal size in this order, i.e., $S_1 = \{r_1, \ldots, r_m\}, \ldots, S_k = \{r_{n-m+1}, \ldots, r_n\}$. We have the following.

Lemma 2. *In the full-compass model, the robots can reach agreement on the order based partition P_{OB} without moving.*

This initial agreement on a partitioning suggests an algorithm in which robots of each subset S_i perform an arbitrary convergence algorithm (e.g., Algorithm Go_to_COG) within their subset, as formalized in Algorithm Part. For a set of n points $P = \{(x_i, y_i) \mid 1 \le i \le n\}$, define the *center of gravity* of P as $Cog(P) = (\sum_i x_i/n, \sum_i y_i/n)$.

Algorithm Part (code for robot r_i)

1. Determine the order-based partition P_{OB} of R.
2. Identify the robots in your subset S.
3. Calculate $Cog(S)$ and set it as your goal point $\tilde{p}$.
4. Move towards $\tilde{p}$.

For the analysis, first define $CH_i[t]$ to be the *convex hull* of S_i and the goal points of the robots of S_i at time t, namely, the convex hull of the points $\{p_l[t] \mid r_l \in S_i\} \bigcup \{\tilde{p}_l[t] \mid r_l \in S_i\}$. Consider the initial state of these convex hulls (at t_0). Note that initially, only the robot positions in S_i affect $CH_i[t_0]$, since the goal points are either not calculated yet or precisely $Cog(S_i)$ (in any case we will treat them as the latter).

We now make the following two observations. The first states that the different convex hulls do not intersect, based on the unique initial position of each robot (the non-overlap assumption).

Observation 1. *For every $i \neq j$, $CH_i[t_0] \bigcap CH_j[t_0] = \emptyset$.*

The second observation states the relation between points in different convex hulls.

Observation 2. *For $i < j$ and points $p \in CH_i[t_0], p' \in CH_j[t_0]$, it holds that $p <_o p'$.*

The following lemma, proven in [6,5], states that the convex hull shrinks during Algorithm Go_to_COG within a set of robots.

Lemma 3. [6, 5] *For robots performing Go_to_COG with $CH[t]$ defined as above, for times $t_2 > t_1$, $CH[t_2] \subseteq CH[t_1]$.*

Our next lemma states that the partition into subsets as determined initially, stays valid during the execution of the algorithm. (Throughout, proofs are omitted but can be found in [8].)

Lemma 4. *During the execution of Algorithm Part, the partition P_{OB} does not change.*

Note that the internal order of robots *within* a subset *may* change during the algorithm.

We conclude with the validity of Algorithm Part.

Proposition 1. *Algorithm Part solves the PARTITION(n, k) problem in the full-compass ASYNC model and is n-fault-tolerant in the crash-fault model.*

2.2 The Half-Compass SSYNC Model

We now turn to a model where the robots share a half compass, i.e., common direction and orientation of one axis only (w.l.o.g the y-axis). This implies that

the x-axis direction is also known, but not its orientation ("positive direction").
We assume the semi-synchronous timing model.

In this model, a solution based on ordering will not work, as implied by the
following lemma.

Lemma 5. *In the half-compass model, the robots cannot always agree on a full
ordering of their positions at time t_0.*

Nevertheless, the following observation shows that an initial partitioning can
always be obtained statically.

Observation 3. *In the half-compass model, the robots can always reach agreement on a partition without moving.*

Unfortunately, the following observation indicates that devising an algorithm
based on an initial partition as in the previous observation is problematic.

Observation 4. *In the half-compass model, there is an initial setting P_I s.t. in
any partition that the robots can agree on without moving (as in Observation 3),
the convex hulls of at least two different subsets intersect.*

A consequence of the last observation is that use of the simple Algorithm
Go_to_COG (such as in Part) will not work here. Besides the possibility of robots
from different subsets crossing one another and maybe switching subsets, there
can be situations in which the centers of gravity of different subsets coincide at
the same point, preventing a minimal distance between the subsets.

Subsequently, for the half-compass model we use Algorithm Part enhanced
with a *tie-break* procedure, described next. The procedure is activated on a set
L of robots sharing the same x coordinate x_0, with at least one robot on each side.

**Procedure Tie-Break (code for a robot r in a set L of l robots with
the same x position x_0)**

- Let $\delta_+ = \min\{x - x_0\}$ over all robots with $x > x_0$.
- Let $\delta_- = \min\{x_0 - x\}$ over all robots with $x < x_0$.
- Let $\delta = \min\{\delta_+, \delta_-\}$.
- Robot $r \in L$ moves a distance of $\delta/2$ in the direction determined as
  follows:
  (a) If r is among the $\lceil l/2 \rceil$ robots with the largest y coordinates in L
       then set dir ← positive, else set dir ← negative.
  (b) If $\delta_+ > \delta_-$ then move towards x direction dir.
  (c) If $\delta_+ < \delta_-$ then move towards the x direction opposite dir.
  (d) If $\delta_+ = \delta_-$ then move towards the positive x direction.

Note that although the robots do not agree on the positive x orientation,
in case $\delta_+ \neq \delta_-$ their choice of direction will be consistent. Also note that in
the SSYNC model it may happen that only some of the robots in L perform

Procedure Tie-Break on a certain cycle, since some may not be active on that cycle. Clearly, δ is updated every cycle.

We now describe Algorithm Part2, which uses Procedure Tie-Break in order to partition the robots. For Algorithm Part2, consider the ordering of the robots by their x positions and the order-based partition P_{OB} induced by it as in Subsection 2.1. Conflicts of robots with the same x position are resolved in the following manner. For a group L of robots with the same x coordinate x_0, define "out" to be the x orientation with fewer robots (relative to x_0), and "in" as the other x orientation. The robots of L can now be partitioned into subsets by their y positions, so that those with larger y coordinates belong to outer subsets and those with smaller y coordinates belong to inner subsets.

By this partitioning method $\hat{P}$, a robot can determine the index of its subset S_i (although not necessarily all of its members) in most cases. The only case in which it cannot do so is if its x position is not unique, and there is an equal number of robots on both x directions.

It follows that there could be at most one group of robots that cannot determine their set in the partition, and these robots are in the same x coordinate as the median robot.

Denote by Y the group of median robots, when the robots are ordered by their x positions (i.e., all robots in Y have the same x coordinate x_0 and there are fewer than $n/2$ robots on either side of Y). Note that if n is even then $|Y| \geq 0$ and if n is odd then $|Y| \geq 1$. Also Y does not necessarily have the same number of robots on both sides. Denote by $\texttt{Same}(Y)$ the event that all robots in the median group Y belong to the same subset in the partition $\hat{P}$.

Algorithm Part2

- If all robots share the same x coordinate, then partition into groups by constructing the y-order-based partition P_{OB} on the y coordinates.
- Determine a partition $\hat{P}$.
- Determine the index i of your subset.
- Identify the subset of robots that are guaranteed to be in your subset S_i.
  Denote this new subset as S_i'.
- State [Same]: Event $\texttt{Same}(Y)$ holds.
  - If $Y \subseteq S_i$, then set your goal point $\tilde{p} \leftarrow Cog(Y)$.
  - Else, set your goal point $\tilde{p} \leftarrow Cog(S_i')$.
- State [Diff]: Event $\texttt{Same}(Y)$ does not hold.
  - If you belong to Y, then invoke Procedure Tie-Break.
  - Else, set $S_i' \leftarrow S_i' \setminus Y$ and set your goal point $\tilde{p} \leftarrow Cog(S_i')$.

Let us sketch the correctness proof of Algorithm Part2. We first examine event $\texttt{Same}(Y)$.

Lemma 6. *If all robots in Y belong to some subset S_i then the only robots that can join Y are robots from S_i.*

Corollary 1. *Once state [Same] is reached, the system will remain in state [Same].*

In state [Diff], the number of robots in Y decreases with time as shown in the next lemma.

Lemma 7. *In state [Diff], if $|Y| = p > 0$ then eventually the system will reach either a configuration with $|Y| < p$ or state [Same].*

Since having zero or one robots in Y implies state [Same], the system will eventually reach it.

Corollary 2. *State [Same] will eventually be reached.*

Combining Corollaries 2 and 1, the system will eventually reach state [Same] and remain in this state. For the convergence of the different subsets in state [Same], a proof similar to that of Algorithm Part yields the following proposition.

Proposition 2. *Algorithm Part2 solves the PARTITION(n, k) problem in the half-compass SSYNC model and is n-fault-tolerant in the crash-fault model.*

The fault-tolerance of the algorithm stems from the fact that all non-faulty robots will converge to the center of gravity of the faulty robots in their subset. Crashed robots in Y cannot prevent other robots from converging, since the non-faulty robots in Y will eventually move, and achieve separation of the median robots. All other robots will converge as in Algorithm Part.

2.3   The No-Compass Model

Finally, we turn to the extreme model where the robots do not share any common direction, orientation or unit distance. As will be shown in Section 3, in this model the partitioning problem is not deterministically solvable. Thus we allow the robots to use randomness, and present a randomized (Las Vegas) algorithm for solving the partitioning problem in the no-compass SSYNC model against an adaptive adversary.

Let $D_{Max} = \max_{i,j}\{dist(p_i, p_j)\}$ denote the maximal distance between a pair of robots. Define Q to be the set of robots that belong to a robot pair of maximal distance, i.e., $Q = \{r_i : \exists r_j \text{ s.t. } dist(p_i, p_j) = D_{Max}\}$. Denote by $\hat{Q}$ the set of positions occupied by the robots in Q. If a robot $r \notin Q$ is allowed to move only within the convex hull of the robot positions then Q may change, but $\hat{Q}$ will not change, as shown in the following lemma.

Lemma 8. *Let $CH_P[t]$ be the convex hull of all robot positions, $Q[t]$ the set of maximal distance robots and $\hat{Q}[t]$ the set of positions of these robots (all at time t). If robots in Q do not move, and all other robots move within $CH_P[t]$, then robots can join Q only in positions of $\hat{Q}[t]$, i.e., $\hat{Q}[t+1] = \hat{Q}[t]$*

A configuration is *unique* if there is only one pair of robot positions, (p_1, p_1'), such that $dist(p_1, p_1') = D_{Max}$, i.e., $\hat{Q} = \{p_1, p_1'\}$. Note that each of these two positions may be occupied by more than one robot, i.e., $|Q| \geq 2$.

In a unique configuration, define the x axis direction to be on the line connecting p_1 and p_1'. Now determine an order-based partitioning of the robots based on this axis and the ordering induced by it. For undecided groups of robots (having the same x coordinate and possibly also the same y coordinate), we can either use the deterministic Tie-Break procedure, or define a new randomized Procedure Rand-Tie-Break presented next, in which the robots choose randomly between one of the two x directions. In both cases an additional restriction must be imposed, to ensure that no pair of robots moves more than D_{Max} apart, thus changing the x axis agreed upon.

Procedure Rand-Tie-Break
(for robot r_i in a subset L of robots with the same x coordinate x_0)

1. Calculate the convex hull of robot positions, CH_P.
2. Set $\delta \leftarrow \min\{|x_j - x_0|\}$ over all robot x coordinates x_j such that $x_j \neq x_0$.
3. If p_i is on CH_P then do:
   (a) Let e be the edge of CH_P containing p_i.
   (b) Choose one of the two directions of the edge e randomly.
   (c) Move a distance of $\delta/4$ in that direction.
4. Else do:
   (a) Choose one of the two x directions randomly.
   (b) Let d be the distance to CH_P in that direction.
   (c) Move a distance of $\min\{\delta/4, d\}$ in the direction chosen.

Notice that a robot r_i cannot have the same x coordinate as p_1, unless it is located at p_1. The same applies to p_1'. Moreover, moving in both directions defined in Procedure Rand-Tie-Break is always possible (from positions different than p_1 and p_1'), and does not affect D_{Max}.

In a non unique configuration, the following Procedure Expand transforms the system into a unique configuration by moving a random robot from Q outwards. All other robots move onto positions in $\hat{Q}$, to handle the case where all robots in Q crash.

Procedure Expand (code for robot r_i)
If $r_i \in Q$ then do:

1. Pick r_j such that $dist(p_i, p_j) = D_{Max}$.
2. With probability $\frac{1}{|Q|}$ move a distance of $D_{Max}/100$ in the direction opposite of p_j.
   With probability $1 - \frac{1}{|Q|}$ do not move.

Else pick $p_j \in \hat{Q}$ such that $dist(p_i, p_j)$ is minimal and set p_j to be your goal point.

Lemma 9. *If Procedure* Expand *is activated at time t in a non unique configuration, and at least one robot in Q is active, then the probability that the configuration at time t + 1 be unique is at least $\frac{1}{n}$.*

The following algorithm uses Procedures Expand and Rand-Tie-Break to solve PARTITION(n, k).

Algorithm Part_Expand (code for robot r_i in position p_i)

1. State [Non-unique]: The configuration is not unique. Invoke Procedure Expand.
2. State [Unique]: The configuration is unique.
   - Define the x axis and determine an order-based partition P_{OB}, with p_1 and p'_1 as the two robot positions of maximal distance.
   - Case [Point]: $p_i \in \{p_1, p'_1\}$.
     - Substate [Point-few]: At most m robots reside at p_i: do not move at this cycle.
     - Substate [Point-many]: More than m robots reside at p_i: invoke Procedure Expand.
   - Case [Middle]: $p_i \notin \{p_1, p'_1\}$.
     - Substate [Middle-tied]: You cannot determine your subset (due to other robots with the same x coordinate): invoke Procedure Rand-Tie-Break.
     - Substate [Middle-untied]: You can determine your subset:
       (a) Identify the subset S'_i of robots guaranteed to be in your subset S_i.
       (b) If there exists $r_j \in S'_i$ such that $p_j \in \{p_1, p'_1\}$, then set goal point $\tilde{p} \leftarrow p_j$.
       (c) Else, set goal point $\tilde{p} \leftarrow Cog(S'_i)$.

The following proposition states the correctness of the algorithm.

Proposition 3. *The randomized Algorithm* Part_Expand *solves* PARTITION(n, k) *in the SSYNC model with probability 1 and is n-fault-tolerant in the crash-fault model.*

Finally, we remark that the randomized method used in order to break symmetry exploits the ability of the robots to decide randomly whether to make a move or not. This ability is effective in the FSYNC and SSYNC models. In the ASYNC model, however, an adversary can delay a robot's move until other robots decide to move as well, so this method is not applicable.

3   Basic Conditions for Partitioning

This section examines the effect of common orientation levels on the ability of the robots to achieve partitioning. The robots are assumed to be failure-free.

We consider a scale of possible levels of common orientation (ranging from full-compass to no-compass) and establish the following.

Proposition 4. *1. In the FSYNC and SSYNC model, having common axes direction is necessary and sufficient for* $\text{PARTITION}(n, k)$ *to be deterministically solvable for all values of n and k.*

2. In the ASYNC model, having common axes direction is necessary, and having common axes direction and one common axis orientation is sufficient, for $\text{PARTITION}(n, k)$ *to be deterministically solvable for all values of n and k.*

Half-compass: In the half-compass model, Proposition 2 states that the partitioning problem is solvable in the SSYNC model. We now claim that it is solvable in the ASYNC model as well.

Proposition 5. *Partitioning is solvable in the half-compass ASYNC model.*

The solution is not fault-tolerant in the crash-fault model, since one crashed robot can delay the algorithm forever. Whether there exists a non-sequential (and fault-tolerant) such algorithm is still an open question.

No-compass: We now examine the *no-compass* model in which the robots do not have any common direction or orientation. Unfortunately, as shown next, in this model no algorithm can solve the partitioning problem.

Proposition 6. *In the no-compass model, for any timing model,* $\text{PARTITION}(n, k)$ *is unsolvable for* $k > 1$.

The proof above holds also for a model in which the robots have some sense of direction. For the sake of refining the scale of different models, define the *axes-only* model in which the robots have common axes directions, but they do not share the positive axes orientations, and moreover, they cannot distinguish between the x and y axes. For some values of n and k $\text{PARTITION}(n, k)$ can be shown to be unsolvable with a similar proof (for example $n = 4, k = 2$ and the robots start out in a symmetric square). This is stated in the following proposition.

Proposition 7. *In the axes-only model, for any timing model,* $\text{PARTITION}(n, k)$ *is unsolvable for some values of n and k.*

Direction-Only: For the *direction-only* model in which the robots share the x and y axes directions but not their positive orientations, the following holds.

Proposition 8. *In the direction-only model and for the SSYNC timing model,* $\text{PARTITION}(n, k)$ *is solvable for all values of n and k.*

For the direction-only model in the ASYNC timing model, we do not yet know if such an algorithm exists. The algorithm above cannot be considered, since Procedure Tie-Break does not work asynchronously when two or more robots are allowed to move at the same time. Nevertheless, we can argue that partitioning is achievable when n is odd (for any k) by the same argument as for the SSYNC model, based on [10].

References

1. Agmon, N., Peleg, D.: Fault-tolerant gathering algorithms for autonomous mobile robots. In: Proc. 15th SODA, pp. 1063–1071 (2004)
2. Cieliebak, M.: Gathering non-oblivious mobile robots. In: Proc. Latin American Conf. on Theoretical Informatics, pp. 577–588 (2004)
3. Cieliebak, M., Flocchini, P., Prencipe, G., Santoro, N.: Solving the robots gathering problem. In: Baeten, J.C.M., Lenstra, J.K., Parrow, J., Woeginger, G.J. (eds.) ICALP 2003. LNCS, vol. 2719, pp. 1181–1196. Springer, Heidelberg (2003)
4. Cieliebak, M., Prencipe, G.: Gathering autonomous mobile robots. In: Proc. 9th SIROCCO, pp. 57–72 (2002)
5. Cohen, R., Peleg, D.: Convergence properties of the gravitational algorithm in asynchronous robot systems. In: Albers, S., Radzik, T. (eds.) ESA 2004. LNCS, vol. 3221, Springer, Heidelberg (2004)
6. Cohen, R., Peleg, D.: Robot convergence via center-of-gravity algorithms. In: Kralovic, R., Sýkora, O. (eds.) SIROCCO 2004. LNCS, vol. 3104, Springer, Heidelberg (2004)
7. Cohen, R., Peleg, D.: Convergence of autonomous mobile robots with inaccurate sensors and movements. In: Durand, B., Thomas, W. (eds.) STACS 2006. LNCS, vol. 3884, Springer, Heidelberg (2006)
8. Efrima, A., Peleg, D.: Algorithms for partitioning swarms of autonomous mobile robots. Technical Report MCS06-08, The weizmann Institute of Science (2006)
9. Efrima, A., Peleg, D.: Distributed models and algorithms for mobile robot systems. In: van Leeuwen, J., Italiano, G.F., van der Hoek, W., Meinel, C., Sack, H., Plášil, F. (eds.) SOFSEM 2007. LNCS, vol. 4362, Springer, Heidelberg (2007)
10. Flocchini, P., Prencipe, G., Santoro, N.: Widmayer P. Distributed coordination of a set of autonomous mobile robots. In: Proc. IEEE IVS, pp. 480–485 (2000)
11. Prencipe, G.: Achievable patterns by an even number of autonomous mobile robots. Technical report, Dipartimento di Informatica, Universita di Pisa (2000)
12. Prencipe, G.: Corda: Distributed coordination of a set of autonomous mobile robots. In: Proc. 4th Europ. Res. Sem. on Advances in Distributed Systems, pp. 185–190 (2001)
13. Prencipe, G.: Instantaneous actions vs. full asynchronicity: Controlling and coordinating a set of autonomous mobile robots. In: Restivo, A., Ronchi Della Rocca, S., Roversi, L. (eds.) ICTCS 2001. LNCS, vol. 2202, pp. 185–190. Springer, Heidelberg (2001)
14. Prencipe, G.: Distributed Coordination of a Set of Autonomous Mobile Robots. PhD thesis, Universita Degli Studi di Pisa (2002)
15. Prencipe, G.: On the feasibility of gathering by autonomous mobile robots. In: Pelc, A., Raynal, M. (eds.) SIROCCO 2005. LNCS, vol. 3499, pp. 246–261. Springer, Heidelberg (2005)
16. Sugihara, K., Suzuki, I.: Distributed algorithms for formation of geometric patterns with many mobile robots. Robotic Systems 13, 127–139 (1996)
17. Suzuki, I., Yamashita, M.: Agreement on a common x-y coordinate system by a group of mobile robots. In: Proc. Dagstuhl Seminar on Modeling and Planning for Sensor-Based Intelligent Robots, pp. 313–330 (1996)
18. Suzuki, I., Yamashita, M.: Distributed anonymous mobile robots - formation and agreement problms. In: Proc. 3rd SIROCCO, pp. 313–330 (1996)
19. Suzuki, I., Yamashita, M.: Distributed anonymous mobile robots: Formation of geometric patterns. SIAM J. on Computing 28(4), 1347–1363 (1999)

Local Edge Colouring of Yao-Like Subgraphs of Unit Disk Graphs

J. Czyzowicz[1], S. Dobrev[2], E. Kranakis[3], J. Opatrny[4], and J. Urrutia[5]

[1] Département d'informatique, Université du Québec en Outaouais, Gatineau,
Québec J8X 3X7, Canada. Research supported in part by NSERC
[2] School of Information Technology and Engineering, University of Ottawa, Ottawa,
Canada, on leave from Slovak Academy of Sciences, Bratislava, Slovakia. Research
supported in part by NSERC
[3] School of Computer Science, Carleton University, 1125 Colonel By Drive, Ottawa,
Ontario, Canada K1S 5B6. Research supported in part by NSERC and MITACS
[4] Department of Computer Science, Concordia University, 1455 de Maisonneuve Blvd
West, Montréal, Qubec, Canada, H3G 1M8. Research supported in part by NSERC
[5] Instituto de Matemáticas, Universidad Nacional Autónoma de México, Área de la
investigación científica, Circuito Exterior, Ciudad Universitaria, Coyoacán 04510,
México, D.F. México. Partially supported by CONACYT of Mexico, Proyecto
SEP-2004-Co1-45876, and PAPIIT (UNAM), Proyecto IN110802

Abstract. The focus of the present paper is on providing a local deterministic algorithm for colouring the edges of *Yao-like* subgraphs of Unit
Disc Graphs. These are geometric graphs such that for some positive integers l, k the following property holds at each node v: if we partition the
unit circle centered at v into $2k$ equally sized wedges then each wedge can
contain at most l points different from v. We assume that the nodes are
location aware, i.e. they know their Cartesian coordinates in the plane.
The algorithm presented is local in the sense that each node can receive
information emanating only from nodes which are at most a constant (depending on k and l, but not on the size of the graph) number of hops away
from it, and hence the algorithm terminates in a constant number of steps.
The number of colours used is $2kl + 1$ and this is optimal for local algorithms (since the maximal degree is $2kl$ and a colouring with $2kl$ colours
can only be constructed by a global algorithm), thus showing that in this
class of graphs the price for locality is only one additional colour.

Keywords and Phrases: Edge Colouring, Geometric Graphs, Unit
Disk Graphs, Local Algorithm, Location Awareness, Wedge, Wireless
Network.

1 Introduction

The problem of graph edge coloring consists of associating colors to the edges of
the graph, so that no two adjacent edges are of the same color. The minimal such
number of colors required is called *edge chromatic number* of the graph. The well-
known theorem of Vizing [1964] implies that the edge chromatic number of a simple

G. Prencipe and S. Zaks (Eds.): SIROCCO 2007, LNCS 4474, pp. 195–207, 2007.

graph is either Δ or $\Delta + 1$, where Δ is the maximum vertex degree of the graph. For arbitrary graphs the problem was shown to be NP-complete by Holyer [1981].

Interest in the wireless networks research community on the edge coloring problem (see Ramanathan [1999]) comes from its applications, e.g. in packet scheduling (Kumar et al. [2004]), channel assignment (Kodialam and Nandagopal [2005]) or link scheduling (Gandham et al. [2005]). Many wireless and ad hoc networks are modeled by *Unit Disk Graphs* (UDGs), where nodes are embedded in the plane and two nodes are adjacent if and only if their Euclidean distance is at most one. In this model it is assumed that wireless nodes possess the same transmission range and two nodes can communicate when they are in each other's transmission range.

To reduce network complexity and enable important network functions, researchers often construct network *spanners*, whereby certain edges of the network are being omitted from consideration while at the same retaining network connectivity. Although many different types of spanners have been considered in the literature, good spanners must possess some useful properties like small node degree, low edge set complexity and stretch factor. Another important property of spanners for wireless and ad hoc networks is that they may enable *local computations*, i.e. distributed algorithms that never need to transmit messages more than a constant number of hops from source to destination. Introduced in the seminal work of Linial [1992] the concept of locality has been explored in different contexts (cf. Peleg [2000]). It follows that in this model, a *local algorithm* is completed in a constant number of steps and thus its complexity is independent of the size of the network. In dynamically changing, in mobile wireless as well as in ad hoc networks locality has proven to be particularly important, because eventual changes are *localized* and need only be performed within the areas affected, without disturbing the entire system.

The most popular spanners of UDGs, which may be computed locally include relative neighborhood graphs, Gabriel graphs and Yao graphs. However among them only Yao graphs achieve *bounded stretch factor* (or dilation), i.e. the distance of two nodes in the spanner is at most a constant factor larger than their distance in the original graph. In this paper we propose local algorithms for edge coloring of (l, k)-edge/wedge graphs. These graphs, besides being generalizations of Yao graphs, include other classes of subgraphs of UDGs, which appear in the literature, like local approximations of minimum spanning trees, Delauney triangulations or half-space proximals (most typically with $l = 1$ and $k = 3$). Notwithstanding proposed heuristics and algorithms for coloring UDGs (e.g. Marathe et al. [1995], Matsui [1998], Barrett et al. [2006]), none of these approaches address edge coloring under the locality conditions defined above. We assume a geometric network with *location aware* nodes, i.e. nodes that know their Cartesian coordinates in the plane.

1.1   Preliminaries and Results of the Paper

Consider a class of graphs for which the following property holds in each node u. For some integer parameters k and l, if we divide all possible edge directions into

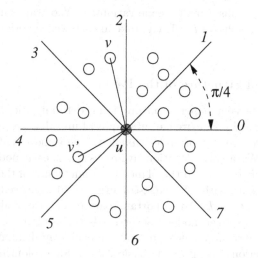

Fig. 1. 8 wedges of width $\pi/4$ around the vertex u. Further, we assume that the horizontal slope is not internal to any wedge. Depicted are the eight wedges numbered $0, 1, \ldots, 7$, an upstream edge (u, v) and a downstream edge (u, v').

$2k$ equally sized wedges (or cones) of width π/k then at most l edges incident to v lie in any fixed wedge. Figure 1 depicts such a partition into eight wedges ($k = 4$) with at most three points per wedge ($l = 3$). We will call such graphs (l, k)-*edge/wedge graphs*. This gives rise to a natural partition of the edges: we partition the edges into k classes $\mathcal{C}_i$ for $i = 0, 1, \ldots, k - 1$ (edges of two opposite wedges–called double-wedge–are put into the same class). Notice that we want an edge to belong to the same class when looking from both its endpoints, which also explains why we do not consider the case of odd number of wedges. More formally, for $i \leq k - 1$, $\mathcal{C}_i$ is the set of all directed edges (u, v) such that the edge's slope lies either within the wedge $\langle i\pi/k, (i+1)\pi/k \rangle$ and the edge's length $|u, v| < 1$, or within the wedge $\langle \pi + i\pi/k, \pi + (i+1)\pi/k \rangle$ and $|u, v| \leq 1$; we call the former *upstream edges* and the latter *downstream edges*. A path p is said to belong to the class $\mathcal{C}_i$, and abusing notation we abbreviate this with $p \in \mathcal{C}_i$, if each of the edges of p is from the class $\mathcal{C}_i$.

Many subgraphs of UDGs that appear in the literature are in fact (l, k)-edge/ wedge graphs, for some l, k. These include Yao graphs as defined in Yao [1982], for k chosen arbitrarily as desired. Similarly, Local Delaunay in Li et al. [2003], Local Minimum Spanning Tree (LMST) Spanners in Chavez et al. [2006b], Wang and Li [2003], and Half-space proximal in Chavez et al. [2006a]) with typical values $l = 1$ and $k = 3$. Interestingly, for many subgraphs of unit disc graphs either there is also a lower bound on the sharpest angle between incident edges (e.g., in LMST as given in Chavez et al. [2006b], Wang and Li [2003] and Half-space proximal Chavez et al. [2006a]) or by construction there is an explicit limit on the number of edges per wedge (e.g. Yao graphs in Yao [1982] and their variants Chavez et al. [2006b]). This means that many spanners of geometric graphs are, in fact,

(l, k)-edge/wedge graphs. Finally, some variants of Yao graphs in Yao [1982] are (l, k)-edge/wedge graphs for $l > 1$ (e.g. take in each wedge both the shortest and the longest edge).

1.2 Results and Structure of the Paper

The focus of the present paper is on providing deterministic, local algorithms for edge-colouring (l, k)-edge/wedge subgraphs of unit disk graphs. The wireless networks considered consist of nodes that have the same *circular* transmission range of size 1. We assume location awareness, i.e., each node of the graph knows its geometric position in the plane. The main result of the paper is a new local, deterministic algorithm for edge-colouring (l, k)-edge/wedge subgraphs of UDGs, for all integers l, k. An important parameter in our algorithms is the *horizon distance* d: a given node u never needs to be aware of the location of nodes beyond its horizon, as measured by the euclidean distance from u. More specifically, in Section 2 we give the basic $2k + 1$ edge colouring algorithm for (l, k)-edge/wedge subgraphs of UDGs using a local horizon distance $7.81 \cdot lk$. Section 3 contains two lower bounds. First, we show that no colouring can be locally constructed in UDGs if there is no geometric information available to the nodes, and second give a lower bound of $d = 2kl$ for the horizon distance used in our algorithm. In Section 4 we give improved constructions for the practically important cases of $k = 3$ and $k = 4$.

2 Basic Construction

In this section we provide the basic construction in detail. To clarify the ideas we first consider the case of at most one point per wedge and later show how to generalize it to multiple points per wedge.

2.1 Virtual Cutting Lines and Grids

Consider first the case $l = 1$ whereby for each node there is at most one point per wedge. The edges of class C_i form a set of vertex-disjoint paths. If we colour each path of class C_i by alternating the colours $2i$ and $2i + 1$, we get a globally consistent edge colouring of the whole graph. However, such an algorithm is inherently global, in fact, even a single path cannot be 2-coloured using a local algorithm since the edges at the endpoints would have to decide on their colour locally, in which case an adversary could adjust the number of edges in the middle segment (which are beyond the horizon of the endpoints) in order to force the use of the third colour.

Therefore the first idea in our algorithm is to use the geometric information to design a tiling of the plane containing the network. This way we can cut the potentially long paths of edges of the same class into path segments of bounded size, each one within a single tile, locally colour the segment edges and use a third colour (per class) to resolve conflicts on the segment boundaries. Let us look more closely at one of the classes C_i, for $i = 0, 1, \ldots, k - 1$. The cutting of paths of class C_i can be achieved by selecting a set of virtual cutting lines (abbreviated VCL) which are normal to the axis of the double-wedge corresponding to C_i and spaced at a distance more than 2. We ensure that each edge $e \in C_i$ intersecting such cutting line will take the third colour, say $2k + i$. The placement of VCLs inside each tile will be identical with respect to the tile boundaries. Hence, the nodes at the endpoints of edges which are cut by the VCLs will use the geometric information to identify the edges which will receive the additional third colour. The fact that the cutting lines are normal to the wedge as well as sufficiently separated ensures that there are no two adjacent edges coloured by the same third colour (if two edges share a node located exactly on a cutting line, only one of them gets assigned the third colour). In this approach the VCLs of different classes do not interfere and can be considered separately, thus resulting in a colouring using overall $3k$ colours.

The main technical contribution of this paper comes from the realization that the VCLs can be designed in such a way that a single conflict-resolution colour can be shared by all classes without interference. The crucial concept is that of d-*cutting the class* C_i, which refers to a set S of line segments that intersects every path $p \in C_i$ with Euclidean diameter at least d. More formally we define

Definition 1 (d-cutting the class C_i). *We say that a set S of line segments in a plane is d-cutting the class C_i if, for each path $p \in C_i$ it is true that*

- *if the Euclidean diameter of p is at least d then p intersects S, and*
- *two consecutive edges of p may intersect S only if their shared endpoint is in S*

Note that S is an infinite set; typically it is a pattern of line segments that is repeated infinitely. Also observe that we can deal with the case $l > 1$ by locally ordering (arbitrarily) the edges of class C_i incident to a vertex v. In this case the edges of the same class and the same rank within this class form a set of vertex disjoint paths. This means that we need as many as lk VCLs (l for each class) which we denote by VCL_i^j for $i = 0, 1, \ldots, k - 1$ and $j = 0, 1, \ldots, l - 1$, with each VCL_i^j d-cutting the class C_i.

The following definition captures the area reachable by edges of C_i which also intersect VCL_i^j. Namely we have

Definition 2 (Reachability by edges of C_i). *Let us denote by MS_i^j the Minkowski sum of VCL_i^j and C_i, that is $MS_i^j = \{v | u \in \mathsf{VCL}_i^j \text{ and } (u, v) \in C_i\}$*

As we require that the VCLs do not interfere with each other, a *Virtual Cutting Grid* is a union of VCLs for each class and each rank such that their Minkowski sums are pairwise disjoint:

Definition 3 (d-Virtual Cutting Grid (VCG)). *S is a d-Virtual Cutting Grid (abbreviated VCG) for (l, k)-edge/wedge graphs if*

- *$S = \bigcup_{i=0}^{k-1} \bigcup_{j=0}^{l-1} \mathsf{VCL}_i^j$,*
- *each VCL_i^j is d-cutting C_i, and*
- *$MS_i^j \cap MS_{i'}^{j'} = \emptyset$, for all i, j, i', j' such that $i \neq i'$ or $j \neq j'$.*

2.2   Local Edge Coloring Algorithm

Before proceeding with the construction of the VCG, let us formally present the colouring Algorithm 1.

Algorithm 1. Local Edge Colouring Algorithm

Input: k and an implicitly known $d - \text{VCG}$
    // *The algorithm at vertex u: Independently process each downstream edge (u, v).*
1: Determine your class $\mathcal{C}_i$; locally sort all edges of class $\mathcal{C}_i$ incident to u in clockwise direction and determine your rank r (from 0 to $l - 1$) – this determines VCL_i^r relevant to you.
    // *Learn your path upstream up to Euclidean distance d.*
2: Broadcast your location to all reachable points downstream along the path of class $\mathcal{C}_i$ and rank r; forward downstream any message received from upstream if its origin is closer than d.
3: **if** intersecting VCL_i^r in a node different then v **then**
4:     select colour $2kl$
5: **else if** e is the first edge of the path, or immediately follows (in its path) an edge coloured $2kl$ **then**
6:     select colour $2(il + r)$
7: **else**
    // *This is local computation using the information obtained in line 2.*
8:     select colour $2(il + r)$ or $2(il + r) + 1$, depending on the parity of the path from an edge coloured by the previous rule, ensuring the colours alternate.
9: **end if**

The following lemma asserts the locality of the algorithm by specifying the length of the required horizon at each node.

Lemma 1. *Each edge decides on its colour using only information from vertices which are at most $2d \cos(\pi/k)$ hops away.*

Proof. According to line 2 of the algorithm, no information is forwarded further than Euclidean distance d from its origin. This may take $2d - 1$ hops if forwarded in a straight line (alternating hops of length 1 and ϵ, where ϵ depends on k and can be chosen arbitrarily small). However the edges of class $\mathcal{C}_i$ can zig-zag only within a wedge of angular width π/k, thus possibly increasing the required distance by a factor of at most $\cos(\pi/k)$. This completes the proof of the lemma.

Assuming that each edge is "waken up" by any activity within its 1-neighbourhood, Lemma 1 also implies that each edge decides on its colour after at most $4d \cos(\pi/k)$ steps since it may need to wake up the upstream path. The correctness of the algorithm follows by construction and from the definition of the VCG.

Theorem 1. *Algorithm 1 computes a $2kl+1$ edge colouring of a (l,k)-edge/wedge subgraph of a UDG.*

Proof. We first prove that each edge can compute its colour. Let us call *start edges* the edges that have determined their colour in lines 4 or 6 of Algorithm 1. From the fact that VCL_i^j is a d-cutting class C_i it follows that each edge is at distance d from a start edge and will therefore compute its colour in line 8 (if it did not do so in lines 4 or 6).

The definition of VCG (second property of each VCL_i^j and the fact that all MS_i^j s are pairwise disjoint) implies that no two edges of colour $2kl$ are incident to each other. From the construction (line 8) and the structure of (l,k)-edge/wedge graphs it follows that no two edges of the same colour are incident.

2.3 Construction of Virtual Cutting Grids

It remains to be shown that given k and l, there indeed exists a $d-\mathsf{VCG}$ for (l,k)-edge/wedge graphs with d depending only on k and l. Without loss of generality, let us rotate all classes $\pi/4$ clockwise. This allows us to partition the classes into three categories as follows:

- *mostly horizontal:* $C_i \bmod \pi \subseteq (-\pi/4, \pi/4)$,
- *mostly vertical:* $C_i \bmod \pi \subseteq (\pi/4, 3\pi/4)$, or
- *straddling the diagonal:* $C_i \bmod \pi \subseteq (\pi/6, \pi/3)$

The basic idea is to use vertical VCLs to cut the *mostly horizontal* classes and horizontal VCLs to cut the *mostly vertical lines.* However, as VCLs must be separated, there must be gaps in the vertical cutting lines to allow placement of the horizontal cutting lines, and vice versa. The gaps are offset by an angle of $\pi/4$ from each other, but because of their width this still allows paths to leak from one gap to another. The square areas between the Minkowski sums of the mesh lines are used for placement of additional cutting lines to stop these leaks. The resulting VCG for $k=2$ and $l=1$ is depicted in Figure 2.

Each leak-stopping square (we call them *leak-stoppers*) contains two vertical and two horizontal line segments, each of them blocking half of the gap width. The leak-stoppers are of two types (called *black* and *white*), placed in chessboard manner, with black leak-stoppers being mirror image of white ones.

This design can be straightforwardly generalized to $l > 1$ by enlarging the basic mesh by a factor of l and having VCL_i^j shifted $2j$ horizontally (for $i=0$) or vertically (for $i=1$). This idea is illustrated in Figure 3, refer also to the first line of Table 1.

Note that if k is even, the classes $C_0, C_1, \ldots, C_{k/2-1}$ are *mostly horizontal* and the remaining $k/2$ classes are *mostly vertical.* This means that we can reuse solution for $k'=2$, $l'=lk/2$, i.e. the pattern from Figure 3 works for $k=6, l=1$ as well (refer to the second line of Table 1).

If k is odd, there will be exactly one class $C_{\lfloor k/2 \rfloor}$ straddling the diagonal, with $\lfloor k/2 \rfloor$ classes being *mostly horizontal* and *mostly vertical,* respectively. For such class, neither vertical, nor horizontal VCL works. Consequently, in this case, we

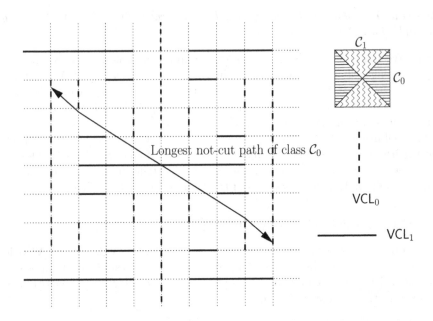

Fig. 2. The design of VCG for $k = 2$, showing also the diameter of the longest not-cut path of class $\mathcal{C}_0$. The side of a basic square is 2.

use the pattern for $k + 1$ classes and let $\mathsf{VCL}^j_{\lfloor k/2 \rfloor}$ be a union of a horizontal and a vertical VCL– this also facilitates uniform construction with the same number of horizontal and vertical VCLs. For example, for $k = 5$ and $l = 1$, this result in the construction illustrated in the last row of Table 1.

Summing up, this means that we can design a VCG for every l and k.

Theorem 2. $d \geq \left(2\sqrt{61(l\lceil k/2 \rceil)^2 + 34(l\lceil k/2 \rceil) + 5}\right)$, *every* (l, k)-*edge/wedge graph, for* $k \geq 2$ *and* $l \geq 1$, *has a* d-*VCG*.

Proof. Firstly, we need to prove that the constructed set is indeed a VCG for l and k. For this purpose we have to show that (1) Minkowski sums are disjoint and that the VCLs satisfy the requirements of Definition 1, in particular that (2) no two consecutive edges are incident to a VCL unless they share an endpoint

Table 1. Possible assignments of VCLs for $lk = 6$ or $k = 5$, $l = 1$

k	l	L0	L1	L2	L3	L4	L5
2	3	VCL^0_0	VCL^1_0	VCL^2_0	VCL^0_1	VCL^1_1	VCL^2_1
6	1	VCL^0_0	VCL^0_1	VCL^0_2	VCL^0_3	VCL^0_4	VCL^0_5
5	1	VCL^0_0	VCL^0_1	VCL^0_2	VCL^0_3	VCL^0_4	VCL^0_2

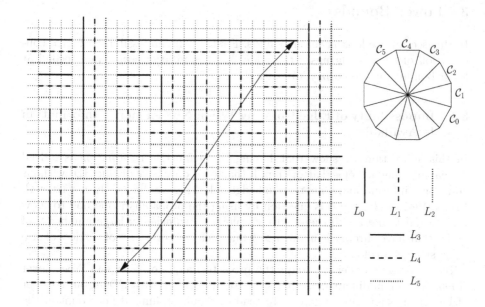

Fig. 3. The design of VCG for $lk = 6$, showing also the longest non-cut path. The side of a basic square is 2.

and (3) there is a finite (depending only on l and k) diameter d such each VCL_i^j d-cuts all paths of class $\mathcal{C}_i$ and rank j.

Proving (1): The MS_i^j's are pairwise disjoint because the closest distance between two points of different VCL_i^j and $\mathsf{VCL}_{i'}^{j'}$ is 2, i.e. two edges of length at most 1 connecting them must lie on the same line – and therefore must lie in the same class. Hence, one is an upstream edge and the other one is a downstream edge. However, by definition $\mathcal{C}_i$ does not contain upstream edges of length 1, i.e. the Minkowski sums do not overlap even when the VCLs are at distance exactly 2.

Proving (2): Note that for each i, no line segment of VCL_i has a slope inside $\mathcal{C}_i$. This is also true for the classes with bounding slopes being multiples of $\pi/4$ (the classes contain only the smaller of their boundary angles, by definition the *mostly horizontal/vertical* classes contain only the larger one). Together with (1) this means that the second property of Definition 1 is satisfied as well.

Proving (3): It is not difficult to observe that the diameter points shown in Figures 2 and 3 are indeed the furthest possible points of a path of class $\mathcal{C}_0$ which is not cut by its VCL. By symmetry, their distance is the diameter distance for other classes as well (the diameter of the *straddling the diagonal* class, if any, is even smaller). As the VCL line separation is 2, the diameter points are offset by $2(5l\lceil k/2\rceil + 1)$ horizontally and $2(6l\lceil k/2\rceil + 2)$ vertically. The statement of the theorem is obtained using the Pythagorean theorem.

3   Lower Bounds

In this section we look at lower bounds. First, at the impossibility of local construction of edge colourings if the nodes are not location aware and second at lower bounds for the horizon distance d.

3.1   Impossibility of Edge Coloring When No Geometric Information Is Available

In this subsection we show that having the geometric information is indeed crucial in achieving a local algorithm. If the nodes only have distinct IDs without additional information, no non-trivial edge colouring can be locally constructed. Erdös [1962] addressed this question by using the probabilistic method to show that for all k and $\epsilon > 0$ there exist graphs with sufficiently large number n of nodes that have chromatic number $> k$ but such that every set of vertices of size at most ϵn is 3 colourable.

In the present paper we adapt the proof technique of Naor and Stockmeyer [1995]. The main difference is that since we are dealing with UDGs, our concept of locality is somewhat different. Instead of requiring limited hop-count in the actual graph, we limit the hop-count to the underlying UDG. The principal result of Naor and Stockmeyer [1995] we use is based on Ramsey's theorem and states that we can limit ourselves to algorithms that do not depend on the actual values of the IDs, but only on their relative ordering.

Theorem 3 (Naor and Stockmeyer [1995]). *Fix a class of graphs $\mathcal{G}$ and a locally checkable labeling problem $\mathcal{L}$. If there is a local algorithm A with time bound t that solves a problem $\mathcal{L}$ in $\mathcal{G}$, then there is an order-invariant local algorithm A' that solves $\mathcal{L}$ in $\mathcal{G}$ in time t.*

Informally, a problem has a *locally checkable labeling* if it can be locally verified whether a given vertex (or edge) labeling satisfies the requirements. Refer to Naor and Stockmeyer [1995] for the formal statement. For our purposes it is sufficient to state that vertex colouring and edge colouring are both locally checkable labellings.

Theorem 4. *For every $l \geq 1$, $k \geq 2$ there is a (l, k)-edge/wedge subgraph of a UDG that is not locally colourable.*

Proof. Consider a mesh of vertices with the distance w between neighboring mesh vertices equal to $w = \sqrt{2}l \tan(\pi/k)$. Define the graph G in which each vertex v of the mesh is connected to the closest l vertices in each wedge. The choice of w ensures that all these edges are of length at most 1, i.e. G is a subgraph of the UDG of these mesh vertices. Since neighboring vertices of the mesh are neighbors in G, for each edge e of the UDG there exists a path in G of length at most $x = \sqrt{2}/w$ connecting the endpoints of e.

Assume now, by contradiction, there exists a local (in the UDG) algorithm A, examining the neighbourhood up to distance t. Therefore, there must exist a constant x and a local (in G) algorithm A' examining the neighbourhood up to distance xt. By Theorem 3 there is also an order invariant algorithm A''.

Consider now a mesh of size $(2xt + 3) \times (2xt + 3)$, with the vertex at location $[i, j]$ labeled $j(2xt + 3) + i$. As A' examines the neighbourhood up to distance at most xt, nodes $v_1 = [xt, xt]$, $v_2 = [xt + 1, xt]$ and $v_3 = [xt + 2, xt]$ all see the same pattern (monotonically increasing from upper left to lower right) and cannot deterministically decide on different colours for edges (v_1, v_2) and (v_2, v_3). This contradicts the fact that A' is a local edge-colouring algorithm.

3.2   Lower Bound on d

Theorem 2 gives a bound for d of about $7.81 \cdot lk + O(1)$. As the construction seems quite ad hoc and not optimized (it gives the same bound for fixed lk, despite the fact that the case $l = 1$ and large k is much more restrictive than $k = 2$ and large l), an obvious question is how for from the optimal it actually is. A somewhat surprising answer is that it is not far at all, namely we have the following result.

Theorem 5. *There is no $d - VCG$ for (l, k)-edge/wedge graphs with $d < 2kl$.*

Proof. The idea is to show that each MS_i^j covers a $1/2d$ fraction of the area. Since all these Minkowski sums must be pairwise disjoint, this would prove the theorem. Consider VCL_i^j. Take a line l whose slope is the axis of the class C_i. Note that l is cut by VCL_i^j into line segments of length at most d. The parts of l at distance less then 1 both downstream and upstream from the cutpoints belong to MS_i^j, i.e. a $2/d$ fraction of l is in MS_i^j. As this is true for every position of l in the plane, (and the behaviour is continuous, as VCLs are sets of line segments) MS_i^j indeed takes $1/2d$ fraction of the area and the statement of the theorem follows.

4   Refined Constructions

As noted above, the construction used in Theorem 2 is based on the case $k = 2$, with the angular width of the wedges being $\pi/2$. This means that the paths have a lot of 'wiggle room' to avoid the VCL s, forcing a conservative construction with relatively large horizon distance d. The horizon distance can be noticeably reduced by more careful analysis for specific k. In particular, the case $k = 3$ warrants special attention, as this is the most relevant case in practical situations (both LMST and Half-space proximal lead to $(1, 3)$-edge/wedge graphs and $k = 3$ is the lowest k for which the Yao graph is connected). An ad-hoc construction of the VCG using hexagonal symmetry and careful analysis of Minkowski sums yields:

Theorem 6. *There exists 15.466-VCG for $(1, 3)$-edge/wedge graphs.*

The case $k = 4$ is practically important as well, as it corresponds to the smallest k for which the Yao graph is connected and has a convenient east/north/west/south symmetry.

Theorem 7. *There exists 26.989-VCG for* $(1, 4)$*-edge/wedge graphs.*

Finally, for the limit case of $k \to \infty$ the wedges are very narrow, which allows the use of several additional techniques to further reduce the horizon distance:

Theorem 8. *There exists* $3.3745lk + o(lk)$*-VCG for* (l, k)*-edge/wedge graphs.*

Details of the proofs of these theorems will appear in a forthcoming complete version of this paper.

5 Conclusions

In this paper we identified a new subclass of unit disk graphs, called (l, k)-edge/wedge graphs, that encompass a variety of well-known and commonly used graphs in the wireless networks literature (Yao graphs, Local Delaunay, Local Minimum Spanning Tree spanners, and Half-Space Proximals). We gave local, deterministic algorithms for edge coloring these graphs and provided several ways for improving the algorithm depending on the density of the point set that determines the graph. Several questions and tradeoffs concerning the optimality of our algorithm remain open and would be interesting to consider in future investigations.

References

Barrett, T.C., Istrate, G., Kumar, A., Marathe, M., Thite, S., Thulasidasan, S.: Strong edge coloring for channel assignment in wireless radio networks. In: IEEE International Workshop on Foundation and Algorithms for Wireless Networking, IEEE, New York (2006)

Chavez, E., Dobrev, S., Kranakis, E., Opatrny, J., Stacho, L., Tejeda, H., Urrutia, J.: Half-space proximal: A new local test for extracting a bounded dilation spanner. In: Proceedings of OPODIS. LNCS, Springer, Heidelberg (2006a)

Chavez, E., Dobrev, S., Kranakis, E., Opatrny, J., Stacho, L., Urrutia, J.: Local construction of planar spanners in unit disk graphs with irregular transmission ranges. In: Correa, J., Hevia, A., Kiwi, M. (eds.) SWSWPC 2004. LNCS, pp. 286–297. Springer, Heidelberg (2006b)

Erdös, P.: On circuits and subgraphs of chromatic graphs. Mathematika 9, 170–175 (1962)

Gandham, S., Dawande, M., Prakash, R.: Link scheduling in sensor networks: distributed edge coloring revisited. In: INFOCOM, pp. 2492–2501. IEEE Computer Society Press, Los Alamitos (2005)

Holyer, I.: The NP-completeness of edge-coloring. SIAM Journal on Computing 10(4), 718–720 (1981)

Kodialam, M.S., Nandagopal, T.: Characterizing achievable rates in multi-hop wireless mesh networks with orthogonal channels. IEEE/ACM Trans. Netw. 13(4), 868–880 (2005)

Kumar, V.S.A., Marathe, M.V., Parthasarathy, S., Srinivasan, A.: End-to-end packet-scheduling in wireless ad-hoc networks. In: Proceedings of the Fifteenth Annual ACM-SIAM Symposium on Discrete Algorithms, SODA 2004, New Orleans, Louisiana, USA, January 11-14, 2004, SIAM, pp. 1021–1030 (2004)

Li, X.-Y., Calinescu, G., Wan, P.-J., Wang, Y.: Localized delaunay triangulation with application in ad hoc wireless networks. IEEE Trans. Parallel Distrib. Syst. 14(10), 1035–1047 (2003)

Linial, N.: Locality in distributed graph algorithms. SIAM Journal on Computing 21(1), 193–201 (1992)

Marathe, M.V., Breu, H., Hunt III, H.B., Ravi, S.S., Rosenkrantz, D.J.: Simple heuristics for unit disk graphs. Networks 25(1), 59–68 (1995)

Matsui, T.: Approximation algorithms for maximum independent set problems and fractional coloring problems on unit disk graphs. In: Akiyama, J., Kano, M., Urabe, M. (eds.) Discrete and Computational Geometry. LNCS, vol. 1763, pp. 194–200. Springer, Heidelberg (1998)

Naor, M., Stockmeyer, L.: What can be computed locally? SICOMP: SIAM Journal on Computing, vol. 24 (1995)

Peleg, D.: Distributed Computing: A Locality-Sensitive Approach. SIAM (2000)

Ramanathan, S.: A unified framework and algorithm for channel assignment in wireless networks. Wireless Networks 5(2), 81–94 (1999)

Vizing, V.G.: On the estimate of the chromatic class of a p-graph. Diskret. Analiz. 3, 25–30 (1964)

Wang, Y., Li, X.-Y.: Localized construction of bounded degree and planar spanner for wireless ad hoc networks. In: DialM: Proceedings of the Discrete Algorithms and Methods for Mobile Computing & Communications (2003)

Yao, A.: On constructing minimum spanning trees in k-dimensional spaces and related problems. SIAM Journal on Computing 11(4), 721–736 (1982)

Proxy Assignments for Filling Gaps in Wireless Ad-Hoc Lattice Computers*

Tiziana Calamoneri[1], Emanuele G. Fusco[1],
Anil M. Shende[2], and Sunil M. Shende[3]

[1] Dipartimento di Informatica
Università di Roma "La Sapienza"
via Salaria, 113-00198 Rome, Italy
{calamoneri, fusco}@di.uniroma1.it
[2] Dept. of Mathematics, Computer Science & Physics
Roanoke College
Salem, Virginia 24153, USA
shende@roanoke.edu
[3] Dept. of Computer Science
Rutgers University
Camden, NJ 08102, USA
shende@camden.rutgers.edu

Abstract. The proliferation of cheap portable, wireless computing devices (e.g., cell phones and PDAs) promises the availability of a large number of computing devices in a relatively small geographic region. Researchers have proposed using such an ensemble of wireless devices to create a wireless ad-hoc lattice computer (WAdL) to harness the collective computing capabilities of the devices for the common cause of scientific computing via analogical simulations. Faulty devices or lack of wireless coverage leads to "gaps" in a WAdL, rendering it ineffective for analogical simulations.

In this paper we discuss our soultion to the problem of bridging gaps in WAdLs by assigning active devices on the perimeter of the gap as proxies for the defective devices in the gap. We establish lower bounds on the communication dilation witnessed by such proxy assignments for single-row gaps and general row-column convex gaps, and present dilation-optimal, constant time algorithms for computing proxy assignments for single-row gaps and gaps that are rectangular in shape.

1 Introduction

The proliferation of cheap portable, wireless computing devices (e.g., cell phones and PDAs) promises the availability of a large number of computing devices in

---

* This research was supported, in part, by the European Research Project *Algorithmic Principles for Building Efficient Overlay Computers* (AEOLUS). Most of the work reported here was performed while Professor A. Shende visited the Department of Computer Science, University of Rome "La Sapienza". Support through a Visiting Fellowship from the University of Rome "La Sapienza" is gratefully acknowledged.

G. Prencipe and S. Zaks (Eds.): SIROCCO 2007, LNCS 4474, pp. 208–221, 2007.

a relatively small geographic region [9]). Based on the foundations of cellular automata and lattice computers [2,8,11], Gupta et al. [6,7] propose the use of such an ensemble of wireless devices to create a *wireless ad-hoc lattice computer (WAdL)*. The goal of a WAdL is to harness the collective computing capabilities of the devices for the common cause of scientific computing via analogical simulations. The methodology, formalised by earlier work on lattice computers, is (1) to represent, by computing devices logically arranged as a part of a lattice, the region of euclidean space in which a phenomenon unfolds, and (2) to have the computing devices *analogically* simulate the unfolding of the phenomenon in this representation of euclidean space. In analogical simulations on a lattice computer (and hence on a WAdL), the motion of an object across euclidean space is carried out as a sequence of steps, in time proportional to real time, where in each step the representation of the object may move from one processing element to a *neighbour*, as defined by the underlying lattice of the lattice computer [13]. (See Section 2 for a brief discussion of analogical simulations.)

Clearly, the accuracy of the results of these simulations is directly dependent on the resolution of the underlying lattice of a WAdL. Moreover, since the devices in a WAdL are logically, but not necessarily *physically*, at lattice points (in the underlying lattice), the communication distance between devices is not proportional to the physical distance between lattice points represented by those devices. To ensure analogical simulations that unfold in time proportional to the real time unfolding of the phenomenon being simulated, additional adjustments have to be made (see [6,7]) to the lattice computer algorithms described in the literature [2]. Faulty devices or lack of wireless coverage leads to "gaps" in the collection of devices, further exacerbating this problem.

In this paper, we focus on the problem of bridging "gaps" in the underlying lattice of a WAdL. We adopt the approach proposed by Moore et al. [12] – to assign, for each faulty device, x, an active device, $l(x)$, in the WAdL to serve as a proxy for x. As a consequence, the communication between two neighbouring faulty devices x and y will be carried out as a communication between the two proxy devices $l(x)$ and $l(y)$. These proxy devices, though, may not be neighbours, and thus analogical simulations being carried out on such a WAdL will run slower than on a WAdL with no faulty devices. Moore et al. [12] present assignment schemes for square gaps. We extend that work and study gaps of more general shapes. As in Moore et al., our primary goal is to devise assignment schemes for bridging gaps so that the communication time between proxies for any pair of neighbouring faulty devices is minimised.

In Section 3 we introduce some useful concepts, and then formally define the problem we are addressing in this paper. It is convenient to discuss our solution for the problem separately for two different classes of gaps – we discuss our treatment of the two classes in Sections 4 and 5. We conclude in Section 6 with some open problems.

2 Analogical Simulations

A physical phenomenon is a development in a region of Euclidean space over a period of time. At each instant in time (in a given time period), the set of objects participating in the phenomenon, together with their attribute values (such as speed, spin, etc.) at that time, completely describes a snapshot in the unfolding of the phenomenon. Most problems in scientific computing are about phenomena whose unfolding involves the motion of participating objects in Euclidean space. Solutions to these phenomena usually involve determining (predicting) the attribute values of objects over time. Some phenomena can be solved analytically using closed form functions of time. On the other hand, there are phenomena where the only apparent method for predicting the attribute values of participating objects at any instant in time, is to *simulate* the unfolding of the phenomenon up through that instant of time [5].

When carried out on a digital computer, such simulations, *necessarily*, develop in a discretized representation of a region of Euclidean space, and over discrete time units. Moreover, such simulations must use, at any given instant of simulation time, only information available *locally*, at a discrete point in the represented Euclidean space, to compute the attribute values of participating objects at the next instant of simulation time. Cellular automaton machines [8] and lattice computers [2] provide the necessary framework for a discretized representation of Euclidean space in which to carry out such simulations. Several physical phenomena, including spherical wavefront propagation and fluid flow [4,14], have been successfully simulated on such a framework where the simulation algorithms do *not* use the traditional analytical models for the phenomena.

3 Preliminaries

Consider a collection of devices arranged on a subset of the cells of a square grid, one device per cell.

Definition 1. *A collection of devices is said to be **row convex** (respectively, **column convex**) if every row (resp. column) of devices forms a contiguous interval of devices within that row (resp. column). The collection is said to be **row-column convex** if it is simultaneously row convex and column convex. Moreover, if the collection is such that there are at least two devices in each row and each column, then the collection is said to be **thick row-column convex**.*

Figure 1 shows examples of such collections of devices to illustrate the notions in Definition 1. The notion of row-column convex collections is similar to the notion of *hv-convex polyominoes* discussed in the literature [3,10].

As discussed in earlier work on WAdLs [7], the expectation is to harness the collective power of mobile devices in a given geographic region, say a few blocks in the downtown area of a city. It is quite likely that some buildings in that area do not have wireless coverage, or that the office occupying some building is closed on a particular day, and thus there are no mobile devices in the physical

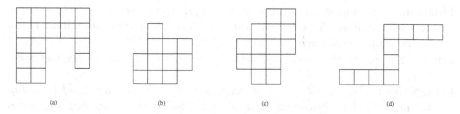

Fig. 1. Examples of collections of devices: (a) column convex but not row convex, (b) row-column convex (not thick), (c) thick row-column convex, (d) row-column convex (not thick)

space occupied by that office. Such situations will lead to gaps in a WAdL in that area, and such gaps can be faithfully modeled by a row-column convex collection of devices. Indeed, in this paper we assume that the collection of faulty devices in a WAdL is row-column convex. We will refer to such a collection of faulty devices in a WAdL as a *gap* or a *hole* in the WAdL.

Thus the overall picture is of a row-column convex device collection consisting of a bunch of *active* devices surrounding a row-column convex gap of defective devices. The set of active devices that are immediately adjacent to the defective ones form a contiguous contour that we will call the *active perimeter* of the gap.

As mentioned in Section 1, to simulate lattice computations completely, we now seek to assign to every defective device, a corresponding device on the active perimeter that will serve as its *proxy* in the computation; for convenience, we will think of active perimeter devices as being trivially their own proxies. Consider a device x and its proxy $l(x)$. If device x were active, it would have been able to communicate directly with adjacent devices along its row and its column. However, if device x is defective, communication for x is now handled instead by the proxy $l(x)$. Additionally, communication takes place along paths whose individual hops (edges) only connect active devices.

Accordingly, in the presence of a hole, the distance, $d(x, y)$, between any two active devices at positions x and y in the square grid is the length of the shortest path between x and y that is composed entirely of horizontal hops (along a row) or vertical hops (along a column) between adjacent active devices. In general, we expect that the communication between two proxies $l(x)$ and $l(y)$ takes place along the shortest path on the active perimeter between $l(x)$ and $l(y)$. Indeed, if the active perimeter encloses a row-column convex gap, then the following folklore result can be easily shown by induction:

Lemma 1. *Given a row-column convex gap in a WAdL, the shortest distance between any two active devices x and y on the active perimeter equals the length of the shortest path that only uses the perimeter edges.*

Definition 2. *For any proxy assignment that maps a device x to a proxy $l(x)$ on the active perimeter, we define the **dilation**, **load** and **congestion** of the assignment as follows:*

Dilation: *The maximum distance, $d(l(x), l(y))$, taken over all pairs of adjacent devices x and y. (Note that, in the literature, the notion of dilation is not restricted to neighbouring devices.)*

Load: *The maximum number of devices assigned to any active perimeter device.*

Congestion: *The maximum, over every active perimeter edge e, of the number of distinct proxy-communication paths that simulate one-hop communications between two devices, at least one of which is defective.*

Henceforth, for convenience, we will use *proxy assignment* to mean "proxy assignment of defective devices to the active perimeter devices". Our problem can be stated as follows: for any row-column convex collection, $\mathcal{C}$, of defective devices in a WAdL, find a proxy assignment such that the dilation, load and congestion of the assignment are minimised. If the dilation of a proxy assignment is the best possible we will call that assignment a *dilation-optimal* assignment (similarly, *load-optimal* assignment and *congestion-optimal* assignment).

Without loss of generality,

Definition 3. *For any row-column convex collection $\mathcal{C}$, $R(\mathcal{C})$ is the unique, smallest rectangle of cells such that $R(\mathcal{C})$ contains $\mathcal{C}$ and each border row and column of $R(\mathcal{C})$ contains at least one element of $\mathcal{C}$. The number of rows and columns of $R(\mathcal{C})$ are denoted by $r(\mathcal{C})$ and $c(\mathcal{C})$, respectively. The number of active perimeter devices around $\mathcal{C}$ is denoted by $p(\mathcal{C})$.*

When the context is clear, we will refer to the above quantities as simply r, c and p. Without loss of generality, we will assume that the top left corner cell of $R(\mathcal{C})$ has co-ordinates $(1, 1)$, and the bottom right corner cell of $R(\mathcal{C})$ has co-ordinates (r, c). We will refer to the devices by the co-ordinates of the cell that the device occupies.

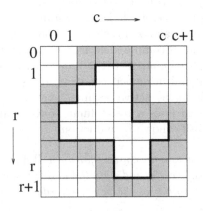

Fig. 2. A row-column convex gap. $r = 6$, $c = 6$, $p = 28$

Figure 2 illustrates the above definition. Figure 2 also illustrates our reference co-ordinate system for naming devices. The active perimeter devices are shown shaded.

It is, then, easy to verify the following two lemmas.

Lemma 2. *For any row-convex collection C,*

$$p(C) = p(R(C)) = 2(r(C) + c(C)) + 4.$$

Lemma 3. *For any gap, the load of any proxy assignment is at least the ratio of the number of defective devices to the number of active perimeter devices.*

4   Single Row Gaps

In this section we will deal with the special case of connected gaps where $r = 1$. Clearly, such gaps are row-column convex. In Section 4.1 we will establish a lower bound on the dilation of any proxy assignment for such a gap. Then, in Section 4.2 we discuss a dilation-optimal and load-optimal assignment scheme for single-row gaps.

4.1   Lower Bound on Dilation

If a gap contains a single row that has one defective device, i.e., $r = c = 1$, then it is easy to verify that the dilation of any proxy assignment must be at least 3. The following theorem establishes a lower bound on the dilation of proxy assignments for single-row gaps with more than one defective device.

Theorem 1. *In a WAdL containing a collection C of defective devices such that $r(C) = r = 1$ and $c(C) = c \geq 2$, every proxy assignment of active perimeter devices to the defective ones has dilation at least $D = \lceil (2c+5)/3 \rceil$.*

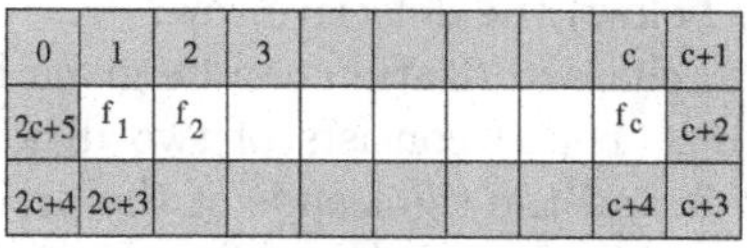

0	1	2	3				c	c+1
2c+5	f_1	f_2					f_c	c+2
2c+4	2c+3						c+4	c+3

Fig. 3. Numbering of devices for a single-row gap

Proof. Given a collection C of defective devices such that $r(C) = r = 1$ and $c(C) = c \geq 2$, the active perimeter $p(C) = p = 2c + 6$. We denote the defective devices as $f_1, f_2, \ldots, f_c$ proceding from left to right and the active perimeter devices as $0, 1, \ldots, 2c + 5$ starting from cell $(0,0)$ and proceding clockwise (see Figure 3 for an example). For convenience, for every $x \in [1, c]$, we denote the devices $2c + 4 - x$ as $b(x)$; with this notation, the defective device f_x has device x as its neighbor above and device $b(x)$ as its neighbor below. Thus, in any proxy assignment l that assigns active device $l(x)$ to the defective device f_x, the dilation will be lower bounded by the maximum taken over the following four sets of distances:

1. $\{d(l(x), x) \mid x \in [1, c]\}$,
2. $\{d(l(x), b(x)) \mid x \in [1, c]\}$,
3. $\{d(l(x), l(x+1)) \mid x \in [1, c-1]\}$,
4. $\{d(2c+5, l(1)), d(l(c), c+2)\}$.

Let $D = \lceil (2c+5)/3 \rceil$ (about a third of the active perimeter), and assume, by way of contradiction, that L has dilation at most $D-1$. Let $A(x)$ denote the possible range of values that $l(x)$ can take if $d(l(x), x)$ and $d(l(x), b(x))$ both respect the assumed dilation bound, i.e. if both these distances are no more than $D-1$. Consider the specific defective devices f_x and f_y where x and y are given by:

$$y = D - 1$$
$$= \lceil (2c+5)/3 \rceil - 1 \text{ and}$$
$$x = (c+1) - y$$
$$= \lfloor (c+1)/3 \rfloor.$$

Note that x and y are equi-distant from the left and the right ends of the defective row. From the definition of y, it is easy to see that the distance to $b(y)$ from any device in the range $[0, y-1]$ is greater than D. Similarly, the distance to y from any device in the range $[b(y)+1, 2c+5]$ is at least D. Consequently, $A(y)$, the allowable range for $l(y)$, is the interval $[y, b(y)]$ that consists of the devices to the right of (and including) y (respectively, $b(y)$) in the row above (respectively, row below) the defective row. Similar reasoning yields the complementary fact that $A(x)$, the admissible range for $l(x)$, is the interval $[b(x), x]$ (in the circular ordering in clockwise order around the perimeter). Note that $A(x)$ contains exactly those devices to the left of (and including) x (respectively, $b(x)$) in the row above (respectively, row below) the defective row.

In fact, a more general characterization can be given for $A(x+i)$ as i ranges from 0 through $(y-x)$: $A(x+i)$ consists of two disjoint intervals (not both empty) containing the left and the right ends of the perimeter. Specifically, the proxy $l(x+i)$ either lies in the interval $A_l(x+i) = \{2c+4-(x-i), \ldots, x-i\}$ (in circular clockwise order) or in the interval $A_r(x+i) = \{c+3-i, \ldots, c+1+i\}$; we have

$$A(x+i) = A_l(x+i) \bigcup A_r(x+i)$$

for all $i \in [0, (y-x)]$. For the extreme values of i in its range, we have $A(x) = A_l(x)$ with $A_r(x)$ empty, and have $A(y) = A_r(y)$ with $A_l(y)$ empty. So as one considers the proxies proceeding from f_x towards f_y, there must come a point where $l(x+i)$ is from the left admissible interval $A_l(x+i)$ but the next proxy, $l(x+i+1)$, is from the right admissible interval $A_r(x+i+1)$, i.e. the proxies are from opposite sides of the perimeter.

But we require $d(l(x+i), l(x+i+1))$ to be at most the assumed dilation. However, the shortest perimeter distance between any such pair of proxies is no less than the distance between the closest pair of devices from $A_l(x + i)$ and $A_r(x + i + 1)$ respectively, *viz.* the distance between $x - i \in A_l(x + i)$ and $c + 3 - i \in A_r(x + i + 1)$. In summary,

$$
\begin{aligned}
d(l(x + i), l(x + i + 1)) &\geq d(x - i, c + 3 - i) \\
&= c + 3 - x \\
&= D + 1
\end{aligned}
$$

This contradicts the assumed dilation bound of $D - 1$, and yields the result.   □

Note that the same argument also works for a gap consisting of a single column, and, with minor modifications, for any row-convex gap consisting of connected single rows and columns such that each device in the gap has no more than two neighbours in the gap. (See Figure 1(d) for an example of such a gap.)

4.2   Dilation-Optimal Assignment

The following theorem states that the lower bound for dilation for a gap of one row and $c \geq 2$ columns given in Section 4.1 is tight. The proof is constructive as it provides a dilation-optimal and load-optimal proxy assignment.

Theorem 2. *There exists a dilation-optimal and load-optimal proxy assignment for a gap with one row and $c \geq 2$ columns.*

Proof. Let $d = \lceil (2c + 5)/3 \rceil$ and $x = \lfloor \frac{c}{2} \rfloor + 1 + d$.
Assign $l(\lfloor \frac{c}{2} \rfloor + 1) = x$, and $l(\lfloor \frac{c}{2} \rfloor) = x + d$.
Complete the proxy assignment l as follows:

- $l(\lfloor \frac{c}{2} \rfloor + i + 1) = x - i$ for $1 \leq i \leq \lceil \frac{c}{2} \rceil - 1$;
- $l(\lfloor \frac{c}{2} \rfloor - j) = (x + d + j) \pmod{2c + 6}$ for $1 \leq j \leq \lfloor \frac{c}{2} \rfloor - 1$.

It is easy to see that l is feasible and load-optimal, and its dilation is equal to $\lceil (2c + 5)/3 \rceil$.
From this result and Theorem 1, the claim follows.   □

5   Row-Column Convex Gaps

In this section we first establish, in Section 5.1, a lower bound on the dilation of proxy assignments for general row-column convex gaps with $r, c > 1$. Specifically, we will show that every proxy assignment will have dilation greater than or equal to about one-fourth of the active perimeter (rather than about one-third for the case discussed in the case of single-row gaps (see Theorem 1 above)). Then, in Section 5.2 we restrict our attention to rectangular gaps with $r, c > 1$ and present an algorithm for a dilation-optimal *and* load-optimal proxy assignment for such gaps.

5.1   Lower Bound on Dilation

Theorem 3 establishes a lower bound on the dilation of any proxy assignment for row-column convex gaps. The idea of the proof for the theorem is inspired by the proof of Sperner's lemma in Aigner et al. (see [1], page 148).

Theorem 3. *In a WAdL containing a row-column convex hole with $r, c > 1$ and active perimeter p, every proxy assignment of active perimeter devices to the defective ones in the hole has dilation at least $D = \lceil p/4 \rceil$.*

Proof. Any row-column convex hole of defective devices can be easily shown to have an even number of devices on its active perimeter. Consider a row-column convex hole with active perimeter p. Starting at an arbitrary position on the perimeter and proceeding clockwise, we *color* the active perimeter devices as follows:

- The first $\lceil p/4 \rceil$ devices are colored A.
- The next $\lceil p/4 \rceil$ devices in order are colored B.
- The next $\lfloor p/4 \rfloor$ devices in order are colored C.
- The last $\lfloor p/4 \rfloor$ devices are colored D.

Consider any proxy assignment l that assigns an active perimeter device $l(x)$ to a defective device x within the gap. As usual, this proxy device is now responsible for simulating the communication that would have ordinarily been initiated by the device x, now defunct. The proxy assignment induces - in a natural way - a coloring of the defective devices: the device x is given the same color $(A, B, C$ or $D)$ as that given to its active proxy $l(x)$.

   Given assignment l, let G_l be the corresponding induced subgraph of the mesh consisting of cells corresponding to the active perimeter devices and the defective devices in the row-column convex gap enclosed by the active perimeter. The edges of G_l replicate the mesh connectivity among adjacent devices and each vertex of G_l is assigned the same color $(A, B, C$ or $D)$ as its corresponding device.

   Clearly, G_l is a planar graph with square faces, except for the outer face that traces the active perimeter contour. We now define a special kind of *planar dual* of G_l as follows. Every face f of G_l (including the outer face) corresponds to a unique vertex u_f in the dual graph G_l^d. However, unlike the standard planar dual, vertices u_f and $u_{f'}$ are connected by an edge in G_l^d iff f and f' are adjacent faces and *the common edge between the adjacent faces f and f' has its endpoints colored AB or BC or CD.* (Note that G_l is not the standard planar dual.)

   Figure 4(a) shows a row-column convex gap with $r = 3$ and $c = 2$, and the coloring induced by a proxy assignment. Note that $p = 14$, and so there are 4 active perimeter devices colored A, 4 colored B, 3 colored C and 3 colored D. Figure 4(b) shows the connectivity graph of the devices in dotted lines and the dual graph as defined above in bold lines. The dual graph vertex corresponding to the outer face is labelled u_0.

   Consider the dual vertex u_0 that corresponds to the outer (perimeter) face of G_l. Since the perimeter is colored in sequence with contiguous As, Bs, Cs and Ds, it follows that u_0 has degree 3 witnessed by the unique edges colored

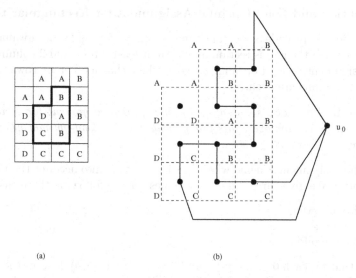

(a)   (b)

Fig. 4. (a) Coloring induced by a proxy assignment. (b) Dual graph.

AB, BC and CD at the color transitions. All other dual vertices correspond to square faces of G_l and hence have degree at most 4. Moreover, by the handshake lemma for graphs, it follows that among the remaining dual vertices (excluding the odd-degree vertex u_0), there *must be an odd number of vertices of odd degree*. In particular, there must be at least one dual vertex u_f that corresponds to a mesh square (face) f and has degree 1 or 3.

It is easy to verify that up to rotational and mirror symmetries, a dual vertex u_f has degree 3 if and only if the four corners of face f in G_l are colored $ABCD$ in cyclic order. Also, a dual vertex u_f has degree 1 if and only if face f in G_l has corners colored $ABBD$, $ABDA$, $ABDD$, $BCAD$, $CDAC$ or $CDDA$ in cyclic order (modulo rotations and mirror symmetries).

However, recall that the colors identify groups of active perimeter devices that serve as proxies for defective devices in the gap. Hence, any edge of G_l whose endpoints are colored AC or BD will correspond to a pair of proxies that must be at least $\lceil p/4 \rceil$ apart along the perimeter. Since the shortest distance between any two perimeter devices is indeed along the perimeter, we immediately see from the preceding paragraph that all the faces f whose dual vertices u_f have degree 1 witness a dilation of at least $\lceil p/4 \rceil$ for the assignment l.

This leaves the only remaining possibility of a face f with corners colored $ABCD$ in cyclic order. Noting that the corners correspond to four proxies $\bar{a}, \bar{b}, \bar{c}$ and $\bar{d}$ colored A, B, C and D, respectively, along the perimeter. Again, noting that the shortest distance between proxies is along the perimeter, we have

$$d(\bar{a}, \bar{b}) + d(\bar{b}, \bar{c}) + d(\bar{c}, \bar{d}) + d(\bar{d}, \bar{a}) = p,$$

and hence, at least one of the distances must be greater than or equal to $\lceil p/4 \rceil$.

$\square$

5.2   Dilation and Load Optimal Assignment for Rectangular Gaps

In this section we provide a constant time algorithm for a dilation-optimal proxy assignment when the gap is rectangular with at least 2 rows and 2 columns. Our proxy assignment is load-optimal as well when the number of rows and the number of columns are even.

Theorem 4. *There exists a dilation-optimal proxy assignment for rectangular gaps with $r, c > 1$. Moreover, this proxy assignment is load-optimal when r and c are even.*

Proof. The assignment scheme when $r > 1$ must take into account the parity of both r and c. We give three different schemes for the following three cases:

1. r and c are even;
2. $r + c$ is odd;
3. r and c are odd.

Recall that for each $0 \leq a \leq r+1$ and $0 \leq b \leq c+1$, (a, b) denotes the devices in the a-th row and b-th column. Of these, the devices (a, b) where $1 \leq a \leq r$ and $1 \leq b \leq c$ are the defective devices. We number the $p = 2(r + c) + 4$ active perimeter devices sequentially from 0 starting with the device $(0, 0)$ and proceeding clockwise (see Figure 5 for an example).

Case 1:
Compute $d = \frac{c+r}{2} + 1$.
For $1 \leq i \leq \frac{c}{2}$ assign:

- defective device $\left(\frac{r}{2}, i\right)$ to active perimeter device i for $1 \leq i \leq \frac{c}{2}$;
- defective device $\left(\frac{r}{2}, \frac{c}{2} + i\right)$ to active perimeter device $\frac{c}{2} + d + 1 - i$;
- defective device $\left(\frac{r}{2} + 1, \frac{c}{2} + i\right)$ to active perimeter device $\frac{c}{2} + 2d + 1 - i$;
- defective device $\left(\frac{r}{2} + 1, i\right)$ to active perimeter device $3d + i$.

For $1 \leq j \leq \frac{r}{2} - 1$ assign:

- defective device $\left(\frac{r}{2} - j, \frac{c}{2}\right)$ to active perimeter device $2c + \frac{3}{2}r + 3 + j$;
- defective device $\left(\frac{r}{2} - j, \frac{c}{2} + 1\right)$ to active perimeter device $\frac{c}{2} + j$;
- defective device $\left(\frac{r}{2} + 1 + j, \frac{c}{2} + 1\right)$ to active perimeter device $c + \frac{r}{2} + 1 + j$;
- defective device $\left(\frac{r}{2} + 1 + j, \frac{c}{2}\right)$ to active perimeter device $\frac{3}{2}c + r + 2 + j$.

The assignemt can be completed by assigning defective devices from the north-west quadrant to active perimeter devices whose number correspond to an A, north-east with B etc. Note that as these assignment can be freely done, it is possible to balance the number of defective devices assigned to active perimeter devices in the contour of the gap, and thus the assignment witnesses an optimal load.

It is easy to see that the assignment can be computed in constant time and has dilation $d = \frac{r+c}{2} + 1$.

An example of applying this scheme for a gap of 8 rows and 14 columns is depicted in Figure 5.

0	1	2	3	4	5	6	7	8	9	10	11	12	13	14	15
47							46	10							16
46			A				45	9			B				17
45							44	8							18
44	1	2	3	4	5	6	7	19	18	17	16	15	14	13	19
43	37	38	39	40	41	42	43	31	30	29	28	27	26	25	20
42							32	20							21
41			D				33	21			C				22
40							34	22							23
39	38	37	36	35	34	33	32	31	30	29	28	27	26	25	24

Fig. 5. Assignment scheme applied to a rectangular gap of 8 rows and 14 columns

For cases 2 and 3 we give a sketch of the assignment that deals only with the most critical positions. The schemes are easy to complete but tedious to describe in details.

Case 2:
Compute $d = \frac{c+r+1}{2} + 1$.

Suppose w.l.o.g. that r is even and c is odd.

For $1 \leq i \leq r$, assign the defective device (i,c) to active perimeter device $c + 1 + i$.

Assign:

- the defective device $\left(\frac{r}{2}, \lfloor\frac{c}{2}\rfloor\right)$ to active perimeter device $\lfloor\frac{c}{2}\rfloor$;
- the defective device $\left(\frac{r}{2}, \lfloor\frac{c}{2}\rfloor + 1\right)$ to active perimeter device $\lfloor\frac{c}{2}\rfloor + d$;
- the defective device $\left(\frac{r}{2} + 1, \lfloor\frac{c}{2}\rfloor + 1\right)$ to active perimeter device $\lfloor\frac{c}{2}\rfloor + 2d - 1$;
- the defective device $\left(\frac{r}{2} + 1, \lfloor\frac{c}{2}\rfloor\right)$ to active perimeter device $\lfloor\frac{c}{2}\rfloor + 3d - 1$.

To complete the assignment, the scheme for case 1 can be applied with minor changes.

Case 3:
Compute $d = \frac{r+c}{2} + 1$.

For $1 \leq i \leq r$, assign the defective device (i,c) to active perimeter device $c + 1 + i$.

For $j \leq 1 < c$ assign the defective device (r,j) to active perimeter device $2c + r + 3 - j$.

Assign:

- the defective device $\left(\lfloor\frac{r}{2}\rfloor, \lfloor\frac{c}{2}\rfloor\right)$ to active perimeter device $\lfloor\frac{c}{2}\rfloor$;
- the defective device $\left(\lfloor\frac{r}{2}\rfloor, \lfloor\frac{c}{2}\rfloor + 1\right)$ to active perimeter device $\lfloor\frac{c}{2}\rfloor + d$;
- the defective device $\left(\lfloor\frac{r}{2}\rfloor + 1, \lfloor\frac{c}{2}\rfloor + 1\right)$ to active perimeter device $\lfloor\frac{c}{2}\rfloor + 2d$;
- the defective device $\left(\lfloor\frac{r}{2}\rfloor + 1, \lfloor\frac{c}{2}\rfloor\right)$ to active perimeter device $\lfloor\frac{c}{2}\rfloor + 3d$.

Also in this case, the assignment can be completed by applying the scheme for case 1 with minor changes.

As the dilation obtained by the assignments given by these schemes equals the lower bound given in Theorem 3, the thesis follows.   □

6   Conclusions and Open Problems

In this paper we study the problem of bridging gaps in wireless ad-hoc lattice computers by assigning active devices on the perimeter of the gap as proxies to the defective devices in the gap. We establish lower bounds on the communication dilation witnessed by such proxy assignments for single-row gaps and general row-column convex gaps. We present dilation-optimal, constant time algorithms for computing proxy assignments for single-row gaps and gaps that are rectangular in shape. Moreover, we also show that our proxy assignments are load-optimal.

In Section 3, we introduce the notion of *thick* row-column convex gaps. We conjecture that, for any thick row-column convex gap C, the lower bound provided in Theorem 3 is tight, i.e., there exists a dilation-optimal proxy assignment for C with dilation no more than $\lceil p/4 \rceil$. We are currently working on settling this conjecture. Establishing bounds and studying the optimality of proxy assignments with respect to congestion (defined in Section 3) is an interesting open problem.

References

1. Aigner, M., Ziegler, G.M.: Proofs from the book. Springer, Heidelberg (2004)
2. Case, J., Rajan, D.S., Shende, A.M.: Lattice computers for approximating euclidean space. Journal of the ACM 48(1), 110–144 (2001)
3. Chrobak, M., Dürr, C.: Reconstructing hv-convex polyominoes from orthogonal projections. Information Processing Letters 69, 283–289 (1999)
4. Frisch, U., Hasslacher, B., Pomeau, Y.: Lattice-gas automata for the Navier Stokes equation. Physical Review Letters 56(14), 1505–1508 (1986)
5. Greenspan, D.: Deterministic computer physics. International Journal of Theoretical Physics 21(6/7), 505–523 (1982)
6. Gupta, V., Mathur, G.: Wireless Ad-hoc Lattice computers (WAdL) In: 18th Annual Consortium for Computing Sciences in Colleges: Southeastern Conference (2004)
7. Gupta, V., Mathur, G., Shende, A.M.: Lattice formation in a WAdL Wireless Ad-hoc Lattice computer. In: Workshop on Algorithms for Wireless and Mobile Networks (2004)
8. Hillis, W.D.: The connection machine: A computer architecture based on cellular automata. Physica D. 10, 213–228 (1984)
9. Keidar, I., Schuster, A.: Want scalable computing?: speculate! SIGACT News 37(3), 59–66 (2006)
10. Kuba, A.: The reconstruction of two-directional connected binary patterns from their two orthogonal projections. Computer Vision, Graphics, Image Processing 27, 249–265 (1984)

11. Margolus, N.: CAM-8: a computer architecture based on cellular automata. In: Lawniczak, A., Kapral, R. (eds.) Pattern Formation and Lattice-Gas Automata (1993)
12. Moore, P., Shende, A.M.: Gaps in wireless ad-hoc lattice computers. In: International Symposium on Wireless Communication Systems (2005)
13. Shende, A.M.: Digital analog simulation of uniform motion in representations of physical n-space by lattice-work mimd computer architectures, Ph.D. dissertation, SUNY, Buffalo (1991)
14. Yepez, J.: Lattice-gas dynamics, volume I viscous fluids. Technical Report 1200, Phillips Laboratories, Environmental Research Papers (November 1995)

Location Oblivious Distributed
Unit Disk Graph Coloring*

Mathieu Couture, Michel Barbeau, Prosenjit Bose, Paz Carmi,
and Evangelos Kranakis

School of Computer Science, Carleton University, Herzberg Building
1125 Colonel By Drive, Ottawa, Ontario, K1S 5B6 Canada
{mathieu,jit,paz}@cg.scs.carleton.ca
{barbeau,kranakis}@scs.carleton.ca
Fax: 1-613-520-4334

Abstract. We present the first location oblivious distributed unit disk graph coloring algorithm having a provable performance ratio of three (i.e. the number of colors used by the algorithm is at most three times the chromatic number of the graph). This is an improvement over the standard sequential coloring algorithm since we present a new lower bound of 10/3 for the worst-case performance ratio of the sequential coloring algorithm. The previous greatest lower bound on the performance ratio of the sequential coloring algorithm was 5/2. Using simulation, we also compare our algorithm with other existing unit disk graph coloring algorithms.

Keywords: coloring, unit disk graph, approximation algorithms, distributed algorithms, location oblivious algorithms.

1 Introduction

A unit disk graph is a graph that admits a representation where nodes are points in the plane and edges join two points whose distance is at most one unit. In wireless ad hoc networks, communicating nodes are sometimes assumed to have the same communication range. For this reason, unit disk graphs are used to model wireless ad hoc networks. Breu and Kirkpatrick [1] showed that determining if an abstract graph is a unit disk graph is an NP-hard problem, which implies that finding a unit disk graph representation is also NP-hard. This difficulty has led to the development of two varieties of algorithms on unit disk graphs depending on how the graphs are represented. If the unit disk graph representation is given (i.e. vertices are points in the plane and edges join pairs of points whose distance is at most one unit) then this situation is referred to as

---

* The authors graciously acknowledge the support received from the following organizations: Natural Sciences and Engineering Research Council of Canada (NSERC), Mathematics of Information Technology and Complex Systems (MITACS) and High Performance Computing Virtual Laboratory (HPCVL).

G. Prencipe and S. Zaks (Eds.): SIROCCO 2007, LNCS 4474, pp. 222–233, 2007.

location-aware since each node is aware of its geometric location. On the other hand, if one is simply given an abstract graph (i.e. that a valid representation exists), then this situation is referred to as *location oblivious*.

A *coloring* of a graph G is a function c mapping vertices of G to a set of colors (which can be thought of as a set of integers) such that adjacent vertices are assigned different colors. The graph coloring problem is to find a coloring which uses the minimum number of colors. The minimum number of colors needed to color a graph G is called its *chromatic number* and is denoted by $\chi(G)$. The graph coloring problem is NP-complete [6], even for unit disk graphs [5]. The *performance ratio* of a coloring algorithm is defined as the ratio of the number of colors it uses over the chromatic number of the input graph. Approximation algorithms have been proposed to address the unit disk graph coloring problem (see Erlebach and Fiala [4] for a survey), but there exists no coloring algorithm that is 1) distributed, 2) location oblivious, and 3) has a performance ratio of three. In this paper, we introduce the first distributed unit disk graph coloring algorithm that has all these three properties.

A standard approach used in the context of coloring graphs is the *sequential coloring algorithm*. The sequential coloring algorithm is the algorithm that colors the nodes of a graph in an arbitrary order, assigning to each node the lowest color that has not been assigned to one of its neighbors. In the literature, the greatest lower bound on the worst-case performance ratio of the sequential coloring algorithm over unit disk graphs is $5/2$, by Caragiannis *et al.* [2]. Therefore, it was unclear whether one slightly more complex algorithm with a performance ratio of three is better than the trivial sequential algorithm. In this paper, we show that algorithms having a performance ratio of three outperform the sequential coloring algorithm in the worst-case by providing an example where the performance ratio of the sequential coloring algorithm is exactly $10/3$.

The rest of this paper is organized as follows: In Section 2, we review related work on coloring unit disk graphs. In Section 3, we give our coloring algorithm. We prove its termination, correctness and performance properties in Section 4. In Section 5, we give new lower bounds on the worst-case performance ratio of sequential coloring of unit disk graphs. In Section 6, using simulation, we compare the average performance ratio of our algorithm with other algorithms. In Section 7, we discuss some optimization techniques we used to speed-up the simulation. We conclude in Section 8.

2   Related Work

A *sequential* coloring algorithm takes a graph as input, computes some ordering on the nodes, and greedily assigns colors to nodes according to that order. Each node is assigned the lowest color that has not been assigned to any of its neighbors. We denote the maximum degree of a graph G by $\Delta(G)$, and the size of the largest clique in G (the *clique number* of G) by $\omega(G)$. Since the number of colors used by a sequential coloring algorithm cannot exceed $\Delta(G) + 1$, we have that $\chi(G) \leq \Delta(G) + 1$. On the other hand, since no two nodes in a clique can

Table 1. Summary of Unit Disk Graph Coloring Algorithms Properties

	distributed	location oblivious	worst-case perf. ratio
sequential	yes	yes	5
lexicographic	yes	no	3
smallest-last	no	yes	3
our algorithm	yes	yes	3

have the same color, we have that $\chi(G) \geq \omega(G)$. For unit disk graphs, Marathe et al. [7] pointed out the following relation: $\Delta(G) \leq 6\omega(G) - 6$ (see Figure 1). This implies that all sequential unit disk graph coloring algorithms have a performance ratio of at most six. In fact, a minor adjustment of that proof shows that the performance ratio is no greater than five [4].

What distinguishes sequential coloring algorithms from each other is the order in which they color the nodes. When an arbitrary order is used, we will simply refer to it as *the* sequential coloring algorithm. For graphs embedded in the plane, the *lexicographic* ordering is the one induced by the (x, y) coordinates of the nodes (nodes with smaller x-coordinate are colored first, with ties broken according to the y-coordinate). For the case of unit disk graphs, Peeters [10] showed that the lexicographical ordering achieves a performance ratio of three. Note that this approach can be easily implemented in a distributed manner provided the nodes are aware of their location. The *smallest-last* coloring algorithm [9] computes the following ordering over the nodes of a graph G: a node v of minimum degree is colored last (ties are broken arbitrarily). The rest of the ordering is computed recursively on the graph $G \setminus \{v\}$. Gräf et al. [5] showed that the smallest-last coloring algorithm achieves a performance ratio of at most three over unit disk graphs. However, this algorithm is not distributed. Table 1 summarizes unit disk graph coloring algorithms properties. As one can see, there seems to be a trade-off between being distributed, location oblivious, and having a worst-case performance ratio of three. We show that in fact, no such trade-off exists.

3   Location Oblivious Distributed Algorithm

Lexicographic coloring achieves a performance ratio of three because for every node u, no more than $3\omega(G) - 3$ neighbors of u will choose their color before u (see Figure 1). The key of our algorithm is to show how to compute an ordering

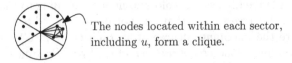

The nodes located within each sector, including u, form a clique.

Fig. 1. The neighborhood of a node does not contain more than $6\omega(G) - 6$ nodes

that has this property in a distributed manner when the nodes do not know their position in the plane (i.e. in a location oblivious manner). The main observation is the following: in every unit disk graph G, there is at least one node that has at most $3\omega(G) - 3$ neighbors.

We denote by $\omega(u)$ the size of a largest clique in which node u belongs. If the neighborhood of a node u has size at most $3\omega(u) - 3$, we say that it has the *small neighborhood* property. Lexicographic coloring exploits the fact that the leftmost node has this property. In fact, all nodes on the convex hull of the nodes also have this property. Since the size of a maximum clique in a unit disk graph can be computed in polynomial time, even without the unit disk representation [11], each node can locally determine whether or not it has the small neighborhood property. Notice that since $\omega(u) \leq \omega(G)$ for every node u, if a node has the small neighborhood property, then it also has at most $3\omega(G) - 3$ neighbors.

The intuition behind our algorithm is the following: in order to reach a performance ratio of three, nodes having the small neighborhood property can pick their colors *after* their neighbors. We then remove all these nodes from the graph, recursively color the remaining subgraph, put the removed nodes back in, and then sequentially color them. Recursion is guaranteed to make progress because there are always nodes having the small neighborhood property. What remains to be shown is how this can be done in a distributed manner.

The distributed algorithm works in two phases. In the first phase, the nodes establish a local order by each selecting a rank. The ranks, together with the identifier, determine the local order in which they will decide their color. The second phase is the actual coloring. Each phase is event-driven, i.e. the nodes do not need synchronous clocks.

Algorithm 1. RankingPhase(id, G_{id})

Input: id, a node identifier
  N, a list containing the identifiers of the neighbors of node id
  G_{id}, the subgraph of G induced by N
Output: $ranks$, a table containing the neighbors ranks
1: $max\_clique \leftarrow \omega(G_{id})$
2: **while** $\{u \in N : ranks[u] = 0\} \neq \emptyset$ **do**
3:   **if** $ranks[id] = 0$ and $|\{u \in N : ranks[u] = 0\}| \leq 3 * max\_clique - 3$ **then**
4:     $ranks[id] \leftarrow max\{u \in N : ranks[u]\} + 1$
5:     **send** RANK($id, ranks[id]$)
6:   **else**
7:     **receive** RANK(u, r)
8:     $ranks[u] \leftarrow r$
9:   **end if**
10: **end while**

The underlying idea of the ranking algorithm is the following: we want to make sure that for every node u of a unit disk graph G, no more than $3\omega(u) - 3 \leq 3\omega(G) - 3$ nodes pick their color before u. In order to ensure this, each node u collects the connectivity information of its distance one neighborhood and

---

Algorithm 2. ColoringPhase($id, N, ranks$)

Input: id, a node identifier
  N, a list containing the identifiers of the neighbors of node id
  $ranks$, a table containing the neighbors' ranks
Output: $colors$, a table containing the node's colors (initial values are all 0)
1: **while** $\{u \in N : colors[u] = 0\} \neq \emptyset$ **do**
2:   **if** $colors[id] = 0$ and
      $\nexists u \in N : colors[u] = 0$ and $\langle ranks[id], id \rangle < \langle ranks[u], u \rangle$ **then**
3:     $colors[id] \leftarrow min\{i > 0 : \{u \in N : colors[u] = i\} = \emptyset\}$
4:     **send** COLOR($id, colors[id]$)
5:   **else**
6:     **receive** COLOR(u, c)
7:     $colors[u] \leftarrow c$
8:   **end if**
9: **end while**

---

computes $\omega(u)$. A node u having a total number of neighbors less than or equal to $3\omega(u) - 3$ (i.e. having the small neighborhood property) selects rank one and informs its neighbors of its decision. A node u having more than $3\omega(u) - 3$ neighbors must wait. Ranking information from neighbors is recorded in a table. When the number of neighbors of a node u with undetermined rank becomes less than or equal to $3\omega(u) - 3$, node u takes a rank that is one more than the maximum rank among its neighbors. Node u then informs its neighbors about its decision. A node u terminates the ranking phase when all its neighbors have chosen their rank. Algorithm 1. gives the details of the ranking phase.

When all neighbors have chosen their ranks, a node may start the coloring phase. Note that two neighbors may have chosen the same rank. Locally, nodes then choose their color according to the order induced by the pair $\langle rank, id \rangle$. Nodes with higher rank pick their color first, and ties are broken according to their identifier. Algorithm 2. gives the details of the coloring phase.

4   Theoretical Properties

Proposition 1. *After Algorithm 1. terminates, all nodes have selected a rank.*

Proof: Suppose after Algorithm 1. terminates, there is a set of nodes S of G which have not yet chosen their rank. This means that every node $u \in S$ has more than $3\omega(u) - 3$ neighbors which have not yet chosen their rank (i.e. which are in S). In particular, this is true for a node v which is on the convex hull of S. Also, since v is on the convex hull of S, all of its neighbors which are in S are located on a half-plane whose boundary passes through v. As for lexicographic coloring, v cannot have more than $3\omega(v) - 3$ neighbors in S, which is a contradiction. Therefore, when no more messages are being sent, all nodes have chosen their ranks. □

Proposition 2. *Algorithm 2. produces a valid coloring.*

Proof: First of all, Algorithm 2. terminates. The reason for this is that, among the nodes which have not yet chosen their color, there is always a node which is a global maximum according to the ordered pair $\langle rank, id \rangle$. In particular, this node is a local maximum which will pick its color. Also, no two neighbors can pick their color at the same time. This is because the ordered pair $\langle rank, id \rangle$ induces a total order on the nodes. Therefore, of two neighbor nodes which have not picked their color, at most one of them can satisfy the condition on line 2. Finally, no two neighbors can pick the same color. This is because the second one will only pick a color which is still available (line 4). □

Lemma 1. *For a node u of a unit disk graph G, let $h(u)$ denote the number of neighbors of u with higher rank than the rank of u. Then $|h(u)| \leq 3\omega(u) - 3$.*

Proof: In Algorithm 1., a node u will choose its rank only when fewer than $3\omega(u) - 3$ of its neighbors have undetermined rank (line 3). Also, when u chooses its rank, it chooses it such that it is greater than all ranks that have been chosen in its neighborhood. Therefore, only less than $3\omega(u) - 3$ nodes could potentially choose rank greater than the one chosen by u. □

Proposition 3. *Using the order computed by Algorithm 1., the color chosen by a node u in Algorithm 2. is less than or equal to $3\omega(u) - 2$.*

Proof: In the neighborhood of u, only nodes with rank greater than u can choose their color before u. By Lemma 1, there are no more than $3\omega(u) - 3$ such nodes. Therefore, the color chosen by u is no greater than $3\omega(u) - 2$. □

Theorem 1. *Using the order computed by Algorithm 1., the number of colors used by Algorithm 2. to color a unit disk graph G is no greater than three times the optimal. During the execution of these algorithms, each node sends exactly two messages.*

Proof: By Proposition 3, all nodes u are assigned color at most $3\omega(u) - 2 \leq 3\omega(G) - 2$. The performance ratio follows from the fact that at least $\omega(G)$ colors are needed to color G. In Algorithm 1., each node sends exactly one RANK message. In Algorithm 2. each node sends exactly one COLOR message. Therefore, each node sends exactly two messages. □

5   Lower Bounds

We now give new lower bounds on the worst-case performance ratio of the sequential coloring algorithm for unit disk graphs. The currently greatest lower bound is $5/2$, given by Caragiannis *et al.* [2]. To prove a lower bound of b, we have to show that there exists a unit disk graph G for which there exists an ordering $<$ of the nodes such that the number of colors used by the sequential coloring algorithm is at least $b \cdot \chi(G)$. The construction of such a unit disk graph proceeds as follows: first, decide what the chromatic number of the graph will be. Then, at least one node must pick color $b \cdot \chi(G)$. In order to ensure this, it must

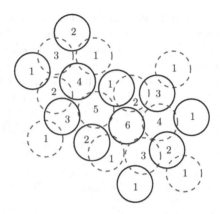

Fig. 2. Lower Bound of 3

have at least $b \cdot \chi(G) - 1$ neighbors, picking all colors ranging from 1 to $b \cdot \chi(G) - 1$. The construction of the graph then continues recursively in order to force these nodes to pick these colors. To force the sequential algorithm to use many colors, one needs to construct a graph with vertices of high degree. The difficulty lies in increasing the degree of vertices without increasing the chromatic number of the graph.

We first prove a lower bound of three. A case where a performance ratio of three can be reached is shown in Figure 2.[1] This graph is bipartite (no two dashed nodes touch each other, and no two solid nodes touch each other). Hence, it can be colored with two colors. However, it is also possible to order the nodes in such a way that the sequential coloring algorithm uses six colors: simply order them in non-decreasing order of labels. Using this ordering, the sequential coloring algorithm will then use six colors whereas only two colors are necessary, which means that in that case, it achieves a performance ratio of three. Thus, since the graph shown in Figure 2 is triangle-free, we have proved the following:

Proposition 4. *For triangle-free unit disk graphs, the worst-case performance ratio of the sequential coloring algorithm is at least three.*

The reason why it is interesting to restrict the preceding proposition to triangle-free unit disk graphs is that the bound is tight for that class of graphs, as shown below.

Proposition 5. *For triangle-free unit disk graphs, the worst-case performance ratio of the sequential coloring is at most three.*

Proof: Suppose that there exists a node u of a unit disk graph G such that the color attributed to u by the sequential coloring algorithm is 7. This means that u has degree at least 6. Since no node of a unit disk graph can have more than

---

[1] The exact positions of the points generating the unit disk graphs of Figure 2 and Figure 3 can be found in the technical report version of this paper [3].

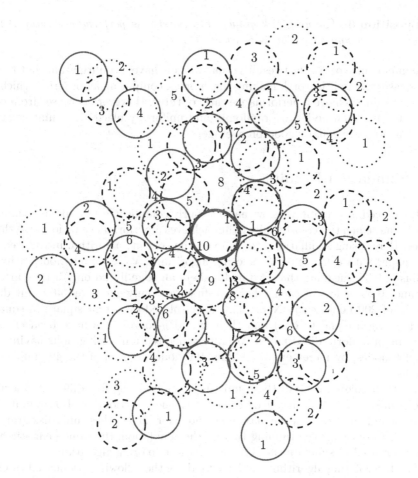

Fig. 3. Lower Bound of 10/3

five independent neighbors [7], at least two of these six neighbors, say v and w, are neighbors of each other. Therefore, v, w and u form a triangle, which means that G is not triangle-free. □

For graphs that are not necessarily triangle-free, we show a worst-case lower bound of 10/3. The construction is depicted in Figure 3. As one can see, this graph can be colored using only three colors (solid, dashed and dotted nodes form a 3-partition of the graph). However, there exists an ordering of the nodes such that sequentially coloring the graph in that order uses ten colors, leading to a performance ratio of 10/3. In order to force the sequential coloring algorithm to use ten colors, the solid bold node which has degree nine is colored last. Its nine neighbors are forced to take all colors ranging from one to nine thereby forcing color ten on the solid bold node. The coloring of its neighbors is forced in a similar fashion. The existence of this graph allows us to conclude the following:

Proposition 6. *For unit disk graphs, the worst-case performance ratio of the sequential coloring algorithm is at least 10/3.*

The importance of this last result is that we now have a confirmation that there exist cases where it is worth making the effort of computing an ordering which is guaranteed to achieve a performance ratio of three. However, as we see from our simulations, when nodes are randomly and uniformly placed in a unit square, all strategies are equally good on average.

6   Simulation Results

In the preceding section, we saw that it is fairly complicated to build an example where the sequential coloring algorithm achieves a performance ratio worse than three. Here, using simulation, we compare the coloring algorithm introduced in this paper with other existing coloring algorithms. We gave our nodes a fixed radius of 0.05, meaning that two nodes share an edge if and only if they are at distance 0.05 or less of each other. We first randomly generated 400 unit disk graphs of 200 nodes each. Nodes have been placed on a unit square (a square of side length equal to one) and their x and y-coordinates have been chosen following a uniform distribution. We also generated unit disk graphs having up to 2000 nodes, by incrementally adding 100 nodes to each of the 400 unit disk graphs.

We then colored each of these unit disk graphs using five different coloring algorithms. Using the heuristic described in the next section, we also computed a lower bound on the size of the maximum clique for each of these unit disk graphs. In order to optimize the running time of the simulation, the same heuristic has also been used to simulate the three-cliques-last coloring algorithm.

The five coloring algorithms we have used are the following: sequential (nodes are colored in the order induced by their identifier), three-cliques-last (the algorithm introduced in this paper), lexicographic (nodes are colored from left to right), smallest-last (nodes of small degree are colored last) and largest-first (nodes of large degree are colored first).

The difference between smallest-last and largest-first is the following: in smallest-last, a node u with minimum degree in a graph G is colored last, and the order in which the other nodes are colored is computed recursively on the graph $G \setminus \{u\}$. In largest-first, a node u with maximum degree is colored first, and the order in which the other nodes are colored is computed recursively on the same graph.

Figure 4 shows the simulation results we obtained. It displays the average number of colors used by each algorithm as a function of the number of nodes in the graph. It also plots the average estimated value of the size of a maximum clique. As explained in the next section, this estimated value is a lower bound on the actual size of a maximum clique. Therefore, it is also a lower bound on the chromatic number. The exact values of our simulation results can be found in the technical report version of this paper [3].

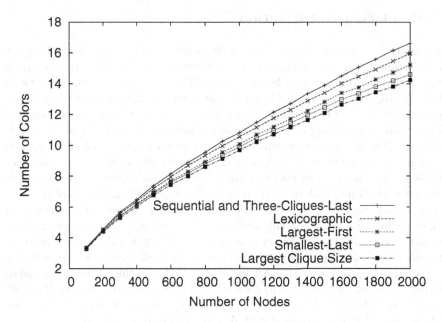

Fig. 4. Simulation Results

The first observation that can be made by looking at the simulation results is that the algorithm we proposed in this paper (three-cliques-last) provides almost no significant improvement over sequential coloring. In fact, the difference between values obtained for the two algorithms is less than the width of the 95% confidence interval. However, this does not really mean that our algorithm performs badly. What it really means is that sequential coloring performs better than expected.

Also, it is not surprising to see that the algorithm which performed the best is the smallest-last coloring. The *span* of an ordering is defined as the maximum, over all nodes u, of the number of neighbors of u that have smaller index than u. Matula [8] showed that smallest-last ordering attains minimum span. Since the span of an ordering provides an upper bound on the number of colors that will be used, smallest-last coloring can be expected to provide good results.

What is really interesting is to see is that largest-first coloring provided better results than both three-cliques-last and lexicographic. There is no known proof that largest-first has a performance ratio better than five, and still it performs better than algorithms which have an upper bound of three on the performance ratio. Since largest-first is distributed, location oblivious and simpler to implement than three-cliques-last, looking at the simulation results allows us to conclude that it is preferable to use largest-first even though there is no proof that it performs better.

7   Simulation Optimization

Since computing the maximum clique in the neighborhood of a node can be quite
time consuming for simulation purposes, we used some heuristics to compute a
lower bound on the size of a largest clique. The main idea of our heuristic is
the following: the size of the largest clique is the maximum number of nodes
contained in a subset of the plane whose diameter is at most one. Since the geometric shape maximizing an area of fixed diameter is the circle, it is reasonable
to expect that the maximum number of nodes contained in a disk of radius one
is a good approximation of the size of a maximum clique. Since it is sufficient
to look at disks having two nodes on their boundaries, the maximum number of
points contained in a disk of radius one can be computed in time $O(n^3)$.

For a node u, let $C(u)$ be the maximum number of nodes contained in a
disk of radius one which also contains u, $\omega(u)$ be the size of a maximum clique
containing u, and $N(u)$ be the set containing u and its neighbors. The heuristic
we used is the following: if $|N(u)| \leq 3C(u) - 3$, then use $C(u)$ as an estimate for
$\omega(u)$. Otherwise, compute the exact value of $\omega(u)$. Using this estimate instead of
computing the exact value of $\omega(u)$ does not affect the simulation results. If a node
u is such that $|N(u)| \leq 3C(u) - 3$, then it is also the case that $|N(u)| \leq 3\omega(u) - 3$
and therefore it will be assigned rank one in Algorithm 1. anyway.

Table 2. Percentages of nodes u such that $|N(u)| \leq 3C(u) - 3$

nodes	%	nodes	%	nodes	%	nodes	%
100	0.9999	600	0.9977	1100	0.9968	1600	0.9957
200	0.9996	700	0.9976	1200	0.9966	1700	0.9953
300	0.9991	800	0.9975	1300	0.9964	1800	0.9950
400	0.9987	900	0.9971	1400	0.9962	1900	0.9946
500	0.9982	1000	0.9971	1500	0.9960	2000	0.9943

Table 2 shows the proportion of nodes u which were such that $|N(u)| \leq
3C(u) - 3$. The 95% confidence interval for these values is at most ± 0.0002.
The first observation to be made is that the heuristic allowed us to accelerate
the simulation in more than 99% of the cases. This means that the heuristic
was worth using it. The second observation to be made is that the percentages
diminish as the graph becomes denser. This makes sense, because the area of a
disk of diameter one is only 1/4 the area of the unit disk around a node.

The most important observation to be made is that all nodes such that
$|N(u)| \leq 3C(u) - 3$ are assigned rank one in Algorithm 1.. Therefore, Table 2
also gives a lower bound on the proportion of nodes which are assigned rank one.
Since this proportion is always higher than 99%, the order used by Algorithm 2.
in the second phase is almost the same as the one used by the sequential algorithm, and this gives an intuition of why the simulation results are so similar for
these two algorithms.

8 Conclusion

We presented the first distributed location oblivious coloring algorithm which achieves a performance ratio of three on unit disk graphs. However, simulation results showed that this algorithm does not provide a significant improvement over the algorithm which sequentially colors the nodes in an arbitrary order. Simulation results also showed that, in the average case, largest-first (which is also distributed and location oblivious) performs better than the algorithm we proposed. It also performs better than lexicographic coloring, which also has a worst-case performance ratio of at most three. However, no one has shown whether largest-first has a better worst-case performance ratio than five. In fact, it is also an open question whether coloring the nodes of a unit disk graph in an arbitrary order can, on the worst case, use less than five or more than 10/3 times the minimum number of colors that are necessary.

References

[1] Breu, H., Kirkpatrick, D.G.: Unit disk graph recognition is np-hard. Comput. Geom. Theory Appl. 9(1-2), 3–24 (1998)

[2] Caragiannis, I., Fishkin, A.V., Kaklamanis, C., Papaioannou, E.: bound for online coloring of disk graphs. In: Pelc, A., Raynal, M. (eds.) SIROCCO 2005. LNCS, vol. 3499, pp. 78–88. Springer, Heidelberg (2005)

[3] Couture, M., Barbeau, M., Bose, P., Carmi, P., Kranakis, E.: Location oblivious distributed unit disk graph coloring. Tech. Rep. TR-07-01, School of Computer Science, Carleton University (2007)

[4] Erlebach, T., Fiala, J.: Independence and coloring problems on intersection graphs of disks. In: Bampis, E., Jansen, K., Kenyon, C. (eds.) Efficient Approximation and Online Algorithms. LNCS, vol. 3484, Springer, Heidelberg (2006)

[5] Gräf, A., Stumpf, M., Weißenfels, G.: On coloring unit disk graphs. Algorithmica 20(3), 277–293 (1998)

[6] Karp, R.M.: Reducibility among combinatorial problems. In: Miller, R.E., Thatcher, J.W. (eds.) Complexity of Computer Computations, pp. 85–103. Plenum Press, New York (1972)

[7] Marathe, M., Breu, H., Ravi, S., Rosenkrantz, D.: Simple heuristics for unit disk graphs. Networks 25, 59–68 (1995)

[8] Matula, D.: k-components, clusters, and slicings in graphs. SIAM Journal of Applied Mathematics 22(3), 459–480 (1972)

[9] Matula, D.W., Beck, L.L.: Smallest-last ordering and clustering and graph coloring algorithms. J. ACM 30(3), 417–427 (1983)

[10] Peeters, R.: On coloring j-unit sphere graphs. Tech. Rep. FEW 512, Department of Economics, Tilburg University, Tilburg, The Netherlands (1991)

[11] Raghavan, V., Spinrad, J.: Robust algorithms for restricted domains. J. Algorithms 48(1), 160–172 (2003)

Edge Fault-Diameter of Cartesian Product of Graphs

Iztok Banič[1,2] and Janez Žerovnik[1,2]

[1] University of Maribor,
Smetanova 17, Maribor 2000, Slovenia
[2] Institute of Mathematics, Physics and Mechanics,
Jadranska 19, Ljubljana
iztok.banic@uni-mb.si,janez.zerovnik@imfm.uni-lj.si

Abstract. Let G be a k_G-edge connected graph and $\overline{\mathcal{D}}_c(G)$ denote the diameter of G after deleting any of its $c < k_G$ edges. We prove that if G_1, G_2, ..., G_q are k_1-edge connected, k_2-edge connected,..., k_q-edge connected graphs and $0 \leq a_1 < k_1$, $0 \leq a_2 < k_2$,..., $0 \leq a_q < k_q$ and $a = a_1 + a_2 + \ldots + a_q + (q-1)$, then the edge fault-diameter of G, the Cartesian product of G_1, G_2, ..., G_q, with a faulty edges is $\overline{\mathcal{D}}_a(G) \leq \overline{\mathcal{D}}_{a_1}(G_1) + \overline{\mathcal{D}}_{a_2}(G_2) + \ldots + \overline{\mathcal{D}}_{a_q}(G_q) + 1$.

Keywords: Cartesian graph products, edge fault-diameter, interconnection network.

1 Introduction

In the design of large interconnection networks several factors have to be taken into account. A usual constraint is that each processor can be connected to a limited number of other processors and the delays in communication must not be too long. Extensively studied network topologies in this context include graph products and bundles. For example the meshes, tori, hypercubes and some of their generalizations are Cartesian products. It is less known that some well-known topologies are Cartesian graph bundles, i.e. some twisted hypercubes [5,8] and multiplicative circulant graphs [16]. Other graph products, sometimes under different names, have been studied as interesting communication network topologies [16,15,4].

Furthermore, an interconnection network should be fault-tolerant. Since nodes and edges of a network do not always work, if some nodes or edges are faulty, some information may not be transmitted through these nodes and by these edges. Usually, it is assumed that either only nodes or only edges are faulty and hence either node fault-diameter (or, simply, fault diameter) or edge fault-diameter is studied. The fault diameter has been determined for many important networks recently [7,6,14,18]. The concept of fault diameter of Cartesian product graphs was first described in [13], but the upper bound was wrong, as shown by Xu, Xu and Hou who corrected the mistake [18]. An upper bound for the fault diameter of Cartesian graph bundles was given in [1] and for arbitrary

G. Prencipe and S. Zaks (Eds.): SIROCCO 2007, LNCS 4474, pp. 234–245, 2007.

Cartesian products in [2]. When a preliminary version of [1,2] was presented at the conference [3], a question was asked whether similar results can be proved for the edge fault-diameter.

In this paper we study the diameter of a graph after deleting some of the edges. As a k-edge connected graph remains connected if up to $k - 1$ edges are missing, we study the diameter of a graph with any permitted number of edges deleted. In this paper, we prove the three theorems listed below.

Theorem 1. *Let* G_1, G_2, $\ldots$, G_q *be* k_1-*edge connected,* k_2-*edge connected,$\ldots$,* k_q-*edge connected graphs. Let* $0 \leq a_1 < k_1$, $0 \leq a_2 < k_2, \ldots$, $0 \leq a_q < k_q$ *and* $a = a_1 + a_2 + \ldots + a_q + (q-1)$. *Then the edge fault-diameter of a graph* G, *the Cartesian product of* G_1, G_2, $\ldots$, G_q, *with* a *faulty edges is*

$$\overline{\mathcal{D}}_a(G) \leq \overline{\mathcal{D}}_{a_1}(G_1) + \overline{\mathcal{D}}_{a_2}(G_2) + \ldots + \overline{\mathcal{D}}_{a_q}(G_q) + 1.$$

In fact, Theorem 1 implies a more precise result for an upper bound of the edge fault-diameter of a Cartesian product of graphs:

Theorem 2. *Let* G_1, G_2, $\ldots$, G_q *be* k_1-*edge connected,* k_2-*edge connected,$\ldots$,* k_q-*edge connected graphs, and* G *the Cartesian product of* G_1, G_2, $\ldots$, G_q. *Let* $0 \leq a < k_1 + k_2 + \ldots + k_q$. *Then*

$$\overline{\mathcal{D}}_a(G) \leq \min\{\overline{\mathcal{D}}_{a_1}(G_1) + \overline{\mathcal{D}}_{a_2}(G_2) + \ldots + \overline{\mathcal{D}}_{a_q}(G_q) + 1 \mid$$

$$a_1 + a_2 + \ldots + a_q = a - (q-1), 0 \leq a_1 < k_1, 0 \leq a_2 < k_2, \ldots, 0 \leq a_q < k_q\}.$$

Furthermore, for cases with a small number of faulty edges we prove the exact formula for computing the edge fault-diameter:

Theorem 3. *Let* $G_1, G_2, \ldots, G_q$ *be connected graphs, and* $G = \square_{i=1}^{q} G_i$. *Then*

1. $\overline{\mathcal{D}}_a(G) = \sum_{i=1}^{q} \overline{\mathcal{D}}_0(G_i) = \overline{\mathcal{D}}_0(G)$ *for* $0 \leq a < q - 1$;
2. $\overline{\mathcal{D}}_0(G) \leq \overline{\mathcal{D}}_{q-1}(G) \leq \overline{\mathcal{D}}_0(G) + 1$.

Note that $\overline{\mathcal{D}}_0(G)$ is just the diameter of G. In fact we prove more than stated in Theorem 3, namely: if all the factors are trees and there is a factor that is a complete graph (i.e. K_2), then $\overline{\mathcal{D}}_{q-1}(G) = \overline{\mathcal{D}}_0(G) + 1$; and if none of the factors is a complete graph, then $\overline{\mathcal{D}}_{q-1}(G) = \overline{\mathcal{D}}_0(G)$.

While the results proven here are very similar to the results for the node fault version [2] and the methods used are similar, we were not able to find a faster proof, for example a theorem which would "translate" the results from node fault version to the edge fault version of the theorems. It may be an interesting task to look for such a tranformation or to find reasons why this seemingly is difficult. On the positive side, we believe that the same method can be used to prove the edge fault version of the result [1] for graph bundles. Another interesting question is whether the approach can be extended to prove analogous bounds for the mixed problem, in which both nodes and edges may be faulty.

2   Preliminaries

A *simple graph* $G = (V, E)$ is determined by a *vertex set* $V = V(G)$ and a set $E = E(G)$ of (unordered) pairs of vertices, called the set of *edges*. As usual, we will use the short notation uv for edge $\{u, v\}$. Two graphs G_1 and G_2 are *isomorphic*, $G_1 \simeq G_2$, if there is a bijection between the vertex sets that preserves adjacency.

Let G_1 and G_2 be graphs. The *Cartesian product* of graphs G_1 and G_2, $G = G_1 \square G_2$, is defined on the vertex set $V(G_1) \times V(G_2)$. Vertices (u_1, v_1) and (u_2, v_2) are adjacent if either $u_1 u_2 \in E(G_1)$ and $v_1 = v_2$ or $v_1 v_2 \in E(G_2)$ and $u_1 = u_2$. For $u \in V(G_1)$ we define the *layers* $G_2(u) = \{(u, x) \mid x \in V(G_2)\}$ and for $v \in V(G_2)$ the *layers* $G_1(v) = \{(x, v) \mid x \in V(G_1)\}$. The layers are clearly isomorphic to factors, $G_1(u) \simeq G_1$ and $G_2(v) \simeq G_2$. For further reading on graph products we recommend [9].

A *walk* between x and y is a sequence of vertices and edges $v_0, e_1, v_1, e_2, v_2, \ldots, v_{k-1}, e_k, v_k$ where $x = v_0$, $y = v_k$, and $e_i = v_{i-1} v_i$ for each i. A walk with all vertices distinct is called a *path*. The *length* of a path P, denoted by $\ell(P)$, is the number of edges in P. The *distance* between vertices x and y is the length of a shortest path between x and y in G. The *diameter* of a graph G, $\mathbf{d}(G)$, is the maximum distance between any two vertices in G. A path P in G, defined by a sequence $x = v_0, e_1, v_1, e_2, v_2, \ldots, v_{k-1}, e_k, v_k = y$ can alternatively be seen as a subgraph of G with $V(P) = \{v_1, v_2, \ldots, v_k\}$ and $E(P) = \{e_1, e_2, \ldots, e_k\}$.

Let G be a graph, $x, y \in V(G)$ distinct vertices, P a path from x to y in G, and $z \in V(P) \setminus \{x, y\}$. We will use $x \xrightarrow{P} z$ to denote the subpath $\tilde{P} \subseteq P$ from x to z. If z is adjacent to x in P, we will simply use $x \to z$.

Let $G = G_1 \square G_2$, P a path in G_2, and v a vertex of G_1. For simplicity of notation, we will also use P to denote the path $\{v\} \square P$ in the layer $G_2(v)$.

Let G be a graph and $X \subseteq V(G)$. A path P from a vertex x to a vertex y *avoids* X in G, if $V(P) \cap X = \emptyset$, and it *internally avoids* X, if $(V(P) \setminus \{x, y\}) \cap X = \emptyset$.

The *edge connectivity* of a graph G, $\lambda(G)$, is the minimum cardinality over all edge-separating sets in G. A graph G is said to be *k-edge connected*, if $\lambda(G) \geq k$. An *x, y-edge cut* is an edge-separating set that separates x and y, i.e. where the nodes x and y are in different connected components.

To state that G is 1-edge connected we will sometimes simply say it is edge connected, or, even shorter, connected.

One of the Menger's theorems (see, for example, [17], page 167) reads:

Theorem 4. (Menger) *If x and y are vertices of a graph G and $(x, y) \notin E(G)$, then the minimum size of an x, y-edge cut equals the maximum number of pairwise edge disjoint x, y-paths.*

The following well-known corollary easily follows

Corollary 1. *Let G be a k-edge connected graph and δ_G be its minimum degree. Then $\delta_G \geq k$.*

For Cartesian product graphs, there is a well-known bound for the edge connectivity of the product (see for example [7]).

Corollary 2. *Let G_1 and G_2 be k_1 and k_2-edge connected graphs, respectively. Then $G_1 \square G_2$ is at least $(k_1 + k_2)$-edge connected.*

Let G be a graph and $x \in V(G)$ a vertex. The *neighborhood* of the vertex x in the graph G, $N_G(x)$, is the set of all vertices in G that are adjacent to x.

Let G be a k-edge connected graph and $0 \leq a < k$. Then we define the *a-edge fault-diameter* of G as

$$\overline{\mathcal{D}}_a(G) = \max\{d(G \setminus X) \mid X \subseteq E(G), |X| = a\}.$$

Note that $\overline{\mathcal{D}}_a(G)$ is the largest diameter among subgraphs of G with a edges deleted, hence $\overline{\mathcal{D}}_0(G)$ is just the diameter of G. For $a \geq k$, the edge fault-diameter of k-edge connected graph does not exist. In this case we write $\overline{\mathcal{D}}_a(G) = \infty$ as some of the graphs are not edge connected.

3 Product of q Factors

Before proving Theorem 1, let us prove Lemma 1, which will be useful in the proof of Theorem 1.

Lemma 1. *Let $G_1, G_2, \ldots, G_q$ be 1-edge connected graphs, and $G = \square_{i=1}^q G_i$. Then*

1. $\overline{\mathcal{D}}_a(G) = \sum_{i=1}^q \overline{\mathcal{D}}_0(G_i) = \overline{\mathcal{D}}_0(G)$ *for $0 \leq a < q - 1$;*
2. $\overline{\mathcal{D}}_0(G) \leq \overline{\mathcal{D}}_{q-1}(G) \leq \overline{\mathcal{D}}_0(G) + 1$;
3. *if none of the factors G_i is a complete graph, then $\overline{\mathcal{D}}_{q-1}(G) = \overline{\mathcal{D}}_0(G)$;*
4. *if all of the factors G_i are trees and at least one of G_i is a K_2, then $\overline{\mathcal{D}}_{q-1}(G) = \overline{\mathcal{D}}_0(G) + 1$.*

Proof. Let x and y be two distinct vertices of G. For each $i = 1, 2, 3, \ldots, q$, let $p_i : G \to G_i$ be the projection on G_i.

Case 1. $p_i(x) \neq p_i(y)$ for all i. Then there are at least q edge-disjoint paths P of length $\ell(P) \leq \sum_{i=1}^q \overline{\mathcal{D}}_0(G_i) = \overline{\mathcal{D}}_0(G)$. As $a < q$, at least one of them avoids faulty edges. Therefore there is a path P in G such that P avoids faulty edges, and $\ell(P) \leq \overline{\mathcal{D}}_0(G)$. Therefore $\overline{\mathcal{D}}_a(G) \leq \overline{\mathcal{D}}_0(G)$.

Case 2. $p_i(x) = p_i(y)$ for at least two indices i. Without loss of generality, assume that $p_i(x) \neq p_i(y)$ for $i=1,2,\ldots,k$ and $p_i(x) = p_i(y)$ for $i=k+1,\ldots, q$. There are at least k edge disjoint shortest paths between x and y within the first k factors. The length of these paths is at most $\ell = \sum_{i=1}^k \overline{\mathcal{D}}_0(G_i)$. We can construct additional $q - k$ edge disjoint paths from x to y of length $\ell + 2$ as follows. Take any of the shortest paths P, choose a neighbor in the i-th factor (for $i = q, q-1, \ldots, k+1$) and construct a new path.

$$x \to u \xrightarrow{P} v \to y$$

More precisely, for $i = q$, take $u = (x_1, \ldots, x_k, x_{k+1}, x_{k+2}, \ldots, x_{q-1}, u_q)$, a neighbor of x. Then $v = (y_1, \ldots, y_k, x_{k+1}, x_{k+2}, \ldots, x_{q-1}, u_q)$ is a neighbor of y and there is a path of length $1+\ell+1$ from x to y. Clearly $\ell+2 \leq \sum_{i=1}^{k} \overline{\mathcal{D}}_0(G_i)+q-k \leq \sum_{i=1}^{q} \overline{\mathcal{D}}_0(G_i)$.

Case 3. $p_i(x) = p_i(y)$ for exactly one i. Say $p_q(x) = p_q(y)$. Then there are at least $q-1$ edge disjoint paths P from x to y in the layer $L(x) = p_q^{-1}(p_q(x))$ with length $\ell(P) \leq \sum_{i=1}^{q-1} \overline{\mathcal{D}}_0(G_i) < \sum_{i=1}^{q} \overline{\mathcal{D}}_0(G_i)$. If $a < q-1$, then there is a path P in G that avoids faulty edges and $\ell(P) \leq \sum_{i=1}^{q} \overline{\mathcal{D}}_0(G_i) = \overline{\mathcal{D}}_0(G)$. Therefore $\overline{\mathcal{D}}_a(G) \leq \overline{\mathcal{D}}_0(G)$. If $a = q-1$, then either one of the paths has no faulty edges or all the faulty edges appear in $L(x)$. In the worst case (if all the faulty edges appear in $L(x)$) there is a path P from x to y with length $\ell(P) \leq 1 + \sum_{i=1}^{q-1} \overline{\mathcal{D}}_0(G_i) + 1 \leq \overline{\mathcal{D}}_0(G) + 1$. Therefore $\overline{\mathcal{D}}_a(G) \leq \overline{\mathcal{D}}_0(G) + 1$.

Summarizing, we have $\overline{\mathcal{D}}_a(G) \leq \overline{\mathcal{D}}_0(G)$ for $a < q-1$. As $\overline{\mathcal{D}}_a(G) \geq \overline{\mathcal{D}}_0(G)$ for each a, hence $\overline{\mathcal{D}}_a(G) = \overline{\mathcal{D}}_0(G) = \sum_{i=1}^{q} \overline{\mathcal{D}}_0(G_i)$ for all a, $0 < a < q-1$. If $a = q-1$, then we have $\overline{\mathcal{D}}_a(G) \leq \overline{\mathcal{D}}_0(G) + 1$ and hence $\overline{\mathcal{D}}_0(G) \leq \overline{\mathcal{D}}_a(G) \leq \overline{\mathcal{D}}_0(G) + 1 = \sum_{i=1}^{q} \overline{\mathcal{D}}_0(G_i) + 1$. Furthermore, if there is an integer i such that G_i is a complete graph, then $\mathbf{d}(G_i) < 2$, and it is easy to construct examples showing that $\overline{\mathcal{D}}_{q-1}(G) = \overline{\mathcal{D}}_0(G) + 1$. If none of the factors is a complete graph, i.e. $\mathbf{d}(G_i) \geq 2$ for all integers i, then $\overline{\mathcal{D}}_{q-1}(G) = \overline{\mathcal{D}}_0(G)$.

Now assume that all graphs G_i are trees and at least one factor is a K_2. Let $G_q = K_2$ and let $v = (v_1, v_2, \ldots, v_{q-1}, 1)$ and $u = (u_1, u_2, \ldots, u_{q-1}, 1)$ be two vertices at maximal distance in $(\square_{i=1}^{q-1} G_i) \square \{1\}$, i.e. such that v_i and u_i are at maximal distance in G_i for each $i = 1, 2, \ldots q-1$. As $(\square_{i=1}^{q-1} G_i) \square \{1\}$ is $(q-1)$-edge connected, it is clear that $q-1$ faulty edges may cut all the paths between u and v in $(\square_{i=1}^{q-1} G_i) \square \{1\}$. Hence $\overline{\mathcal{D}}_{q-1}(G) = \overline{\mathcal{D}}_0(G) + 1$. $\quad\square$

Example 1. Let $G = Q_q = \square_{n=1}^{q} K_2$. Then G is q-edge connected. It follows from Lemma 1 that $\overline{\mathcal{D}}_{q-1}(G) = q + 1$, and $\overline{\mathcal{D}}_a(G) = \overline{\mathcal{D}}_0(G) = q$ for $0 \leq a < q-1$. (See Fig. 1.)

In the proof of Lemma 1 we only used the assumption that the factors are edge connected, therefore essentially the same proof gives

Theorem 3. *Let $G_1, G_2, \ldots, G_q$ be edge connected graphs, and $G = \square_{i=1}^{q} G_i$. Then*

1. $\overline{\mathcal{D}}_a(G) = \sum_{i=1}^{q} \overline{\mathcal{D}}_0(G_i) = \overline{\mathcal{D}}_0(G)$ *for $0 \leq a < q-1$;*
2. $\overline{\mathcal{D}}_0(G) \leq \overline{\mathcal{D}}_{q-1}(G) \leq \overline{\mathcal{D}}_0(G) + 1$.

Now we recall and prove Theorem 1.

Theorem 1. *Let $G_1, G_2, \ldots, G_q$ be k_1-edge connected, k_2-edge connected,..., k_q-edge connected graphs. Let $0 \leq a_1 < k_1$, $0 \leq a_2 < k_2$,..., $0 \leq a_q < k_q$ and $a = a_1 + a_2 + \ldots + a_q + (q-1)$. Then the edge fault-diameter of G, the Cartesian product of $G_1, G_2, \ldots, G_q$, with a faulty edges is*

$$\overline{\mathcal{D}}_a(G) \leq \overline{\mathcal{D}}_{a_1}(G_1) + \overline{\mathcal{D}}_{a_2}(G_2) + \ldots + \overline{\mathcal{D}}_{a_q}(G_q) + 1.$$

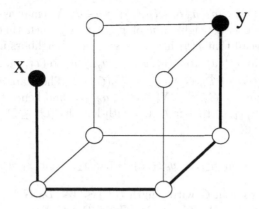

Fig. 1. $\overline{\mathcal{D}}_2(Q_3) = \overline{\mathcal{D}}_0(Q_3) + 1 = 3 + 1 = 4$

Proof. We prove the result by induction on the number of factors. The assertion holds for $q = 1$, trivially.

Let $G = \square_{i=1}^{q} G_i$, $X \subseteq E(G)$, such that $|X| = a$, and let $x = (x_1, x_2, \ldots, x_q)$ and $y = (y_1, y_2, \ldots, y_q) \in V(G)$. We shall construct a path P from x to y in $G \setminus X$ such that the length $\ell(P) \leq \overline{\mathcal{D}}_{a_1}(G_1) + \overline{\mathcal{D}}_{a_2}(G_2) + \ldots + \overline{\mathcal{D}}_{a_q}(G_q) + 1$. We assume that the theorem holds for less than q factors and show that it holds also for $q \geq 2$ factors.

In the rest of the proof below we also assume that there is at least one factor which is more than 1-edge connected. (Recall that we proved the claim of Theorem 1 for a product of 1-edge connected graphs by proving Lemma 1.)

Let for each $i = 1, 2, 3, \ldots, q$, $p_i : G \to G_i$ be the i-th projection on G_i.

Case 1. $p_i(x) = p_i(y)$ for some i. Without loss of generality, we can say $i = q$. For each $u \in V(G_q)$ let $w(u) = |X \cap p_q^{-1}(u)|$ and for each $e \in E(G_q)$ let $w(e) = |X \cap p_q^{-1}(e)|$. If $w(p_q(x)) < a - a_q$ then there is a path P in $p_q^{-1}(p_q(x)) \subseteq G$ from x to y with the required length, by induction. Assume $w(p_q(x)) \geq a - a_q$. The vertex $p_q(x)$ has at least $a_q + 1$ neighbors $z_1, z_2, z_3, \ldots, z_{a_q+1}$ in G_q because G_q is at least $(a_q + 1)$-edge connected. As $w(p_q(x)) \geq a - a_q$, there is an index $i \in \{1, 2, 3, \ldots, a_q + 1\}$ such that $w(e_i) = 0$ and $w(z_i) = 0$, where $e_i = p_q(x)$, $z_i \in E(G_q)$. Therefore there is a path $\tilde{P}$ in $p_q^{-1}(z_i)$ from $(x_1, x_2, \ldots, x_{q-1}, z_i)$ to $(y_1, y_2, \ldots, y_{q-1}, z_i)$ with length $\ell(\tilde{P}) \leq \overline{\mathcal{D}}_0(G_1) + \overline{\mathcal{D}}_0(G_2) + \ldots + \overline{\mathcal{D}}_0(G_{q-1})$ and

$$P : x \to (x_1, x_2, \ldots, x_{q-1}, z_i) \overset{\tilde{P}}{\to} (y_1, y_2, \ldots, y_{q-1}, z_i) \to y$$

is a path from x to y in G with length $\ell(P) \leq 1 + \overline{\mathcal{D}}_0(G_1) + \overline{\mathcal{D}}_0(G_2) + \ldots + \overline{\mathcal{D}}_0(G_{q-1}) + 1 \leq \overline{\mathcal{D}}_{a_1}(G_1) + \overline{\mathcal{D}}_{a_2}(G_2) + \ldots + \overline{\mathcal{D}}_{a_q}(G_q) + 1$.

Case 2. $p_i(x) \neq p_i(y)$ for all i. Let G_q be one of the factors which is more than 1-edge connected. As before, let $w(u) = |X \cap p_q^{-1}(u)|$ for each $u \in V(G_q)$ and $w(e) = |X \cap p_q^{-1}(e)|$ for each $e \in E(G_q)$. We distinguish two subcases.

Subcase 2.1. $w(p_q(x)) \geq a - a_q$ or $w(p_q(y)) \geq a - a_q$. We may assume $w(p_q(x)) \geq a - a_q$. Then there is a neighbor u of $p_q(y)$ in G_q such that $w(u) = 0$ and $w(p_q(x)u) = 0$ (recall that $p(y)$ has at least $a_q + 1$ neighbors in G_q). Therefore there is a path $\tilde{P}$ in $p_q^{-1}(u)$ from $(y_1, y_2, \ldots, y_{q-1}, u)$ to $(x_1, x_2, \ldots, x_{q-1}, u)$ with length $\ell(\tilde{P}) \leq \overline{\mathcal{D}}_0(G_1) + \overline{\mathcal{D}}_0(G_2) + \ldots + \overline{\mathcal{D}}_0(G_{q-1})$. There are at most a_q faulty edges in the layer of x, $|G_q(x) \cap X| \leq a_q < a_q + 1$, and hence there is a path Q from $(x_1, x_2, \ldots, x_{q-1}, u)$ to x in $G_q(x)$ with length $\ell(Q) \leq \overline{\mathcal{D}}_{a_q}(G_q)$. Therefore the path (see Fig. 2)

$$P : y \rightarrow (y_1, y_2, \ldots, y_{q-1}, u) \xrightarrow{\tilde{P}} (x_1, x_2, \ldots, x_{q-1}, u) \xrightarrow{Q} x$$

is a path from y to x in G with length $\ell(P) \leq 1 + \overline{\mathcal{D}}_0(G_1) + \overline{\mathcal{D}}_0(G_2) + \ldots + \overline{\mathcal{D}}_0(G_{q-1}) + \overline{\mathcal{D}}_{a_q}(G_q) \leq \overline{\mathcal{D}}_{a_1}(G_1) + \overline{\mathcal{D}}_{a_2}(G_2) + \overline{\mathcal{D}}_{a_3}(G_3) + \ldots + \overline{\mathcal{D}}_{a_q}(G_q) + 1.$

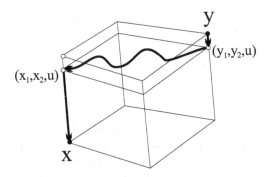

Fig. 2. The construction of a path in subcase 2.1. for $q = 3$

Subcase 2.2. $w(p_q(x)) < a - a_q$ and $w(p_q(y)) < a - a_q$.

Assume first $w(p_q(x)) + w(p_q(y)) > a - a_q$. In $E(G_q)$ there are at most $a_q - 1$ edges e, such that $w(e) > 0$. As G_q is at least $(a_q + 1)$-edge connected, there is a neighbor v of $p_q(x)$ with $w(v) = 0$ and $w(p_q(x)v) = 0$. We construct a path $\tilde{P}$ in G_q from v to $p_q(y)$ with length $\ell(\tilde{P}) \leq \overline{\mathcal{D}}_{a_q}(G_q)$, such that $w(e) = 0$ for each edge e of $\tilde{P}$. As $w(v) = 0$, there is a path Q from $(x_1, x_2, \ldots, x_{q-1}, v)$ to $(y_1, y_2, \ldots, y_{q-1}, v)$ in $p_q^{-1}(v)$ with length $\ell(Q) \leq \overline{\mathcal{D}}_0(G_1) + \overline{\mathcal{D}}_0(G_2) + \ldots + \overline{\mathcal{D}}_0(G_{q-1})$. Therefore the path (see Figure 3)

$$P : x \rightarrow (x_1, x_2, \ldots, x_{q-1}, v) \xrightarrow{Q} (y_1, y_2, \ldots, y_{q-1}, v) \xrightarrow{\tilde{P}} y$$

is the path from x to y in G with length $\ell(P) \leq 1 + \overline{\mathcal{D}}_0(G_1) + \overline{\mathcal{D}}_0(G_2) + \ldots + \overline{\mathcal{D}}_0(G_{q-1}) + \overline{\mathcal{D}}_{a_q}(G_q) \leq \overline{\mathcal{D}}_{a_1}(G_1) + \overline{\mathcal{D}}_{a_2}(G_2) + \ldots + \overline{\mathcal{D}}_{a_q}(G_q) + 1.$

Now assume $w(p_q(x)) + w(p_q(y)) \leq a - a_q$. If there are less than a_q edges e in G_q, such that $w(e) > 0$, then it is easy to construct a required path P from x to y in G. Otherwise, we claim that there is a path $\tilde{P}$ from $p_q(x)$ to $p_q(y)$ in

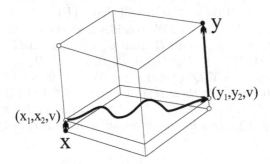

Fig. 3. The construction of the path P for $q = 3$ in subcase 2.2

G_q such that $w(\tilde{P}) \leq a - a_q$. This is easily seen as follows: In the subgraph G_q, choose a set Y of a_q edges with maximal w. Then $w(G_q \setminus Y) \leq a - a_q$ and from the $(a_q + 1)$-edge connectivity of G_q it follows that there is a path $\tilde{P}$ of length at most $\ell(\tilde{P}) \leq \overline{\mathcal{D}}_{a_q}(G_q)$ and with $w(\tilde{P}) \leq a - a_q$, which proves the claim. Let $p :$ $G_1 \square G_2 \square \ldots \square G_{q-1} \square \tilde{P} \rightarrow G_1 \square G_2 \square \ldots \square G_{q-1}$ be the projection, defined with $p(x_1, x_2, \ldots, x_{q-1}, x_q) = (x_1, x_2, \ldots, x_{q-1})$. For each $u \in V(G_1 \square G_2 \square \ldots \square G_{q-1})$ let $W(u) = |p^{-1}(u) \cap X|$ and for each $e \in E(G_1 \square G_2 \square \ldots \square G_{q-1})$ let $W(e) = |p^{-1}(e) \cap X|$. We consider next two possibilities.

– $W(p(x)) = 0$ or $W(p(y)) = 0$. Say $W(p(x)) = 0$. Therefore there is a path Q from x to $(x_1, x_2, \ldots, x_{q-1}, y_q)$ in $p^{-1}(x)$, such that $\ell(Q) \leq \ell(\tilde{P}) \leq \overline{\mathcal{D}}_{a_q}(G_q)$. As $w(p_q(y)) < a - a_q$, there is a path $\tilde{Q}$ in $p_q^{-1}(p(y))$ from $(x_1, x_2, \ldots, x_{q-1}, y_q)$ to y with length $\ell(\tilde{Q}) \leq \overline{\mathcal{D}}_{a_1}(G_1) + \overline{\mathcal{D}}_{a_2}(G_2) + \ldots + \overline{\mathcal{D}}_{a_{q-1}}(G_{q-1}) + 1$. Hence

$$P : x \xrightarrow{Q} (x_1, x_2, \ldots, x_{q-1}, y_q) \xrightarrow{\tilde{Q}} y$$

is a path from x to y in G, such that $\ell(P) \leq \overline{\mathcal{D}}_0(G_q) + \overline{\mathcal{D}}_{a_1}(G_1) + \overline{\mathcal{D}}_{a_2}(G_2) + \ldots + \overline{\mathcal{D}}_{a_{q-1}}(G_{q-1}) + 1 \leq \overline{\mathcal{D}}_{a_1}(G_1) + \overline{\mathcal{D}}_{a_2}(G_2) + \ldots + \overline{\mathcal{D}}_{a_q}(G_q) + 1$ (see Figure 4).

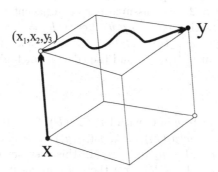

Fig. 4. The construction of the path P for $q = 3$ when $W(p(x)) = 0$

- $W(p(x)) > 0$ and $W(p(y)) > 0$. Recall that $w(\tilde{P}) \leq a - a_q$, i.e. $|(G_1 \square G_2 \square \ldots \square G_{q-1} \square \tilde{P}) \cap X| \leq a - a_q$. We consider next two possibilities.
    - If $q > 2$, there is a path Q from $p(x)$ to $p(y)$ in $G_1 \square G_2 \square \ldots \square G_{q-1}$ with length $\ell(Q) \leq \overline{\mathcal{D}}_{a_1}(G_1) + \overline{\mathcal{D}}_{a_2}(G_2) + \ldots + \overline{\mathcal{D}}_{a_{q-1}}(G_{q-1}) + 1$, such that for each $u \in V(Q) \setminus \{p(x), p(y)\}$, $W(u) = 0$, and for each $e \in E(Q)$, $W(e) = 0$. As $q > 2$, there is such a vertex $u \in V(P)$ (here we need the fact that $G_1 \square G_2 \square \ldots \square G_{q-1}$ is not isomorphic to K_2, which is trivially true if $q > 2$). Let u be the vertex adjacent to $p(x)$ in Q. As $W(u) = 0$,

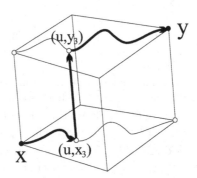

Fig. 5. The construction of the path P for $q = 3$ when $W(p(x)) > 0$ and $W(p(y)) > 0$

there is a path $\tilde{Q}$ from (u, x_q) to (u, y_q) in $p^{-1}(u)$ with length $\ell(\tilde{Q}) \leq \ell(\tilde{P}) \leq \overline{\mathcal{D}}_0(G_q)$. Finally we may construct the required path

$$P : x \to (u, x_q) \overset{\tilde{Q}}{\to} (u, y_q) \overset{Q}{\to} y$$

from x to y, such that $\ell(P) \leq 1 + \overline{\mathcal{D}}_0(G_q) + \overline{\mathcal{D}}_{a_1}(G_1) + \overline{\mathcal{D}}_{a_2}(G_2) + \ldots + \overline{\mathcal{D}}_{a_{q-1}}(G_{q-1}) + 1 - 1 \leq \overline{\mathcal{D}}_{a_1}(G_1) + \overline{\mathcal{D}}_{a_2}(G_2) + \ldots + \overline{\mathcal{D}}_{a_q}(G_q) + 1$ (see Figure5).
    - If $q = 2$, we consider the following possible cases. If $p(x)$ is not adjacent to $p(y)$, the required path P from x to y can be found as in previous case, when $q > 2$. Now assume $p(x)$ is adjacent to $p(y)$, and let e be the edge connecting $p(x)$ to $p(y)$. Let $w(\tilde{P}) = k \leq a - a_2 = a_1 + 1$. If $w(p_2(x)) = k$ or $w(p_2(y)) = k$ (say $w(p_2(x)) = k$), then there (x_1, y_2) is adjacent to y in $p_2^{-1}(p_2(y))$ and the edge is not faulty. Hence the path

$$P : x \overset{\tilde{P}}{\to} (x_1, y_2) \to y$$

is a path from x to y in G with length $\ell(P) \leq \overline{\mathcal{D}}_{a_2}(G_2) + 1 \leq \overline{\mathcal{D}}_{a_1}(G_1) + \overline{\mathcal{D}}_{a_2}(G_2) + 1$. If $w(p_2(x)) < k$, $w(p_2(y)) < k$ and $k < a_1 + 1$, then $w(p_2(x)) < a_1$ and $w(p_2(y)) < a_1$. In this case, let the edge e be one of the faulty edges. In $p_2^{-1}(p_2(x))$ there are at most a_1 faulty edges, and we may construct a path Q from $p(x)$ to $p(y)$ in $p_2^{-1}(p_2(x))$ with length $\ell(Q) \leq \overline{\mathcal{D}}_{a_1}(G_1)$, which contains a vertex $u \neq p(x), p(y)$ with $W(u) = 0$, such that for each edge e on Q, $W(e) = 0$. Now we finish the proof in the same way as in the case when $q > 2$. If $w(p_2(x)) = a_1$, then from $W(p(x)) > 0$ we have $W(p(y)) = 0$, which is a contradiction with the assumption that $W(p(x)) > 0$ and $W(p(y)) > 0$. □

In the proof of Theorem 1 we have assumed that each G_i is at least $a_i + 1$ edge connected, and we only needed that $a_i + 1 \leq k_i$. Given G and X we may read the proof with arbitrary choice of a_i which satisfies the conditions $a_1 + a_2 + \ldots + a_q = a - (q-1), 0 \leq a_1 < k_1, 0 \leq a_2 < k_2, \ldots, 0 \leq a_q < k_q\}$. This argument proves

Theorem 2. *Let G_1, G_2, $\ldots$, G_q be k_1-edge connected, k_2-edge connected,$\ldots$, k_q-edge connected graphs, and G the Cartesian product of G_1, G_2, $\ldots$, G_q. Let $0 \leq a < k_1 + k_2 + \ldots + k_q$. Then*

$$\overline{\mathcal{D}}_a(G) \leq \min\{\overline{\mathcal{D}}_{a_1}(G_1) + \overline{\mathcal{D}}_{a_2}(G_2) + \ldots + \overline{\mathcal{D}}_{a_q}(G_q) + 1 \mid$$

$$a_1 + a_2 + \ldots + a_q = a - (q-1), 0 \leq a_1 < k_1, 0 \leq a_2 < k_2, \ldots, 0 \leq a_q < k_q\}.$$

Example 2. Let $G = C_6 \square C_{100}$. From Theorem 2,

$$\overline{\mathcal{D}}_2(G) \leq \min\{\overline{\mathcal{D}}_1(C_6) + \overline{\mathcal{D}}_0(C_{100}), \overline{\mathcal{D}}_0(C_6) + \overline{\mathcal{D}}_1(C_{100})\} + 1 =$$

$$= \min\{55, 102\} + 1 = 56.$$

Clearly, the bound from Theorem 2 improves the upper bound from Theorem 1. For example, for $a = 0$ and $b = 1$, Theorem 1 gives $\overline{\mathcal{D}}_{a+b+1}(G) \leq \overline{\mathcal{D}}_a(C_6) + \overline{\mathcal{D}}_b(C_{100}) + 1 = 103$. On the other hand, one can easily check that $\overline{\mathcal{D}}_2(G) = 53$, therefore Corollary 2 does not give the exact formula for computing the fault diameter. In fact, the bound can be far from the exact value. For example, by Theorem 2, $\overline{\mathcal{D}}_3(G) \leq 105$, while the exact value is $\overline{\mathcal{D}}_3(G) = 53$.

Example 3. We have computed the fault diameters of the hypercube using Theorem 3 in Example 1. Hypercube Q_q can be represented also as $(\square_{i=1}^{r} C_4) \square K_2$ if $q = 2r + 1$ or $\square_{i=1}^{r} C_4$ if $q = 2r$. Let us apply the Theorem 2.

First, let $a = q - 1$. The only 'legal partitions' of a are $a = \underbrace{1 + 1 + \ldots + 1}_{r} + 0 + r$, if $q = 2r + 1$, and $a = \underbrace{1 + 1 + \ldots + 1}_{r} + r - 1$, if $q = 2r$. By Theorem 2,

$\overline{\mathcal{D}}_a(Q_q) \leq \underbrace{3 + 3 + \ldots + 3}_{r} + 1 + 1 = 3r + 2$ for $q = 2r + 1$. For $q = 2r$, $\overline{\mathcal{D}}_a(Q_q) \leq$

$\underbrace{3 + 3 + \ldots + 3}_{r} + 1 = 3r + 1$.

Second, let $a < q - 1$. The same way as in the first case we get $\overline{\mathcal{D}}_a(Q_q) \leq q + 1$. For example, if $a = q - 2$, we have $a = \underbrace{1 + 1 + \ldots + 1}_{r-1} + 0 + 0 + r = 2r - 1$, if $q = 2r + 1$, and $a = \underbrace{1 + 1 + \ldots + 1}_{r-1} + 0 + r - 1 = 2r - 2$, if $q = 2r$. In the first case we have $\overline{\mathcal{D}}_a(Q_q) \leq \underbrace{3 + 3 + \ldots + 3}_{r-1} + 2 + 1 + 1 = 3r + 1$. In the second case

we have $\overline{\mathcal{D}}_a(Q_q) \leq \underbrace{3 + 3 + \ldots + 3}_{r-1} + 2 + 1 = 3r$.

Hence Theorem 2 gives only an upper bound for $\overline{\mathcal{D}}_a(Q_q)$ which is around 50% too large while Theorem 3 gives the exact result. The example shows that it is profitable to use the Cartesian product structure of the network studied. It may be worth mentioning that the recognition (i.e. factorization) with respect to the Cartesian product can be done efficiently, in time $O(mn)$ [9,10]. Furthermore, it is also possible to efficiently reconstruct a Cartesian product graph with not too many nodes missing under mild assuptions [11,12]. Loosely speaking, if we know that the original structure was a Cartesian product of k factors with at least k vertices each, and at most $k-1$ nodes are missing, then we can reconstruct the original graph, up to isomorphism.

Acknowledgements. The authors would also like to thank anonymous referees for careful reading and valuable comments and suggestions.

This work was supported in part by the Slovenian research agency, grants L2-7207-0101, P1-0294-0101 and P1-0285-0101.

References

1. Banič, I., Žerovnik, J.: Fault-diameter of Cartesian graph bundles. Inform. Process. Lett. 100, 47–51 (2006)
2. Banič, I., Žerovnik, J.: Fault-diameter of Cartesian product of graphs. to appear in Advances in Applied Mathematics
3. Banič, I., Žerovnik, J.: Fault-diameter of generalized Cartesian products. Proceedings of 26th IEEE International Conference on Distributed Computing Systems. (Workshops.) IEEE Computer Society, Los Alamitos (2006) http://doi.ieeecomputersociety.org/10.1109/ICDCSW.2006.51
4. Bermond, J.-C., Comellas, F., Hsu, D.: Distributed loop computer networks: a survey. J. Parallel Distrib. Comput. 24, 2–10 (1995)
5. Cull, P., Larson, S.M.: On generalized twisted cubes. Inform. Process. Lett. 55, 53–55 (1995)
6. Du, D.Z., Hsu, D.F., Lyuu, Y.D.: On the diameter vulnerability of kautz digraphs. Discrete Math. 151, 81–85 (2000)
7. Day, K., Al-Ayyoub, A.: Minimal fault diameter for highly resilient product networks. IEEE Trans. Parallel Distrib. Syst. 11, 926–930 (2000)
8. Efe, K.: A variation on the hypercube with lower diameter. IEEE Trans. Comput. 40, 1312–1316 (1991)
9. Imrich, W., Klavžar, S.: Products Graphs, Structure and Recognition. Wiley, New York (2000)
10. Imrich, W., Žerovnik, J.: J. graph theory, let. 18, t. 6, str, pp. 557–567 (1994) Žerovnik, J.: Factoring Cartesian-product graphs. J. Graph Theory 18, 557–567 (1994)
11. Imrich, W., Žerovnik, J.: On the weak reconstruction of Cartesian-product graphs. Discrete math. 150, 167–178 (1996)

12. Imrich, W., Zmazek, B., Žerovnik, J.: Weak k-reconstruction of Cartesian products. Discuss. Math. Graph Theory 21, 273–285 (2003)
13. Krishnamoorthy, M., Krishnamurty, B.: Fault diameter of interconnection networks. Comput. Math. Appl. 13(5/6), 577–582 (1987)
14. Liaw, S.C., Chang, G.J., Cao, F., Hsu, D.F.: Fault-tolerant routing in circulant networks and cycle prefix networks. Ann. Comb. 2(5-6), 165–172 (1998)
15. Munoz, X.: Asymptotically optimal (δ, d', s)-digraphs. Ars. Combin. 49, 97–111 (1998)
16. Stojmenović, I.: Multiplicative circulant networks: Topological properties and communication algorithms. Discrete Applied Math. 77, 281–305 (1997)
17. West, D.B.: Introduction to Graph Theory. Prentice Hall, Upper Saddle River, NJ 07458 (2001)
18. Xu, M., Xu, J.-M., Hou, X.-M.: Fault diameter of cartesian product graphs. Inform. Process. Lett. 93, 245–248 (2005)

Rapid Almost-Complete Broadcasting in Faulty Networks*

Rastislav Královič[1] and Richard Královič[1,2]

[1] Department of Computer Science, Comenius University,
Mlynská dolina, 84248 Bratislava, Slovakia
[2] Department of Computer Science, ETH Zurich, Switzerland

Abstract. This paper studies the problem of broadcasting in synchronous point-to-point networks, where one initiator owns a piece of information that has to be transmitted to all other vertices as fast as possible. The model of fractional dynamic faults with threshold is considered: in every step either a fixed number T, or a fraction α, of sent messages can be lost depending on which quantity is larger.

As the main result we show that in complete graphs and hypercubes it is possible to inform all but a constant number of vertices, exhibiting only a logarithmic slowdown, i.e. in time $O(D \log n)$ where D is the diameter of the network and n is the number of vertices.

Moreover, for complete graphs under some additional conditions (sense of direction, or $\alpha < 0.55$) the remaining constant number of vertices can be informed in the same time, i.e. $O(\log n)$.

1 Introduction

Fault tolerance has been a crucial issue in the distributed computing since its beginnings [3,5,6,10,16,25]. Because a typical distributed system is designed to contain a large number of individual components, attention must be paid to the fact that, even if the failure probability of a single component is negligible, the probability that some components fail may be high. There are numerous ways how to cope with failures, using either probabilistic or deterministic approaches. In the probabilistic setting, it is supposed that a failure probability of each component follows some probability distribution [4,8,11,26,27]. Failures of individual components are usually assumed to be independent random events. The goal is to design algorithms and protocols that perform well with high probability if the failures follow the conjectured distribution.

The deterministic approach, which is pursued also in this paper, copes with failures in a different way. Instead of considering a failure probability distribution for each individual component, algorithms and protocols are designed to perform well in the worst case, under some a-priori constraints on the failure behavior. [1, 2,7,12,13,14,19,20,22,24,28]. These constraints may take the form of considering only computations with a limited overall number of faults [1,19], limited number

---

* The research has been supported by grant APVV-0433-06.

G. Prencipe and S. Zaks (Eds.): SIROCCO 2007, LNCS 4474, pp. 246–260, 2007.

of faults during any single computation step [7, 13, 14, 24, 28], or during any window of first t steps [20], requiring that after some finite time there is a long enough fault-free computation [10, 15] etc. While the probabilistic model is analyzed with respect to the expected behavior, the deterministic models have been mostly analyzed for the worst case scenario.

We shall focus our attention on synchronous point-to-point distributed systems, i.e. systems in which the communication is performed by sending messages along links connecting pairs of vertices. Moreover, the vertices are synchronized by a common clock, and the delivery of every message takes exactly one time unit. This model has been widely considered [7,8,12,13,14,20,22,24,26,27,28] not only for its theoretical appeal, but for its practical relevance as well (e.g. many wireless networking standards, like IEEE 802.11, or GSM, operate in discrete time steps). We shall consider only one type of failures: *message loss*.

The oldest deterministic model of faults considered in this setting is the static model [1,3], in which it is assumed that at most a fixed constant number k of messages may be lost in every step, and moreover, the failures are always located on the same links. Later, other models have been considered, too, like the dynamic model [7,13,14,24,28] in which the k failures may be located on arbitrary links in every step, linearly bounded faults [20], fractional faults [22], etc.

We continue in the analysis of the *fractional model with threshold* from [12]. Here, the number of messages lost in one time step is bounded by the maximum of a fixed threshold T and a fixed fraction α of sent messages. This restriction implies that if, in a given step, fewer than T messages are sent they may all be lost. On the other hand, if there are many messages sent, at least a fixed fraction $1 - \alpha$ of them is delivered. The threshold T is always assumed to be one less than the edge connectivity, since this is the largest value under which the network stays connected. This model has been developed in order to avoid some unrealistic special cases of static and dynamic models (the number of faults is independent on the actual network traffic), as well as those of fractional model (if just one message is being sent, its delivery is always guaranteed).

The broadcasting problem is a crucial communication task in the study of distributed systems (e.g. [21]). One vertex, called initiator, has a piece of information that has to be distributed among all remaining vertices. The broadcasting has not only been used as a test-bed application for the study of the complexity of communication in various communication models, but has served as a building stone of many applications (e.g. [29]) as well.

We analyze the broadcasting in complete graphs and hypercubes. The broadcasting time in these graphs has been studied in the static [19], dynamic [13, 14, 24], and simple threshold [12][1] models, and the results are summarized in Table 1.

We address a natural relaxation of the broadcasting problem in which we allow a small constant number of vertices to stay uninformed in the end (a

---

[1] If the number of messages sent in a given time step is less than the edge connectivity $c(G)$ in the simple threshold model, all of them may be lost. Otherwise at least one of them is delivered.

Table 1. Known time complexities of the complete broadcasting in various models

Model	K_n, chordal sense of direction	K_n unoriented	Q_d, $n = 2^d$
static	$\Theta(1)$	$\Theta(1)$	$d + 1$ [19]
dynamic	$\Theta(1)$	$\Theta(1)$ [24]	$d + 2$ [13]
fractional	$\Theta(\log n)$	$\Theta(\log n)$ [22]	$O(d^3)$ [22]
simple threshold	$\Omega(n)$, $O(n^2)$ [12]	$\Omega(n^2)$, $O(n^3)$ [12]	$O(n^4 d^2)$ [12]

Table 2. Results for the complete and almost complete broadcasting in the fractional model with threshold

Scenario	Almost complete broadcasting	Complete broadcasting
K_n, unoriented	$O(\log n)$	$\Omega(\log n)$ [22], $O(n^3)$ [12]
K_n, chordal sense of direction	$O(\log n)$	$\Omega(\log n)$ [22], $O(\log n)$
K_n, $\alpha < 0.55$	$O(\log n)$	$\Omega(\log n)$ [22], $O(\log n)$
Q_d	$O(d^2)$	$\Omega(d)$, $O(n^4 d^2)$ [12]

problem called *almost complete broadcasting*), and analyze the worst case time needed to solve the problem. Our main motivation to study almost complete broadcasts is the fact that in large faulty networks it is often vital to finish a communication task fast, even subject to some small error. In the probabilistic setting, this is modelled by allowing a failure probability that tends to zero with increasing network size: in the worst case the task is not successful but this worst case scenario has a small probability. Since in our deterministic setting we study the worst case, another model of allowed error must be chosen. If we look at the broadcast as an optimization problem where the task is to inform as many vertices as possible, it is natural to introduce a constant additive error by allowing a constant number of vertices to stay uninformed[2].

For complete graphs and hypercubes, we show that the problem can be solved in time $O(D \log n)$, where D is the diameter of the graph and n is the number of its nodes.

Moreover, we show that if the complete graph is equipped with the chordal sense of direction, complete broadcasting can be performed in time $O(\log n)$. This is asymptotically optimal since the broadcasting time in the fractional model is a lower bound for the fractional model with threshold. Similarly we show that the broadcasting can be completed in time $O(\log n)$ for values $\alpha < 0.55$. The overview of the results can be found in Table 2.

2 Definitions

We consider a synchronous, point-to-point distributed system with a coordinated start-up. The system consists of a number of nodes and a number of commu-

---

[2] So that the uninformed vertices comprise at most an $O(1/n)$ fraction of all vertices.

nication links connecting some pairs of nodes. The system is modelled by an undirected graph, in which vertices correspond to nodes and edges correspond to communication links. In this respect, we shall use the terms "node" and "vertex" interchangeably. Sometimes we need to argue about outgoing and incoming links; in this cases we consider a directed graph obtained from the undirected one by replacing each edge by two opposite arcs.

At the beginning of the computation all nodes are active and start performing the given protocol. The computation consists of a number of steps: at the beginning of each step, messages sent during the previous step are delivered to their destinations, then each vertex performs some local computation, possibly sending some messages[3], and the next step begins.

The failure model we consider is the *fractional dynamic faults with threshold* from [12], which can be described as a game between the algorithm and an adversary: in a time step t the algorithm sends m_t messages and the adversary may destroy up to

$$F(m_t) = \max\{c(G) - 1, \lfloor \alpha \, m_t \rfloor\}$$

of them, where $c(G)$ is the edge connectivity of the graph and α is a known, fixed constant $0 < \alpha < 1$. There is no built-in mechanism of acknowledgements, so the sender node is not informed whether a particular message was delivered or destroyed.

We consider the problem of broadcasting, where an initiator has a piece of information to be transmitted to all remaining vertices. We call a broadcast *complete* if all vertices have the information after the termination of the algorithm. A broadcast is called *almost-complete* if there is a fixed constant c (independent on the network size) such that after the termination there are at most c uninformed vertices. Hence, to prove the existence of an almost-complete broadcasting algorithm for a family of graphs $\mathcal{G}$, one has to prove that there exists a constant c such that for each $G \in \mathcal{G}$ the broadcasting algorithm informs all but c vertices of G.

In all presented algorithms only the informed vertices send messages. Arcs (i.e. directed edges) leading from an informed vertex can be classified as being either active, passive or hyperactive during the computation:

Definition 1. *Let e be an arc leading from an informed vertex. We call e active if it leads to an uninformed vertex. We call an arc e passive, if some message has been delivered via the opposite arc of e. Finally, we call an arc e hyperactive if it leads to an informed vertex, and is not passive.*

If the arc e is passive, the source vertex of e is aware of the fact that the destination vertex of e has already been informed. The main idea of our algorithms is to perform appropriate number of *simple rounds* defined as follows:

Definition 2. *A simple round consists of two time steps. In the first step, every informed vertex sends a message along each of its incident arcs, excluding*

---

[3] i.e. a vertex may send different message to each of its neighbors in one step.

the passive ones.[4] *In the second step, all vertices that have received a message send an acknowledgement (and mark the arc as passive). Vertices that receive acknowledgement mark the corresponding arc as passive.*

For the remainder of this paper, let $0 < \alpha < 1$ be a known fixed constant, and let us denote

$$X := \frac{1}{\alpha(1 - \alpha)}$$

The rest of the paper is organized as follows. In the next two sections we present algorithms for the almost-complete broadcasting on complete graphs and hypercubes, respectively, that run in time $O(D \log n)$. Then we show how to obtain broadcast in complete graphs equipped with chordal sense of direction, and for unoriented complete graphs for $\alpha < 0.55$, having the same time complexity.

Due to space restrictions some technical parts have been omitted from this paper, and can be found in the technical report [23].

3   Complete Graphs

In a complete graph K_n, all n vertices have degree $n - 1$, and $n - 1$ is also the edge connectivity. Hence, in each step t the adversary can destroy up to $\max\{n-2, \lfloor \alpha m_t \rfloor\}$ messages, where m_t is the number of messages sent in the step t. In this section we present an algorithm that informs all but a constant number of vertices in logarithmic time. The idea of the algorithm is very straightforward – just repeat simple rounds sufficiently many times. However, the arguments given in the analysis of a simple round below hold only if there are enough informed vertices participating in the round. To satisfy this requirement two steps of a simple greedy algorithm are performed, during which each informed vertex just sends the message to all vertices. After two steps of this algorithm, the number of informed vertices is as shown in Lemma 1.

Lemma 1. *After two steps of the greedy algorithm, at least*

$$1 + \min\left\{\frac{n}{2}, (n - 1)(1 - \alpha)\right\}$$

vertices are informed.

After these two steps, the algorithm performs a logarithmic number of simple rounds. To show that logarithmic number of simple rounds is sufficient to inform all but one vertex we first provide a lower bound on the number of acknowledgements delivered in each round, and then we show that each delivered acknowledgement decreases a certain measure function.

---

[4] In this step, a message is sent via all active and hyperactive arcs. The former can inform new vertices, the latter exhibit only useless activity. However, the algorithm can not distinguish between active and hyperactive arcs.

Theorem 1. *Let $\varepsilon > 1$ be an arbitrary constant. For large enough n it is possible to inform all but at most $X\varepsilon$ vertices in logarithmic time. Moreover, the number of remaining hyperactive arcs is at most $X(n-2)$.*

Proof. At the beginning, two steps of the greedy algorithm are executed. Then, a logarithmic number of simple rounds is performed. Now consider the situation at the beginning of the i-th round. Let k_i be the number of uninformed vertices, and h_i the number of hyperactive arcs. We claim that if $k_i > X\varepsilon$ or $h_i > X(n-2)$ then at least $[k_i(n-k_i) + h_i](1-\alpha)^2$ acknowledgements are delivered in this round. Since there are $k_i(n-k_i)+h_i$ messages sent in this round, in order to prove the claim it is sufficient to show that $\alpha(1-\alpha)[k_i(n-k_i)+h_i] \geq n-2$. Obviously, if $h_i > X(n-2)$ the inequality holds, so consider the case $k_i > X\varepsilon$. We prove that in this case $k_i(n-k_i) \geq X(n-2)$, i.e. $k_i^2 - nk_i + X(n-2) \leq 0$. Let $f(n) :=$ $1/2\left(n - \sqrt{n^2 - 4X(n-2)}\right)$; the roots[5] of the equation $k_i^2 - nk_i + X(n-2) = 0$ are $f(n)$ and $n - f(n)$, so we want to show that $f(n) \leq k_i \leq n - f(n)$. Since $\lim_{n\mapsto\infty} f(n) = X$, we get that $k_i > X\varepsilon > f(n)$ holds for large enough n. Hence, the only remaining step is to show the inequality $k_i \leq n - f(n)$. From Lemma 1 it follows that $n - k_i > \min\{n/2, (n-1)(1-\alpha)\}$. Since $f(n) < n/2$, if $n - k_i > n/2$ it holds $k_i < n - f(n)$. So let us suppose that $n - k_i > (n-1)(1-\alpha)$, i.e. $k_i < 1 + \alpha(n-1)$. Let $n \geq \frac{\varepsilon + \alpha(1-\alpha)^2}{\alpha(1-\alpha)^2}$. Then it holds for large enough n that

$$k_i < 1 + \alpha n - \alpha \leq n - \frac{\varepsilon}{\alpha(1-\alpha)} = n - \varepsilon X \leq n - f(n).$$

We have proved that if $k_i > X\varepsilon$ or $h_i > X(n-2)$ then at least

$$[k_i(n-k_i) + h_i](1-\alpha)^2$$

acknowledgements are delivered in round i.

To conclude the proof we show that after logarithmic number of iterations we get $k_i \leq X\varepsilon$ and $h_i \leq X(n-2)$. Let $M_i := 2(n-1)k_i + h_i$; then every delivered acknowledgement decreases M_i by at least one: indeed, if the acknowledgement was delivered over a hyperactive arc, h_i decreases by 1. If, on the other hand, the acknowledgement was delivered over an active arc, the number of uninformed vertices is decreased by at least one, and the number of hyperactive arcs is increased by at most $2n-3$ (new hyperactive arcs are between the newly informed vertex and any other vertex, with the exception of the arc that delivered the acknowledgement which is passive).

From Lemma 1 it follows that either $n - k_i > n/2$ or $n - k_i > (n-1)(1-\alpha)$. In the first case it follows that at least $(1-\alpha)^2[k_i(n-k_i)+h_i] > (1-\alpha)^2[k_in/2 + h_i] \geq \frac{(1-\alpha)^2}{4}M_i$ acknowledgements are delivered. In the second case we get that at least $(1-\alpha)^2[k_i(n-k_i)+h_i] > (1-\alpha)^2[k_i(n-1)(1-\alpha)+h_i] \geq \frac{(1-\alpha)^3}{2}M_i$ acknowledgements are delivered. Let $c := \min\{\frac{(1-\alpha)^2}{4}, \frac{(1-\alpha)^3}{2}\}$, then obviously every iteration decreases the value of M_i at least by factor c. Since

---

[5] Assume that n is large enough such that $f(n)$ is real number.

the value of M at the beginning of the algorithm is $M_1 = O(n^2)$, $\log_{1/c} M_1 = O(\log n)$ steps are sufficient to inform all but a constant number (at most $X\varepsilon$) of vertices and to ensure that the number of remaining hyperactive arcs is linear (at most $X(n-2)$).

4   Hypercubes

In this section we consider d-dimensional hypercubes. The hypercube Q_d has 2^d vertices, and both diameter and edge connectivity are d. We present an algorithm that informs all but a constant number of vertices in time $O(d^2)$.

The general idea is the same as for complete graphs: first we perform two initialization steps to make sure there are enough informed vertices for the subsequent analysis to hold. Next, simple rounds are repeated for a sufficient number of times. The analysis, however, is more complicated in this case.

The next lemma covers the initialization steps. In the first step, the initiator sends a message to all its neighbors, and at least one of these messages is delivered. In the second step, the initiator sends a message to all its neighbors again; moreover, each of the vertices informed in the first step sends a message to all its neighbors except the initiator.

Lemma 2. *After the first two steps of the algorithm, at least $\frac{1-\alpha}{2}(2d-1)$ vertices are informed.*

For the rest of this section we suppose that there are at least $\frac{1-\alpha}{2}(2d-1)$ informed vertices. We show that after $O(d^2)$ simple rounds all but some constant number of vertices are informed, and there are only linearly many hyperactive arcs. At the end of this section, we shall be able to prove the following theorem.

Theorem 2. *Let $\varepsilon \in (0,1)$ be an arbitrary constant. For large enough d it is possible to inform all but at most $X/(1-\varepsilon)$ vertices of Q_d within $O(d^2)$ time steps. Moreover, the number of remaining hyperactive arcs is at most $X(d-1)$.*

In our analysis we need to assert that enough acknowledgements are delivered, given the number of informed vertices. To bound the number of sent messages, we rely heavily upon the following isoperimetric inequality due to Chung et. al. [9]:

Claim. [9] Let S be a subset of vertices of Q_d. The size of the edge boundary of S, denoted as $\partial(S)$ is defined as the number of edges connecting S to $Q_d \setminus S$. Let $\partial(k) = \min_{|S|=k} \partial(S)$, and let lg denote the logarithm of base 2. It holds that

$$\partial(k) \geq k(d - \lg k)$$

The first step in the analysis is to prove that if there are enough uninformed vertices, or enough hyperactive arcs at the beginning of a round i, then sufficiently many acknowledgements are delivered in this round:

Lemma 3. *Consider a d-dimensional hypercube with k non-informed vertices and h hyperactive arcs. Let $\varepsilon \in (0,1)$ be an arbitrary constant, and let $k > X/(1-\varepsilon)$ or $h > X(d-1)$. Then in the second step of a simple round at least $\beta(h + \partial(k))$ acknowledgements are delivered, where $\beta = (1-\alpha)^2$.*

Sketch of the proof. Let S be the set of informed vertices. In the first step of the round, $h + \partial(S)$ messages are sent. Since the edge boundary of informed and uninformed vertices is the same, at least $h + \partial(k)$ messages are sent in the first step of the round. The idea of the proof is to show that $\alpha(h + \partial(k)) \geq d-1$, so in the first step at most $\alpha(h+\partial(k))$ messages are lost, and at least $(1-\alpha)(h+\partial(k))$ of them are delivered. Next we prove that $\alpha(1-\alpha)(h + \partial(k)) \geq d-1$, so in the second step at least $(1-\alpha)^2(h+\partial(k))$ messages are delivered. Since $1-\alpha < 1$, it is sufficient to prove that $\alpha(1-\alpha)(h + \partial(k)) \geq d-1$. If $h > X(d-1)$ then clearly $h + \partial(k) \geq X(d-1)$ and the statement holds. Hence, the main goal of the proof is to show that for $k > X/(1-\varepsilon)$, it holds $\partial(k) \geq X(d-1)$. To do so, the inequality $2^d - k \geq \frac{1-\alpha}{2}(2d-1)$, which is granted by Lemma 2, is used.

In the rest of the proof of Theorem 2 we show that $O(d^2)$ simple rounds are sufficient to inform almost all vertices. The analysis is divided into two parts. In the first part we prove that within $O(d^2)$ rounds at least $2^d/3$ vertices are informed. In the second part we show that another $O(d^2)$ rounds are sufficient to finish the algorithm.

Lemma 4. *After performing $O(d^2)$ simple rounds on Q_d at least $2^d/3$ vertices are informed.*

Sketch of the proof. Let $l := 2^d - k$ be the number of informed vertices. From Lemma 3 it follows that at least $\beta\partial(k)$ acknowledgements are delivered in one simple round. Since the edge boundary of informed vertices is also the boundary of uninformed vertices, the number of delivered acknowledgements is at least $\beta\partial(l)$. Furthermore, every delivered acknowledgement adds one passive arc, so the number of passive arcs grows at least by $\beta\partial(l)$ each round, which we show to be at least a factor of $\left(1 + \frac{1}{\frac{d}{\beta \lg 3}}\right)$. Because the number of passive arcs cannot grow over $d2^d/3$ without informing at least $2^d/3$ vertices, we get the statement of the lemma.

Lemma 5. *Let $\varepsilon \in (0,1)$ be an arbitrary constant, and let $k_i \leq (2/3)2^d$ be the number of uninformed vertices and h_i the number of hyperactive arcs of an d-dimensional hypercube at the beginning of round i. Then after $O(d^2)$ simple rounds there are at most $X/(1-\varepsilon)$ uninformed vertices and at most $X(d-1)$ hyperactive arcs.*

Sketch of the proof. Similarly to the proof of Theorem 1, let us consider the measure $M_i := 2dk_i + h_i$ which decreases with every acknowledgement delivered. We show that as long as the requirements of Lemma 3 hold, M_i decreases in each round by a factor $\left(1 + \frac{\beta \lg(2/3)}{d}\right)$. Since $M_i \leq (5/3)d2^d$, we get the statement of the lemma.

Combining Lemma 2 with Lemma 4 and Lemma 5 completes the proof of Theorem 2.

5   Complete Broadcast in Complete Graphs

In Section 3 we have shown how to inform all but some constant number of vertices in a complete graph K_n in time $O(\log n)$. A natural question is to ask if it is possible to inform also the remaining vertices in the same time complexity. In this section we partially answer this question. In particular, we show in the following subsection that if the graph is equipped with a chordal sense of direction, then the complete broadcasting can be performed in time $O(\log n)$. In the subsequent subsection, we show that if the constant $\alpha < 0.55$, complete broadcast can be performed in time $O(\log n)$ without the sense of direction, too.

5.1   Chordal Sense of Direction

Let us consider a complete graph with a fixed Hamiltonian cycle $\mathcal{C}$ (unknown to the vertices). We say that the complete graph has a chordal sense of direction if in every vertex the incident arcs are labeled by the clockwise distance on $\mathcal{C}$ (see Figure 1). The notion of a sense of direction has been defined formally for general graphs, and it has been known to significantly reduce the complexity of many distributed tasks (e.g. [17, 18]).

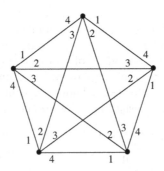

Fig. 1. K_5 with the chordal sense of direction

We show how to perform a complete broadcast on a complete graph with the sense of direction in time $O(\log n)$. The process consists of three steps. First, using Theorem 1, all but a constant number of vertices are informed. In the second phase the information is delivered to all but one vertex. In the last phase the remaining single vertex is informed.

The sense of direction is essential to our algorithm. Since there is a unique initiator of the broadcasting, all vertices can derive unique identifiers defined as their distance on $\mathcal{C}$ from the initiator. Furthermore, the sense of direction allows each vertex to know the identifier of a destination vertex of any of its incident arcs.

Lemma 6. *It is possible to inform all vertices but one on complete graphs with chordal sense of direction in time $O(\log n)$. Furthermore, after finishing the algorithm vertex 0 or vertex 1 knows a constant number of candidates for the uninformed vertex.*

Proof. The outline of the algorithm is as follows: At first the algorithm from Theorem 1 is performed, which ensures that all but a constant number of vertices are informed. Afterwards a significant group of vertices negotiate a common set U of *candidates* for uninformed vertices, such that all uninformed vertices are in U and the size of U is constant. The vertices then cooperate to inform all vertices in U but one. As a side effect, the set U will be known to vertex 0 or vertex 1, hence satisfying the second claim of the lemma. Now we present this algorithm in more detail:

Phase 1. Run the algorithm from Theorem 1. This phase takes $O(\log n)$ time and ensures that there are at most $X\varepsilon$ uninformed vertices and at most $X(n-2)$ hyperactive arcs.

Phase 2. Each vertex v that has at most $3X(1+\varepsilon)$ non-passive (i.e. active or hyperactive) links leading to the set of vertices U_v sends a message containing U_v to vertices with number 0 and 1.

Now we show that at least one of these messages is delivered. It is easy to see that there are at least $2n/3$ vertices satisfying the above-mentioned condition, otherwise there would be more than $n/3$ vertices with at least $3X(1+\varepsilon)$ non-passive links, so there would be more than $nX(1+\varepsilon)$ active or hyperactive arcs. But since the number of uninformed vertices is at most $k \leq X\varepsilon \leq n/2$ for large n, there are $k(n-k) \leq X\varepsilon(n-X\varepsilon)$ active arcs. So the total number of active or hyperactive arcs is at most $X\varepsilon(n-X\varepsilon)+X(n-2) \leq Xn(1+\varepsilon)$, which is a contradiction.

The rest of the algorithm will be time-multiplexed into two parts. In even time steps, the case that the vertex 0 received a message in phase 2 is processed. In odd time steps, the case that the vertex 1 received a message is processed analogously. Hence, we can restrict to the first case in the rest of the algorithm description. As there are only two cases the asymptotic complexity of the algorithm is unaffected by the multiplexing.

Phase 3. The vertex 0 received at least one message containing a set of possibly uninformed vertices. It is obvious that the set of uninformed vertices is a subset of every received message. Hence the set U can be defined as the intersection of the received messages: Indeed, every uninformed vertex is in U and the size of U is at most $3X(1+\varepsilon) = O(1)$. The set U is then distributed using the algorithm in Theorem 1 among at least $n - X\varepsilon$ vertices in time $O(\log n)$.

Phase 4. There are at least $n - X\varepsilon$ vertices aware of the set U. In this phase they cooperate to inform all but one vertex in U, using an idea similar to Lemma 2 in [12]: every vertex aware of the set U iterates through all pairs $[i,j]$ $(i,j \in U)$ in lexicographical order; in each time step it sends the original message to both vertices i and j. Since in each time step at least $2n - X\varepsilon$ messages are sent, at least one of them is delivered (for large enough n). As all vertices process the same pair $[i, j]$ in every time step, this ensures that a new vertex is informed whenever both i and j were uninformed. Hence, at the end of this phase all vertices but one are informed. The time complexity of this phase is $O(|U|^2) = O(1)$.

It is obvious that after finishing the Phase 4 the claim of the Lemma holds.

Finally, we show how to inform the last remaining vertex, thus proving the following theorem:

Theorem 3. *It is possible to perform broadcasting on complete graphs with chordal sense of direction in time $O(\log n)$.*

Sketch of the proof. Suppose that after performing the algorithm from Lemma 6 all vertices with the exception of some vertex v are informed and vertex 0 knows a set U of constant size containing candidates for v. The algorithm from Lemma 6 is used again to broadcast U with two possible outcomes: either v was informed during the broadcast, or all other vertices have the same set of candidates, which they try to inform one by one.

5.2   Without Sense of Direction

As a last result in this paper we show that it is possible to perform broadcasting on complete graphs in time $O(\log n)$ for small values of α (i.e. $\alpha \lesssim 0.55$) even without the sense of direction. The idea is to use the algorithm from Theorem 1 to inform all but constantly many vertices. Next, instead of repeating 2-step simple rounds, some $\log n$-step extended rounds are repeated, such that each extended round informs a yet uninformed vertex. During an extended round messages are sent for $O(\log n)$ steps in such a way that in every step the number of hyperactive arcs is decreased by some factor[6] unless a new vertex is informed.

Theorem 4. *Let $1 - \alpha - 2\alpha^2 + \alpha^3 > 0$. Then it is possible to perform broadcasting on complete graphs without sense of direction in time $O(\log n)$.*

Proof. The algorithm is described as Algorithm 1..

At first, the algorithm from Theorem 1 is performed, ensuring that there are at most $k \leq X\varepsilon$ uninformed vertices and at most $h \leq X(n-2)$ hyperactive arcs (X and ε have the same meaning as in Theorem 1). The purpose of one iteration of the loop on lines 3–14 is to inform at least one uninformed vertex. Taking $L_1 := X\varepsilon = O(1)$ ensures that all vertices will be informed.

The loop on lines 4–10 reduces the number of hyperactive arcs to zero unless a new vertex is informed. One iteration of this loop either informs a new vertex or reduces the number of hyperactive arcs from h to $(1 - Y/2)h$, where $0 < Y < 1$ is a constant (depending on α) defined later. Hence the number of hyperactive arcs decreases exponentially with number of iterations of the loop and $\log_{1/(1-Y/2)} h$ iterations are sufficient to eliminate all hyperactive arcs. Since the condition

---

[6] In this part we need the assumption that α is small enough.

Algorithm 1. Complete graphs without sense of direction

1: perform almost-complete broadcast according to Theorem 1
2: let k denote the number of uninformed vertices, let h denote the number of hyper-active arcs
3: **loop** L_1 times   // Perform L_1 extended rounds
4:     **loop** $L_2(n)$ times   // In each iteration h decreases by a constant factor
5:         $E :=$ set of all currently active or hyperactive arcs; $P := \emptyset$
6:         **loop** L_3 times
7:             send the message via all arcs in $E \cup P$
8:             $P := P \cup \left\{ e \mid \begin{array}{l} \text{a message has been delivered in this step} \\ \text{via the opposite arc of } e \end{array} \right\}$
9:         **end loop**
10:     **end loop**
11:     **loop** $L_4(n)$ times   // Inform new vertex and decrease a
12:         perform one simple round
13:     **end loop**
14: **end loop**
The values of L_1, $L_2(n)$, L_3 and $L_4(n)$ are specified in the analysis of the algorithm, such that $L_1, L_3 = O(1)$ and $L_2(n), L_4(n) = O(\log n)$.

$h \leq X(n-2)$ holds before every execution of the loop (this is provided either directly by Theorem 1 or by the loop on lines 11–13), we can define $L_2 := \log_{1/(1-Y/2)}(X(n-2)) = O(\log n)$.

Now we describe one iteration of the loop on lines 4–10. We distinguish two types of arcs that are hyperactive at the beginning of the considered iteration: An arc e is a *single hyperactive arc* if and only it is hyperactive and the opposite arc of e is passive at the beginning of the iteration. Otherwise (i.e. if both e and the opposite arc of e are hyperactive at the beginning of the iteration), e is a *double hyperactive arc*.

Let E be the set of all active or hyperactive arcs at the beginning of the iteration, and P be the set of all arcs opposite to arcs through which some message has been delivered in the current iteration. Furthermore, let k' be the number of uninformed vertices at the beginning of the current iteration, h' be the number of hyperactive arcs at the beginning of the current iteration and $p = |P \setminus E|$ be number of arcs in P that were passive at the beginning of the current iteration. It clearly holds that $|E| = k'(n-k') + h'$ and that $k'(n-k') + h' + p$ messages are sent on every execution of line 7. Since at least $n-1$ messages are lost (because we may assume that no new vertex is informed), at most $\alpha(k'(n-k') + h' + p)$ of them are lost, i.e. at least $(1-\alpha)(k'(n-k') + h' + p)$ are delivered.

Now assume by contradiction that the number of hyperactive arcs does not decrease below $(1-Y/2)h'$, and no new vertices are informed during the current iteration of the loop on lines 4–10. Consider any message delivered over an arc e which is a double hyperactive arc or an arc in $P \setminus E$; it is easy to see that the opposite arc of e is passive after the delivery and that it was hyperactive at the beginning of the iteration. This fact yields that at most $(Y/2)h'$ messages are delivered over a double hyperactive arc or an arc in $P \setminus E$ on any execution of line 7.

Now we show a lower bound on the number of messages that pass over double hyperactive arcs or arcs in $P \setminus E$ or single hyperactive arcs whose opposite arcs are not in $P \setminus E$. Intuitively, every such message ensures some progress of the algorithm, since either an arc is made passive (in the first two cases) or a new arc is added to $P \setminus E$ (in the third case). As no messages passes over active arcs by our assumption, and at most p messages pass over single hyperactive arcs whose opposite arcs are in $P \setminus E$, there are at least $(1-\alpha)(k'(n-k')+h'+p)-p$ messages satisfying one of these three cases. Using the inequalities $k'(n-k') \geq n-1$ and $p \leq h'$ yields $(1-\alpha)(k'(n-k')+h'+p)-p \geq (1-\alpha)(n-2)+(1-2\alpha)h'$. Because $h' \leq X(n-2)$ which is equivalent to $(n-2) \geq \alpha(1-\alpha)h'$, we have $(1-\alpha)(k'(n-k')+h'+p)-p \geq (1-\alpha-2\alpha^2+\alpha^3)h'$. Defining $Y := 1-\alpha-2\alpha^2+\alpha^3$, which is positive and less than one by the assumption of the Lemma, we have shown that there are at least Yh' messages satisfying one of the three cases.

However, at most $(Y/2)h'$ of them satisfies the first two cases, hence there are at least $(Y/2)h'$ arcs added to P in every execution of line 8. So taking $L_3 := 2/Y + 1$ ensures that P contains opposite arcs to all single hyperactive arcs at the beginning of the last iteration of the loop on lines 6–9. However, this is a contradiction with the fact that new arcs are added to P at line 8.

We conclude the proof with the analysis of the loop on lines 11–13. In the first iteration of the loop a new vertex is informed, because there are no hyperactive arcs left after the loop on lines 4–10 finished (unless the new vertex has already been informed in that loop). Due to Theorem 1, next $O(\log n)$ iterations are sufficient to ensure that $h \leq X(n-2)$, which is an invariant required by the loop on lines 4–10. Hence putting $L_4(n) := O(\log n)$ (according to Theorem 1) is sufficient to make the algorithm work correctly in time $L_1(L_2(n)L_3 + L_4(n)) = O(\log n)$.

6   Conclusions, Open Problems, and Further Research

We have studied the problem of almost complete broadcast under the model of fractional dynamic faults with threshold. We showed that both in complete graphs and in hypercubes, it is possible to inform all but constantly many vertices in time $O(D \log n)$ where D is the diameter of the graph and n is the number of vertices.

Moreover, we have proved that if the complete graph is equipped with the chordal sense of direction, or the parameter $\alpha < 0.55$, a complete broadcast can be performed in time $O(\log n)$.

This research leaves many open questions and directions for further research, from which we mention at least a few. One obvious question is to ask if it is possible to perform a complete broadcast in complete graphs also for large values of α in polylogarithmic time. The difficulty of broadcast in the fractional dynamic model with threshold stems from the fact that, in order to inform the last few vertices, all informed vertices must cooperate very tightly. In general, the relationship between the almost complete and complete broadcast in various models is worth studying. We have also not considered non-constant values of α. It would be interesting to extend our results to more general classes of graphs.

We finish by noting that there is a lack of any non-trivial lower bounds in the model of fractional faults with threshold.

References

1. Ahlswede, R., Gargano, L., Haroutunian, H.S., Khachatrian, L.H.: Fault-tolerant minimum broadcast networks. Networks, vol. 27 (1996)
2. Bagchi, A., Hakimi, S.L.: Information dissemination in distributed systems with faulty units. IEEE Transactions on Computers 43(6), 698–710 (1994)
3. Berman, K.A., Hawrylycz, M.: Telephone problems with failures. SIAM Journal on Algebraic and Discrete Methods 7(1), 13–17 (1986)
4. Berman, P., Diks, K., Pelc, A.: Reliable broadcasting in logarithmic time with Byzantine link failures. Journal of Algorithms 22(2), 199–211 (1997)
5. Bjork, L.A.: Recovery scenario for a db/dc system. In: ACM'73: Proceedings of the annual conference, pp. 142–146. ACM Press, New York (1973)
6. Chang, J.-M., Maxemchuk, N.F.: Reliable broadcast protocols. ACM Transactions on Computer Systems 2(3), 251–273 (1984)
7. Chlebus, B., Diks, K., Pelc, A.: Broadcasting in synchronous networks with dynamic faults. Networks, vol. 27 (1996)
8. Chlebus, B.S., Diks, K., Pelc, A.: Optimal broadcasting in faulty hypercubes. In: FTCS, pp. 266–273 (1991)
9. Chung, F.R.K., Füredi, Z., Graham, R.L., Seymour, P.: On induced subgraphs of the cube. Journal of Combinatorial Theory Series A. 49, 180–187 (1988)
10. Dijkstra, E.W.: Self-stabilizing systems in spite of distributed control. Commun. ACM 17(11), 643–644 (1974)
11. Diks, K., Pelc, A.: Almost safe gossiping in bounded degree networks. SIAM Journal on Discrete Mathematics 5(3), 338–344 (1992)
12. Dobrev, S., Královič, R., Královič, R., Santoro, N.: On fractional dynamic faults with threshold. In: Flocchini, P., Gasieniec, L. (eds.) SIROCCO 2006. LNCS, vol. 4056, pp. 197–211. Springer, Heidelberg (2006)
13. Dobrev, S., Vrťo, I.: Optimal broadcasting in hypercubes with dynamic faults. Information Processing Letters 71(2), 81–85 (1999)
14. Dobrev, S., Vrťo, I.: Dynamic faults have small effect on broadcasting in hypercubes. Discrete Applied Mathematics 137(2), 155–158 (2004)
15. Dolev, S.: Self-stabilization. MIT Press, Cambridge (2000)
16. Fischer, M.J., Lynch, N.A., Paterson, M.S.: Impossibility of distributed consensus with one faulty process. Journal of the ACM 32(2), 374–382 (1985)
17. Flocchini, P., Mans, B., Santoro, N.: On the impact of sense of direction on message complexity. Information Processing Letters 63(1), 23–31 (1997)
18. Flocchini, P., Mans, B., Santoro, N.: Sense of direction in distributed computing. In: International Symposium on Distributed Computing, pp. 1–15 (1998)
19. Fraigniaud, P.: Asymptotically optimal broadcasting and gossiping in faulty hypercube multicomputers. IEEE Transactions on Computers 41(11), 1410–1419 (1992)
20. Gasieniec, L., Pelc, A.: Broadcasting with linearly bounded transmission faults. Discrete Applied Mathematics 83(1–3), 121–133 (1998)

21. Hedetniemi, S., Hedetniemi, S., Liestman, A.: A survey of broadcasting and gossiping in communication networks. Networks 18, 319–349 (1988)
22. Královič, R., Královič, R., Ružička, P.: Broadcasting with many faulty links. In: Sibeyn, J.F. (ed) SIROCCO, of Proceedings in Informatics, vol. 17, pp. 211–222. Carleton Scientific (2003)
23. Královič, R., Královič, R.: Rapid almost-complete broadcasting in faulty networks. Technical report, arXiv:cs.DC/0703122 (2007)
24. Liptak, Z., Nickelsen, A.: Broadcasting in complete networks with dynamic edge faults. In: Butelle, F. (ed.) OPODIS, Studia Informatica Universalis, pp. 123–142. Suger, Saint-Denis, rue Catulienne, France (2000)
25. Pease, M., Shostak, R., Lamport, L.: Reaching agreement in the presence of faults. Journal of the ACM 27(2), 228–234 (1980)
26. Pelc, A.: Broadcasting in complete networks with faulty nodes using unreliable calls. Information Processing Letters 40(3), 169–174 (1991)
27. Pelc, A., Peleg, D.: Feasibility and complexity of broadcasting with random transmission failures. In: PODC '05: Proceedings of the twenty-fourth annual ACM SIGACT-SIGOPS symposium on Principles of distributed computing, pp. 334–341. ACM Press, New York (2005)
28. Santoro, N., Widmayer, P.: Distributed function evaluation in the presence of transmission faults. In: Asano, T., Imai, H., Ibaraki, T., Nishizeki, T. (eds.) SIGAL 1990. LNCS, vol. 450, pp. 358–367. Springer, Heidelberg (1990)
29. Stanoi, I., Agrawal, D., Abbadi, A.E.: Using broadcast primitives in replicated databases. In: ICDCS '98: Proceedings of the The 18th International Conference on Distributed Computing Systems, pp. 148–155. IEEE Computer Society, Washington (1998)

Design of Minimal Fault Tolerant On-Board Networks: Practical Constructions[*]

Jean-Claude Bermond[1], Frédéric Giroire[2], and Stéphane Pérennes[1]

[1] Mascotte Project, CNRS/I3S/INRIA, 2004 route des Lucioles, B.P. 93, _ F-06902
Sophia-Antipolis Cedex, France
[2] Project Algorithms, INRIA Rocquencourt,F-78153 Le Chesnay, France

Abstract. The problem we consider originates from the design of efficient on-board networks in satellites (also called Traveling Wave Tube Amplifiers). Signals incoming in the network through ports have to be routed through an on-board network to amplifiers. The network is made of expensive switches with four links and subject to two types of constraints. First, the amplifiers may fail during satellite lifetime and cannot be repaired. Secondly, as the satellite is rotating, all the ports are not well oriented and hence not available. Let us assume that we have $p + \lambda$ ports (inputs) and $p + k$ amplifiers (outputs), then a $(p, \lambda, k)-$network is said to be *valid* if, for any choice of p inputs and p outputs, there exist p edge-disjoint paths linking all the chosen inputs to all the chosen outputs. Then, the objective is to design a valid network having the minimum number of switches denoted $N(p, \lambda, k)$. In the special case where $\lambda = 0$, these networks were already studied as *selectors*. Here we present validity certificates from which derive lower bounds for $N(p, \lambda, k)$; we also provide constructions of optimal (or quasi optimal) networks for practical values of λ and k ($1 \leq \lambda \leq k \leq 8$) and a general way to build networks for any k and any λ.

1 Introduction

Motivation. The problem we consider here was asked by Alcatel Space Industries and consists of designing efficient on-board networks in satellites (problem called Traveling Wave Tube Amplifiers). The satellites under consideration are used for TV and video transmission (like for example the Eutelsat or Astra series) as well as for private applications. Signals incoming in a telecommunication satellite through ports have to be routed through an on-board network to amplifiers. A first constraint is that the network is built of switches with four links. But other constraints appear. On the one hand the amplifiers may fail during satellite lifetime and cannot be repaired. On the other hand, as the satellite is rotating, all the ports and amplifiers are not well oriented and hence not available. So more amplifiers and ports are needed than the number of signals which have to be routed. Note that in this context, contrary to classical networks, there

---

[*] This work has been partially funded by the European project IST FET AEOLUS.

G. Prencipe and S. Zaks (Eds.): SIROCCO 2007, LNCS 4474, pp. 261–273, 2007.

are no failure of links nor switches. Indeed the switches are very reliable rotating mechanical systems and links are just big wave-tubes. However two different signals cannot use the same tube.

To decrease launch costs, it is crucial to minimize the network physical weight, i.e. for us, to minimize the number of switches. Each switch weighs about 200g and saving a switch implies a gain of more than 20 000 Euros; therefore it is worth saving even one. Space industries are interested in designing such networks for specific values of the parameters. However the general theory is of interest by itself.

Problem. We consider here *networks*, that is graphs connecting *inputs* to *outputs* and where vertices represent the switches. We define a $(p, \lambda, k)-network$ as a network with $p + \lambda$ inputs and $p + k$ outputs. A $(p, \lambda, k)-$network is said to be *valid*, if, for any choice of p inputs and of p outputs, there exist p edge-disjoint paths linking all the chosen inputs to all the chosen outputs. For symmetry reason, we may assume in the following that $k \geq \lambda$ and we note $n := p + k$. Note that when chosen the paths become directed from the input to the output. But the disjointness condition is undirected as there cannot be two signals in a wave-tube (edge).

We study the case where the switches of the network have degree four (although the theory can be generalized to any degree) which is of primary interest for the applications. The problem is to find $N(p, \lambda, k)$, the *minimum* number of switches in a valid $(p, \lambda, k)-$network and to give *constructions* of such optimal networks (see Figure 4).

Note that finding a minimal network is a challenging problem: the number of possible networks grows exponentially and even testing the validity of a given network is hard. Indeed if we fix the valid inputs and the valid outputs testing the validity reduces to a flow problem but the number of possible choices for inputs and outputs grows exponentially as they are binomial coefficients. Still, the problem is in Co-NP, since one prove that a network is not valid by exhibiting a bad cut. In fact, deciding if a given $(p, \lambda, k)-$network is valid is a Co-NP complete problem, see [1].

In the specific case $\lambda = k$, it is interesting to design networks with a particular property: every switch linked to a port is also linked to an amplifier; indeed if there is no failure the incoming signal is routed directly to the amplifier connected to its entering switch. This minimizes the length of signals and avoids the interferences. These networks are called *simplified networks*. Observe that in that case every switch is linked to either two or four switches.

Related Work. When $\lambda = 0$, a valid network is called a *selector* (For a survey on selectors, see [2] or the seminal work of Pippenger [3]). A general theory of selectors can be found in [4] where several results are obtained for small values of k. For example it is proved that $N(p, 0, 4) = \lceil \frac{5p}{4} \rceil$.

In [5] the case of selectors with switches of degree $2k > 4$ is considered. In [6] the authors consider a variant of selectors where some signals have priority and should be sent to amplifiers offering the best quality of service. In [7] the authors study the case were all the amplifiers are different and where a given input has to be sent to a dedicated output the problem being related to permutation networks.

Results. We first present a simple cut criterion which implies the validity of networks. This criterion will be useful both to prove the validity of the designed networks (giving upper bounds) and also to find lower bounds for the minimal number $N(p, \lambda, k)$ of switches of valid (p, λ, k)−networks.

In Section 3, we present ways to build valid networks close to minimal for small values of λ and k ($1 \leq \lambda \leq k \leq 8$). For instance, for $k \in \{3, 4\}$ and $0 < \lambda \leq 4$

$$N(p, \lambda, k) = \lceil \frac{5n}{4} \rceil$$

For $k \in \{5, 6\}$ and $0 < \lambda \leq 6$

$$N(p, \lambda, k) \leq \lceil \frac{3n}{2} \rceil$$

Examples of $(p, 4, 4)$ and $(p, 6, 6)$−networks are given in Figure 3. In Section 4, we present a general way to build networks for any k and any λ.

2   Preliminaries

In this section, we define more formally the problem and introduce notations used throughout the paper.

We state a cut criterion (Proposition 1): this criterion is fundamental because it characterizes the validity of (p, λ, k)-networks. It is extensively used to prove that networks are valid. In Section 3 we use the cut criterion to detect forbidden patterns leading to lower bounds for the number of switches of valid networks.

Proofs of lower and upper bounds are simplified by the construction theorems given here (Theorems 1 and 2). We also give one way to build any (p, λ, k)−networks from (p, k)−selectors leading us to a linear bound for the number of switches of (p, λ, k)−networks.

Notations. Let $G = (V, E)$ be a graph and let $W \subseteq V$ be a subset of vertices of G. We denote by $\Delta(W)$ the set of edges connecting W and $\overline{W} = V \backslash W$ and by $\Gamma(W)$ the set of vertices of $\overline{W}$ adjacent to vertices of W.

The cardinality of $\Delta(W)$ is denoted by $\delta(W)$. More generally, the convention used in this paper is that, if a set is denoted by an upper case letter, the corresponding lower case letter denotes its cardinality.

For the sake of simplicity, when a subset has a single element, we note $\delta(v)$ instead of $\delta(\{v\})$.

(p, λ, k)−**networks and valid** (p, λ, k)−**networks.** A (p, λ, k)-*network* is a triple $N = \{(V, E), i, o\}$, where (V, E) is a graph. i, o are positive integral functions defined on V, called input and output functions, such that for any $v \in V$,

$i(v) + o(v) + \delta(v) \leq 4$ (for parity reasons, some switches can be linked to a dead-end. The total number of inputs is $i(V) = \Sigma_{v \in V} i(v) = p + \lambda$, and the total number of outputs is $o(V) = \Sigma_{v \in V} o(v) = p + k$. We can see a network as a graph where all vertices but the leaves have degree 4, and where the leaves correspond to inputs or outputs or dead-ends. A *non-faulty output function* is a function o' defined on V such that $o'(v) \leq o(v)$ for any $v \in V$ and $o'(V) = p$. A *used input function* is a function i' defined on V such that $i'(v) \leq i(v)$ for any $v \in V$ and $i'(V) = p$. A (p, λ, k)-network is said *valid* if for any faulty output function o' and any used input function i', there are p edge-disjoint paths in G such that each vertex $v \in V$ is the initial vertex of $i'(v)$ paths and the terminal vertex of $o'(v)$ paths.

Design Problem. Let $N(p, \lambda, k)$ denotes the minimum number of switches of a valid (p, λ, k)-network. The *Design Problem* consists in determining $N(p, \lambda, k)$ and in constructing a minimum (p, λ, k)-network, or at least a valid (p, λ, k)-network with a number of vertices close to the optimal value.

We introduce a variation of the problem: consider networks with $p + \lambda$ switches with exactly one input and one output (we call such a switch a doublon), and with $k - \lambda$ switches with only one output. To find minimum valid network like these is what we call the *Simplified Design Problem*. Networks of this kind are especially good for practical applications, as they simplify the routing process, minimize path lengths and lower interferences between signals. The minimal number of switches of such networks is noted $N'(p, \lambda, k)$.

Excess, Validity and Cut-criterion. We show that, to verify if a network is valid, instead of solving a flow/supply problem for each possible configuration of output failures and of used inputs, it is sufficient to look at an invariant measure of subsets of the network, the *excess*, as expressed in the following proposition.

Definition 1 (Excess $\varepsilon(W)$). *Let $\{(V, E), i, o\}$ be a (p, λ, k)-network and $W \subseteq V$ a subset of vertices. The* excess in inputs *of W is defined as*

$$\varepsilon_i(W) := \delta(W) + o(W) - \min(k, o(W)) - \min(i(W), p).$$

The excess in outputs *of W is defined as*

$$\varepsilon_o(W) := \delta(W) + i(W) - \min(\lambda, i(W)) - \min(o(W), p).$$

Lemma 1. *Let $\{(V, E), i, o\}$ be a (p, λ, k)-network. Consider a subset $W \subseteq V$. We have $\varepsilon_o(\overline{W}) = \varepsilon_i(W)$.*

Proof. As $\delta(\overline{W}) = \delta(W)$, $o(\overline{W}) = p + k - o(W)$, $i(\overline{W}) = p + \lambda - i(W)$,

$$\begin{aligned}
\varepsilon_o(\overline{W}) &= \delta(W) + p + \lambda - i(W) - \min(\lambda, p + \lambda - i(W)) \\
&\quad - \min(p + k - o(W), p) \\
&= \delta(W) + p + \lambda - i(W) - (\lambda - i(W) \\
&\quad + \min(i(W), p)) - (p - o(W) + \min(k, o(w))) \\
&= \delta(W) + o(W) - \min(k, o(w)) - \min(i(W), p) \\
&= \varepsilon_i(W).
\end{aligned}$$

So we can mainly restrict our attention to $\varepsilon(W) := \varepsilon_i(W)$.

Proposition 1 (Cut Criterion). *A* (p, λ, k)-*network is* valid *if and only if, for any subset of vertices* $W \subset V$, *the excess of* W *satisfies* $\varepsilon(W) \geq 0$.

The intuition is that the signals arriving in W (in number at most $\min(i(W), p)$) should be routed either to the valid outputs of W (in number at least $o(W) - \min(k, o(W))$) or to the links going outside (in number $\delta(W)$). The omitted formal proof reduces to a supply/demand flow problem.

Proposition 2 (Cut Criterions and Symmetry). *A* (p,λ,k)-*network is valid if and only if one of the following proposition is true.*

1. *For all* W, *we have* $\varepsilon_i(W) \geq 0$.
2. *For all* W, *we have* $\varepsilon_o(W) \geq 0$.
3. *For all* W, *with* $o(W) \leq \lceil \frac{p+k}{2} \rceil$ *or* $i(W) \leq \lceil \frac{p+\lambda}{2} \rceil$, *we have* $\varepsilon_i(W) \geq 0$ *and* $\varepsilon_o(W) \geq 0$.

Proof. Direct by Proposition 1 and Lemma 1.

Proposition 3 (Cut Criterions and Connectivity). *For all cut criterions of Proposition 2, it is sufficient to consider only connected subsets* W *with connected complements* $\overline{W}$.

This comes from the submodularity of ε (the proof is omitted here). Intuitively, if a W has a negative excess, one of its connected component also has a negative excess. If W has a negative excess, $\overline{W}$ has negative excess in outputs, and so one of its connected component.

Construction Theorems

Fig. 1. Valid symmetric (2,0,2)-network and (2,2,0)-network

Theorem 1. *The problem is symmetric in inputs and outputs, that is* $N(p,\lambda,k) = N(p, k, \lambda)$

Proof. As soon as we have a valid (p, λ, k)-network, we can immediately build a valid (p, k, l)-network which is the same after we have inverted the inputs and the outputs (see example of Figure 1). The validity is given by Proposition 2.

Theorem 2. *A valid* (p, λ, k)-*network* $\mathcal{R}$ *can be built from a valid* (p, k, k)-*network* $\mathcal{R}'$ *by removing* $k - \lambda$ *inputs.*

Proof. $\mathcal{R}$ and $\mathcal{R}'$ have the same set of switches, V. Let W be a subset of V. $i_{\mathcal{R}}(W)$ (resp. $i_{\mathcal{R}'}(W)$) denotes the number of inputs attached to W in $\mathcal{R}$ (resp. $\mathcal{R}'$). As $i_{\mathcal{R}}(W) \le i_{\mathcal{R}'}(W)$, $\varepsilon_{\mathcal{R}}(W) \ge \varepsilon_{\mathcal{R}'}(W)$, finishing the proof.

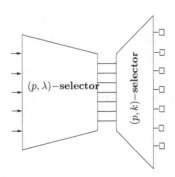

Fig. 2. A $(p, \lambda, k)-$network built with two selectors in series

Links with $(p, k)-$networks or $(p, k)-$selectors. (p, λ, k)-networks where all the inputs are used $(\lambda = 0)$ have been studied in [4] and [8] and are called $(p, k)-$networks or $(p, k)-$selectors. We have

$$N(p, 0, k) = N(p, k, 0) = N(p, k). \tag{1}$$

Definition 2 (2selectors $(p, \lambda, k)-$networks). *A 2selectors $(p, \lambda, k)-$network is formed by a $(p, \lambda)-$selector and a $(p, k)-$selector in serial, as indicated in the following theorem.*

Theorem 3
$$N(p, \lambda, k) \le N(p, \lambda) + N(p, k)$$
$$N(p, \lambda, k) \le O(p + k)$$

Proof. The proof is constructive. We can construct a valid $(p, \lambda, k)-$network with two valid selectors, a $(p, \lambda)-$selector and a $(p, k)-$selector as shown in Figure 2. The idea is to use the first selector in a symmetric way by replacing its $p + \lambda$ outputs by our inputs, then to link the inputs of the two selectors all together. The outputs of the second selector will be our outputs. Any subset of size p of the inputs can be routed to the p central links by the first selector. The second one can route these links to any subset of our outputs. So the network is valid. We call the networks built this way 2selectors $(p, \lambda, k)-$networks (Definition 2).

As it is shown in [4] and [8], that the minimum number of switches of a $(p, k)-$selector is linear, we have the same result for $(p, \lambda, k)-$networks.

3   Constructions for Small Cases

Recall that we may assume here that $1 \leq \lambda \leq k$ (see Theorem 1 and Equation 1). We first present results for $p = 1, 2$ and $k = 1, 2$, then for $p \geq 3$ and $k \geq 3$. We construct valid networks for $1 \leq k \leq 8$, called respectively 2networks, 4networks, 6 networks and 8networks. The number of switches of these networks are optimal for $\lambda = k$. For $\lambda < k$ we use Theorem 2 to obtain networks close to minimal. The proofs are mostly omitted here.

Case $p = 1, 2$ and any λ, k.

Theorem 4. $N(1, \lambda, k) = \lceil \frac{\lambda+k}{2} \rceil$ and $N(2, \lambda, k) = \lceil \frac{\lambda+k}{2} \rceil + 2$

Proof. When $p = 1$, we build a network consisting of a path where one end vertex has 3 inputs and the other 3 outputs. The internal vertices have either 2 inputs or 2 outputs when λ and k are even as shown in Figure 4. One vertex can have only one input or one output or one input and one output according λ or k or both are odd. When $p = 2$, we build similarly a network consisting of a cycle with a maximal number of switches with 2 inputs or outputs, as shown in Figure 4. The numbers of switches of the above networks attain these bounds.

To prove the validity we will use the cut criterion. When $p = 1$, let W be a connected subset of V (see Proposition 3 for this choice of W). As $\delta(W) \geq 1$, we have $\varepsilon(W) \geq 1 + o(W) - \min(o(W), k) - 1 \geq 0$. The cut criterion finishes the proof. When $p = 2$, we have $\delta(W) \geq 2$. So $\varepsilon(W) \geq 2 + o(W) - \min(o(W), k) - 2 \geq 0$. The networks are valid.

When $p = 1$, the network is minimal because a valid minimal network has to be connected. When $p = 2$, if we construct a network with one vertex less, we have a node v with 2 inputs and 1 output (or 2 outputs and 1 input). $\varepsilon(\{v\}) = 1 + 1 - 1 - 2 = -1$ ($\varepsilon_o(\{v\}) = -1$) and the network would not be valid. $\square$

Case $k = 1, 2$ (2networks).

A 2network consists of a cycle of $p + \lambda$ vertices with one input and one output, plus one vertex with one output if $\lambda = 1$ and $k = 2$, connected in a cycle, as shown in Figure 3 A.

Theorem 5. $N(p, \lambda, 2) = p + 2$. $N(p, \lambda, 1) = p + 1$.

Proof. If $W \subseteq V$, we have $\delta(W) = 2$. As $k \leq 2$, $\delta(W) - \min(k, o(W)) \geq 0$. Moreover, as, by construction, $i(W) \leq o(W)$, we have $o(W) - \min(p, i(W)) \geq 0$. So $\varepsilon(W) \geq 0$ and 2networks are valid.

If we construct a network with one vertex less, we obligatary have a vertex with $i(v) = 1$ and $o(v) = 2$. Consider the subset W made of this switch $\varepsilon_o(W) = 1 + 1 - 1 - 2 = -1$. This network would not be valid. So 2networks are minimal. For k= 1 we use a connection of the switches via a path. $\square$

Case $p \geq 3$ and $k \geq 3$

Let $\mathcal{R}$ be a (p, λ, k)-network. Recall that $n := p + k$. In what follows we suppose $p \geq 3$ and $k \geq 3$.

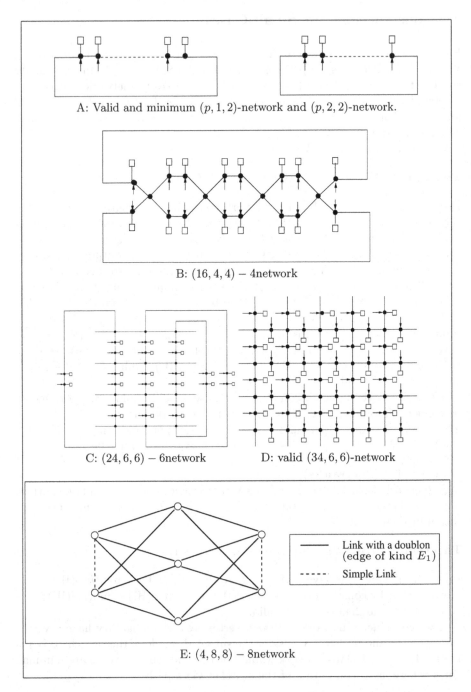

A: Valid and minimum $(p, 1, 2)$-network and $(p, 2, 2)$-network.

B: $(16, 4, 4) - 4$network

C: $(24, 6, 6) - 6$network   D: valid $(34, 6, 6)$-network

———— Link with a doublon (edge of kind E_1)

- - - - Simple Link

E: $(4, 8, 8) - 8$network

Fig. 3. Constructions for small k

Definition 3 (Doublons, R-Switches). *A doublon of $\mathcal{R}$ is a vertex with $i(v) = o(v) = 1$. An R-switch is a vertex that is not a doublon.*

Definition 4 (Edges of kind E_0, E_1 and E_2). *We build a graph G associated to $\mathcal{R}$. Its vertices are the R-switches of $\mathcal{R}$. Its edges are of three kinds, respectively E_0, E_1 and E_2: the edges of $\mathcal{R}$ between two R-switches, the edges corresponding in $\mathcal{R}$ to a path of length 2 with a doublon in the middle and those corresponding to a path of length 3 with 2 doublons in the middle.*

Note that the R-switches and doublons partition $\mathcal{R}$ and that the cut criterion gives immediately that edges of other kinds corresponding to paths of length more than 3 with doublons in the middle are forbidden. Indeed, if we consider the set W consisting of these doublons, we have $\delta(W) = 2$ and $o(W) = i(W) \geq 3$.

Fig. 4. Valid and minimum $(1, \lambda, k)$-network and $(2, \lambda, k)$-network

Case $k \leq 3, 4$, (4networks).
    A 4network is built from blocks made of one node with no inputs nor outputs and 4 doublons on 2 edges of kind E_2. A block is connected to two identical blocks in serial as shown in the $(16, 4, 4) - $4network of Figure 3 B. Each block has 5 nodes, 4 inputs and 4 outputs except one or two if $(p + k) \mod 4 \neq 0$ or $(p + \lambda) \mod 4 \neq 0$. Counting the number of switches and checking that the 4networks are valid gives the following theorem:

Theorem 6. *For $k = 3, 4$, $N(p, \lambda, k) \leq n + \frac{n}{4} + c'_4 - c$, $c'_4 = \lceil \frac{n \mod 4}{4} \rceil$ and $c = \lfloor \frac{k-\lambda}{2} \rfloor$.*

On the other hand we can prove the following lower bound:

Theorem 7. *For $k \geq 3$, $N(p, \lambda, k) \geq n + \frac{n}{4} - c''_4$, where $c''_4 = \frac{k-\lambda}{2} + \frac{k-\lambda}{8}$.*

Notice that the difference between the number of switches of a $(p, \lambda, k) - $4network and the lower bound is at most 1 and that in the case $\lambda = k = 4$ we obtain:

Corollary 1. $N(p, 4, 4) = n + \frac{n}{4}$.

Case $k = 5, 6$, (6networks).
    A 6network is built from blocks made of 3 switches connected in circle, each of them is connected to two doublons on an edge of kind E_2. A block is connected to two identical blocks in serial as shown in the $(24, 6, 6) - $6network of Figure 3 C. Each block has 9 nodes, 6 inputs and 6 outputs except one or two if

$(p+k) \mod 4 \neq 0$ or $(p+\lambda) \mod 4 \neq 0$. Counting the number of switches of a $(p, \lambda, k) - $6network and checking the validity of the 6networks we obtain:

Theorem 8. *For $k = 5, 6$, $N(p, \lambda, k) \leq n + \frac{n}{2} + c_6' - c$, where $c_6' = 3\lceil \frac{n \mod 6}{6} \rceil$ and $c = \lfloor \frac{k-\lambda}{2} \rfloor$.*

Theorem 9. *For $k \geq 5$ and for the Simplified Design Problem*

$$N'(p, \lambda, k) \geq n + \frac{n}{2} - c_6'',$$

with $c_6'' = \frac{k-\lambda}{2} + \frac{k-\lambda}{4}$.

Notice that the difference between the number of switches of a $(p, \lambda, k)-$6network and the lower bound is at most 4. In the general case we have obtained the following bound:

Theorem 10. *For the General Design Problem and for $k \geq 5$*

$$N(p, \lambda, k) \geq n + \frac{3n}{8} - c'',$$

with $c'' = \frac{k-\lambda}{2} + \frac{3(k-\lambda)}{16}$.

We found also an other family of valid networks with the same number of switches. Nodes with no inputs nor outputs are connected on a grid on a sphere with edges of kind E_1, as shown in Figure 3 D.

Case $k = 7, 8$, (8networks).

An 8network is built with n doublons, $\frac{n}{4}$ nodes in N_4 and $\frac{n}{3}$ nodes in N_3. Nodes in N_4 and N_3 have no inputs nor outputs. A node in N_4 is connected to four nodes in N_3 via an edge of kind E_1 (see Definition 4). A nodes in N_3 is connected to three nodes in N_4 via an edge of kind E_1 and to one node in N_3 via an edge in E_0. N_3 is divided in four groups with the condition that two nodes linked with an edge in E_0 are in different groups. Each node in N_4 is connected to there four groups as shown in the $(4, 8, 8)-$8network with 19 nodes of Figure 3 E. Here again counting the number of switches and checking that the 8networks are valid we obtain:

Theorem 11. *For $k = 7, 8$,*

$$N(p, \lambda, k) \leq n + \frac{7n}{12} + c_8' - c.$$

with $c_8' = 7\lceil \frac{n \mod 12}{12} \rceil$ and $c = \lfloor \frac{k-\lambda}{2} \rfloor$.

Theorem 12. *For $k \geq 7$ and for Simplified Design Problem with no edges E_2:*

$$N'(p, \lambda, k) \geq n + \frac{7n}{12} - c_8'',$$

with $c_8'' = \frac{k-\lambda}{2} + \frac{7(k-\lambda)}{24}$.

Notice that the difference between the number of switches of a $(p, \lambda, k)-$8network and the lower bound is at most 8.

4   Constructions for Any k and Any λ. General (p, λ, k)−Networks

We present here General (p, λ, k)−Networks (see Definition 7). Their sizes are close to minimal for small k (see Remark 1). They are built with ν−boxes introduced in Definition 6.

4.1   Preliminaries: ν−Boxes and ν-Permutation Networks

The decisive property of ν−boxes is expressed in Lemma 2.

Definition 5 (ν−graph). *A ν−graph ($\nu \geq 2$) is a pair (G, l), where $G = (V, E)$ is a simple non oriented graph and l a positive integer function defined on V such that $l(V) = 2\nu$ and that, for every vertex x, one has $l(x) + \delta(x) = 4$.*

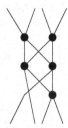

Fig. 5. A minimal 4-box

Definition 6 (ν−box). *A ν−box is a ν−graph such as for all integer function i defined on V with $0 \leq i \leq l$ and $i(V) = \nu$, there exist ν edge-disjoint paths such that every vertex x is the beginning of $i(x)$ paths and the end of $l(x) - i(x)$ paths.*

Examples: For $\nu = 1$ a minimal ν−box is reduced to a vertex. For $\nu = 2$ a minimal ν−box consists of a triangle. A minimal 4−box can be seen in Figure 5. These examples are obtained from permutation networks. A ν−permutation network is a network that can route its ν inputs to any permutation of its ν outputs.

Proposition 4. *A ν−permutation network is a ν−box.*

Proofs of properties of ν−boxes are omitted here. There exist linear asymptotic constructions for ν−boxes. Nevertheless, for small ν, no constructions of ν−box smaller than the corresponding permutation network have been found. For $\nu \leq 6$, it has been proved that minimal permutation networks are optimal ν−boxes. For $\nu \leq 6$, these minimal networks are known as AS-Waksman Permutations networks see [7]. For these reason we choose AS-Waksman permutation networks for our ν−boxes for small k.

An useful tool to prove validity is:

Lemma 2. *In a ν−box, for every subset $X \subseteq V$ we have*

$$|\Gamma(X)| \geq \min(l(X), l(\bar{X})),$$

where $\Gamma(W)$ is the set of vertices of $\overline{X}$ adjacent to a vertex of W.
The proof reduces to a flow problem.

4.2   General (p, λ, k)−Networks

Definition 7 (General (p, λ, k)−Networks). *A General (p, λ, k)−Networks (see Figure 6) is built with $\left\lceil \frac{p+k}{k} \right\rceil \left\lceil \frac{k}{2} \right\rceil$ −boxes connected in circle. These boxes are connected with*

- *a maximal number, $\left\lfloor \frac{p+\lambda}{2} \right\rfloor$, edges of type E_2,*
- *1 edge of kind E_1 if $p + \lambda$ is odd and 0 otherwise,*
- *$\left\lfloor \frac{k-\lambda}{2} \right\rfloor$ edges with one node with 2 outputs on it (edges of kind E_2'),*
- *1 edge with one node with one output on it (edge of kind E_1'), if $k - \lambda$ is odd and 0 otherwise,*
- *the remaining of type E_0 $\left(\left\lceil \frac{k}{2} \right\rceil \left\lceil \frac{p+k}{k} \right\rceil - e_2 - e_1 - e_2' - e_1' \right)$.*

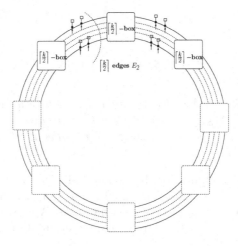

Fig. 6. Scheme of General (p, λ, k)−Networks

Lemma 3. *The number of switches of a general-(p, λ, k)−network is*

$$n + \frac{B_{min}(\lceil \frac{k}{2} \rceil)}{2 \lceil \frac{k}{2} \rceil} n + c_g' - c,$$

with $B_{min}(\nu)$ the number of nodes of a minimal ν−box, with $c_g' = B_{min} \left\lceil \frac{n \mod 2 \lceil \frac{k}{2} \rceil}{2 \lceil \frac{k}{2} \rceil} \right\rceil$ and $c = \left\lfloor \frac{k-\lambda}{2} \right\rfloor$.

k	3,4	5,6	7,8	9,10	13,14
Size	$n + \frac{n}{4}$	$n + \frac{n}{2}$	$n + \frac{5n}{8}$	$n + \frac{8n}{10}$	$2n$

Fig. 7. Size of General (p, λ, k)−networks for small k

Remark 1. *The sizes of General (p, λ, k)−networks for small k using AS-Waksman permutation networks as $\lceil \frac{k}{2} \rceil$−boxes can be seen in Figure 7. General (p, λ, k)−networks for small k are close to minimal networks (To compare with the networks of Section 3).*

Proposition 5. *A General (p, λ, k)−Network is a valid (p, λ, k)−Network.*

5 Conclusion

There remain a lot of open problems in this general issue. For example, the demands of Alcatel include networks with few (six) to around thirty unused inputs that have to tolerate few (ten) to around twenty output failures. We present some networks close to minimal for k from one to a dozen. But it remains to find tighter bounds for them and to explore larger values of k.

References

1. Blum, M., Karp, R., Papadimitriou, C., Vornberger, O., Yannakakis, M.: The complexity of testing whether a graph is a superconcentrator. Inf. Proc. Letters 13, 164–167 (1981)
2. Du, D.Z., Hwang, F.K. (eds.): Notes on the Complexity of Switching Networks. 42. Kluwer Academic Publishers, Boston (2000)
3. Pippenger, N.: Superconcentrators. SIAM Journal on Computing 6, 298–304 (1977)
4. Bermond, J.C., Perennes, S., Tóth, C.D.: Fault tolerant on-board networks with priorities. Manuscrit (To be published)
5. Bermond, J.C., Delmas, O., Havet, F., Montassier, M., Perennes, S.: Réseaux de télécommunications minimaux embarqués tolérants aux pannes. In: Algotel. pp. 27–32 (2003)
6. Bermond, J.C., Havet, F., Tóth, D.: Fault tolerant on board networks with priorities. Networks 47(1), 9–25 (2006)
7. Beauquier, B., Darrot, E.: Arbitrary size Waksman networks and their vulnerability. Parallel Processing Letters 3(4), 287–296 (2002)
8. Havet, F.: Repartitors, selectors and superselectors. Submitted to Journal of Interconnection Networks (2006)

Dynamic Compass Models and Gathering Algorithms for Autonomous Mobile Robots

Yoshiaki Katayama[1], Yuichi Tomida[1], Hiroyuki Imazu[1,2], Nobuhiro Inuzuka[1], and Koichi Wada[1]

[1] Nagoya Institute of Technology, Gokiso-cho, Showa-ku, Nagoya, Aichi, Japan
[2] This work has been done at Nagoya Institute of Technology. Now he is belonging to AISHIN AW CO., LTD., Japan

Abstract. This paper studies a gathering problem for a system of asynchronous autonomous mobile robots that can move freely in a two-dimensional plane. We consider robots equipped with inaccurate (incorrect) compasses which may point a different direction from other robots' compasses. A gathering problem is that the robots are required to eventually gather at a single point which is not given in advance from any initial configuration.

In this paper, we propose several inaccurate compass models and give two algorithms which solve the gathering problem on these models. One algorithm is the first result dealing with the compasses whose indicated direction may change in every beginning of execution cycles of robots. It solves the problem when compasses point different at most $\pi/8$ from the (absolute) north. The other one solves the problem when the compasses never change its pointed direction and their difference is at most $\pi/3$ among robots.

1 Introduction

Background. The system constituted by autonomous mobile robots is one of the autonomous distributed systems. Such a system is really effective under the circumstances like in deep sea or space where it is hard to control the robots because of difficulties of communications (i.e., long delay, bit deficiencies and so on). There are two standpoints in this research area that are practical and theoretical ones. We are focusing on the theoretical point in this paper.

In recent years, the research of the problem which achieves some kind of consensus among autonomous mobile robots has attracted attentions of researchers. One of the interesting problems is a gathering problem, where all robots meet at a single point which is not predefined, with inconsistent coordinate systems. There are some literatures dealing with this problem under various assumptions (i.e., additional capabilities of robots such as a multiplicity, restrictions for compasses' inaccuracies, and so on).

In this paper, we study on inaccuracies of compasses and the gathering problem in asynchronous autonomous mobile robot systems with inaccurate compasses. Prencipe proposed a model of the asynchronous autonomous robot called

G. Prencipe and S. Zaks (Eds.): SIROCCO 2007, LNCS 4474, pp. 274–288, 2007.

CORDA[1,2]. In the asynchronous robot systems, it is known that a very simple problem, a gathering problem, cannot be solved under some assumptions. Prencipe has shown that there exist no deterministic algorithms which solve the gathering problem in a finite time [3]. If robots have a capability of multiplicity, which means that a robot can count the number of robots that is at a same point, it has been shown that the problem is solvable for three or more robots [4]. In [5], it has been introduced that the problem can be solved for any number of robots under the assumption that coordinate systems on the robots are consistent, that is to say, all robots have perfectly accurate compasses. But it is an unrealistic assumption, so Souissi et al.[6,7] and Imazu et al.[8] have proposed some models of inaccurate compasses and algorithms that solve the gathering problem on them.

Our main interests are that (1) what and how many kinds of the inaccuracies of compasses exist? (2) with what kinds of the inaccurate compass the gathering problem can be solved? (3) what is the bound of the angles between inaccurate compasses to solve the problem? Therefore, in this paper, we classify inaccurate compasses and propose their models, and give two algorithms that solve the gathering problem for two robots. Our robot model is CORDA, that is, robots act asynchronously and are oblivious, meaning that, they cannot memorize their previous states, actions, and positions of other robots. This model seems to be too weak (over-restrictive). But it is very interesting to develop algorithms from a theoretical point, because any algorithm which solves a problem correctly in this model, can solve it in other (stronger) models, that is to say, the applicable situation is very wide. Moreover, the algorithm can get to have a fault tolerance, which is called self-stabilizing property, that the system reaches some predefined legal configuration (behavior) in despite of starting the algorithm from any initial configuration.

Our contribution. There are three our main contributions in this paper. The first is classifications and definitions of inaccurate compass models. The second is proposing two algorithms solving the gathering problem for two types of inaccurate compass models. One is for two robots with $\pi/4$-absolute error semi-dynamic compasses. This compass model means that the compass of each robot may change its pointed direction infinitely often during the execution of the algorithm, and its difference from the direction of absolute north is at most $\pi/8$. That is to say, the angle between two pointed directions of two robots' compasses is at most $\pi/4$. This algorithm is the first algorithm with a model where the compass may dynamically change its pointed direction during execution of algorithms. Another one is for two robots with $\pi/3$-relative error fixed compasses. This algorithm solves the problem when a difference of any two robots' compasses is at most $\pi/3$. On relative error fixed compass, there exist two algorithms that solve the problem when a difference is at most $\pi/4$ [7,8]. Our result improves these results with regard to inaccuracy of angle.

Related work. Suzuki and Yamashita [9,10] proposed the first algorithm to solve the gathering problem deterministically under their semi-synchronous robot model. We call their model SYm. They have also proven that it is impossible to solve the gathering problem if oblivious robots have no common orientation (coordinate system) even when the number of robots is 2.

In [9,10], the robots have unlimited visibility. Ando et al. proposed an algorithm with the same model except limited visibility of the robot [11].

Prencipe proposed an asynchronous robot model called CORDA [1,2], and Cielibak et al. provided a deterministic algorithm which solves the gathering problem using CORDA [4]. Their algorithm requires ability to detect the multiplicity of robots at a point. Flocchini et al. proposed an algorithm for the gathering problem using CORDA with limited visibility [5]. But their algorithm requires robots to have perfectly accurate compasses.

Recently, several fault tolerant algorithms are proposed. Agmon et al. studied the gathering problem in the presence of faulty robots [12]. They proposed an algorithm which tolerates one crash faulty robot where there are three or more robots in the system. They also introduced impossibility of solving the gathering problem where a Byzantine robot exists in asynchronous 3 robots system. Their result is improved by Defago et al. in [13]. They showed that impossibility also holds in a stronger model. Imazu et al. [8] and Souissi et al. [7] introduced algorithms which solve the gathering problem using inaccurate compasses. Souissi et al. [6] also proposed a gathering algorithm that solves the problem using SYm with eventually stabilizing compasses. Cohen et al. studied about the effect of errors in robot measurements, calculations and move [14]. One of their main positive results is an algorithm for convergence with bounded errors of measurement, movement and calculation. But they did not consider errors of compasses.

Structure. The remaining parts of this paper are organized as follows. In Section 2, we introduce a model of robots, inaccurate compasses, and some basic terminologies. In Section 3 we describe our first proposed algorithm to solve the gathering problem for two robots under the assumption of absolute $\pi/8$ inaccurate and semi-dynamic compasses, and prove its correctness. In Section 4, our second proposed algorithm solving the problem with relative $\pi/3$ inaccurate and fixed compasses and its proof is described. Finally, we conclude our work in Section 5.

2   System Models and Definitions

In this section, we describe models of systems which consist of autonomous robots, and then, classify and define some types of inaccurate compass models.

2.1   Robot Model

We consider a system which consists of a set of autonomous mobile robots $\mathcal{R} = \{r_1, r_2, \cdots, r_n\}$. We use CORDA model as a model of robots and is defined as follows:

- Each robot has no volume and so we can treat it as a point.
- It can move freely in the two-dimensional plane.

- It is anonymous in the sense that it does not have any kinds of identifiers and cannot be distinguished from other robots.
- It repeats its own cycle asynchronously that consists of the four states, "Look", "Compute", "Move" and "Wait".
- The range of its view (the observation range) is infinite.
- Each robot is oblivious, meaning that it cannot remember any kind of information of the previous cycles, which includes its actions and positions of other robots (observations).
- It has sensorial capabilities to observe positions of all other robots and no direct means of communication capabilities. It can form its local view of the world by observation.
- All robots execute a same algorithm.
- Each robot has its own rectangular coordinate system which may be different from other robots', in other words, their direction of y axis and/or the unit distance is different from other robots. The x and y axes of the coordinate system are given by their own compass. We assume that the direction of y positive axis on the robot is agreed with the north directed by its compass.

As described above, the cycle consists of four states described as follows:

- Look: In this state, a robot observes the world by using its sensor, which will return a snapshot of positions of all other robots as a set of robots' coordinate with respect to its local coordinate system. We consider that the observating (looking) robot exists on the origin of its coordinate system. It is assumed that robots are transparent. It means that any robot does not obstruct the view of other robots.
- Compute: In this state, a robot performs local computation according to its result of observation and the algorithm. The result of the computation is a destination point (a coordinate on its own coordinate system).
- Move: The robot moves on the line from the current position to the destination. If the destination is equal to its current location, it performs a null movement that means the robot does not move actually.
- Wait: In this state, a robot is idle. We assume that all robots are in this state at the initial configuration.

A snapshot of the system is called "configuration". So, in look state, a robot can get configuration at the time when it observes the world by its sensor.

The time that passes between two successive states of the same robot is finite, but unpredictable. And no time assumption within a state can be made. This implies that a destination got in a last "Compute" state may be different from a destination which is computed with the actual (real time) configuration.

The model makes two assumptions on the cycle of the robot.

Assumption 1. A robot travels at least the distance of some constant $\delta(> 0)$ during a single move state. If a distance to the destination is less than or equal to δ, the robot can reach there in one move. Moreover, a robot may stop at any point on its way of destination (of course, the distance between starting and stopping points is greater than or equal to δ).

Assumption 2. A time needed to finish one cycle is at least some constant $\varepsilon(> 0)$ and finite.

An execution of the system is a sequence of configurations. And it is produced by a schedule which consists of a sequence of time that a robot performs its actions and the way of its movement. The way of a movement means all of nondeterministic factors of movements, typically including the traveling distance.

We may take two different models with respect to the behavior of a robot movement as the followings.

(i) Movement *depending* on a local coordinate system: A robot moves toward its destination represented by its local coordinate system. The coordinate of the destination is changed according to the variance of its local coordinate system (i.e., the variance of its compass). If the coordinate system is changed during movement, the robot's trajectory will not be a straight line.

(ii) Movement *not depending* on a local coordinate system: A robot moves toward its destination according to a movement vector for the destination. For instance, the command like "turn right wheel 2, turn left wheel -3" is given to an actuator on the robot. The destination point on the absolute (global) coordinate system (if exists) is never changed even though a local coordinate system is changed.

In this paper, we use a model of (i).

2.2   Compass Models

In this subsection, we introduce new models of compasses with which robots are equipped. These models are classified with regard to inaccuracies of their indicated directions. In general, compasses (magnetic, gyro, GPS and so on) are error-prone because of magnetic interference, limitation of precision and/or mechanical preciseness. Therefore, it is very natural to assume that compasses have inaccuracies.

By standing two points of view, the variance of indicated direction and the difference between compasses, we define inaccurate compass models.

Variance of Compasses. We define four types of inaccurate compasses on the point of variance of indicated direction.

Definition 1. *The full dynamic compass (FDC) is a compass whose indicated direction may vary at any time during the execution.*

Definition 2. *The semi-dynamic compass (SDC) is a compass whose indicated direction may vary at the time between any two cycles, but it is never changed during one cycle.*

Definition 3. *The fixed compass (FXC) is a compass whose indicated direction never varies.*

Definition 4. *The eventually fixed compass (EFC)[6] is a compass whose indicated direction is fixed after some point of a time but may vary before that time.*

FDC is the weakest model and SDC is stronger than FDC. SDC is weaker than FXC. FXC is the strongest model in these four models. It means that if an algorithm solves a problem with FDC, it can solve the problem with SDC. For SDC and FXC, we can make the same discussion too.

EFC can be considered as a property of FDC, SDC and FXC. For instance, FDC-EFC model can be considered that the indicated direction of the compass may be varied at any time but it is eventually fixed. Note that, under the assumption of an infinite visibility, EFC can be considered as the same with FXC in the sense of solvability. That is, if a robot's view is limited, the case may exist such that some robot moves out from other robot's view due to the variance of the compass. On the other hand, if an infinite view is assumed, every robot can always observe each other. And after the compass was fixed, we can look on EFC as FXC.

In this paper, we assume infinite visibility, so EFC is not considered. And FDC and SDC can be considered as the same model (in the sense of solvability) if a movement is not depending on the local coordinate system. Our work is the first work dealing with the compasses which change their indicated direction dynamically.

Difference between Compasses. We define two types of inaccurate compasses on the point of difference of indicated directions between any two compasses equipped on robots.

Definition 5. *(α-absolute error) Existence of an absolute north is assumed. The absolute north is a vector that indicates the fixed (absolute) north direction. It is agreed with the absolute y positive axis on the coordinate system of the world. For compasses on robots, they have α-absolute error iff, for every robot, the each angle which is formed by the indicated direction of its compass and the absolute north direction is at most $\alpha/2$.*

Definition 6. *(α-relative error) For compasses on robots they have α-relative error iff the angle which is formed by indicated directions of any two robots' compasses is at most α.*

Now, we consider the variance of a compass is a part of schedule.

From the definitions of the models, we can get the first lemma about the robots' behavior.

Lemma 1. *Consider a robot r with SDC or FXC. If there is a time t such that the robot r computes the same goal (point) p in any compute state after t, it can reach the point p in a finite time after t.*

Proof. If the robot exists on the point p, lemma holds.

According to the robots model, one cycle is done in a finite time and the robot r moves at least δ during any move. And robots can move only straight, therefore, r can reach the point p after at most $\lceil x/\delta \rceil$ cycles.   □

Combination of Variances and Differences. Our inaccurate compass models can be considered as combinations with three types of variances and two types of differences. For instance, we can consider an inaccurate compass α-absolute error SDC in which the compass' indicated direction may vary at the time between any two cycles and the angle which is formed by the indicated direction of each robot and the absolute north direction is at most $\alpha/2$.

Note that, it is easy to understand, the compass model of α-relative error FXC is the same one as that of α-absolute error FXC in the sense of problem solvability. But we can not do the same discussion on SDC.

2.3   Gathering Problem

The gathering problem is a problem to require robots to gather at a non-predefined single point. An algorithm solves the gathering problem iff, from any initial configuration, under any schedules, all robots can gather at a point in a finite time when the robots travel according to the algorithm.

The problem cannot be solved under the assumption of FDC for α-absolute ($\alpha > 0$) and α-relative ($\alpha \geq 0$) error. Consider the last move before two robots gathering. Intuitively, a robot cannot arrive at a point where other robot exists because the direction of movement can be changed by an adversarial scheduler.

2.4   Notations

In this paper, the following notations are used. Let p, q and r be vertices.

- $\Delta(p, q, r)$ indicates the triangle formed by vertices p, q, and r.
- $[p, q]$ indicates the segment starting at p and terminating at q, including p and q.
- $[p, q)$ indicates the half line starting at p and getting through q.
- $\widehat{pqr}$ indicates the acute angle formed by two segments $[p, q]$ and $[r, q]$.

3 $\pi/4$-Absolute Error SDC Algorithm

The algorithm introduced in this section has been published by Imazu et al. in [8]. They provided it as a $\pi/4$-relative error FXC algorithm for two robots. We show that the algorithm solves the gathering problem with $\pi/4$-absolute error SDC. Souissi et al. also proposed a $\pi/4$-relative error FXC algorithm [7], but it cannot solve the problem with $\pi/4$-absolute error SDC.

```
1. If (no robot is observed) then
     // gathering has been finished.
2.    no move;
3. else if (other robot is observed in
                (1), (2) or (3)) then
4.    p := the coordinate where
          the other robot exists;
5.    Move(p); //the robot travels toward p.
6. else if (other robot is observed in
                (4), (5) or (6)) then
7.    no move;
8. else if (other robot is observed in
                (0) or (7)) then
9.    p := the coordinate which is right above
          (same x-coordinate) and I can observe
          the other robot in (6);
10.   Move(p);
```

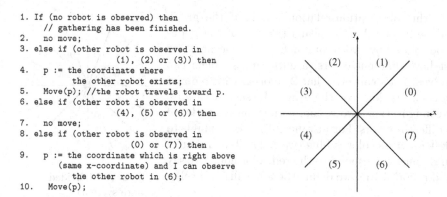

Fig. 1. Imazu's algorithm($\pi/4$-absolute error SDC algorithm)

Fig. 2. An 8 divided coordinate system

3.1   Outline of Imazu's Algorithm

Fig.1 shows the algorithm. In this algorithm, the world (a view of a robot) is divided into 8 sectors by the same angle $\pi/4$ from positive x axis (Fig.2). Each border is included in the sector adjacent on the counter-clockwise side. Every robot decides its action according to the sectors in which the other robot is observed.

We explain the idea of this algorithm. Two robots r1 and r2 are considered. In the case that r1 observes r2 in sector (1), if their compasses are correct, r2 observes r1 in (5). Actually, the compasses differ by at most $\pi/4$, and so r2 may observe r1 in (4), (5), or (6). That is, if r1 observes r2 in a sector, r2 observes r1 in the origin-symmetric sector or its neighboring sectors. In this fact, we get the following observation.

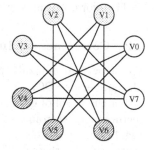

Observation 1. Let r1 and r2 be two robots. In the case that r1 observes r2 in sector (i) ($0 \leq i \leq 7$) on its own coordinate system, r2 observes r1 in sector (j), where $j = i + 3, i + 4$ or $i + 5 \pmod 8$.

Fig. 3. An observation relation graph $G_{\pi/4}$

From Observation 1, "observation-relation graph" can be defined such that we consider sectors as nodes, and observation relations as edges. Fig.3 is the observation-relation graph $G_{\pi/4}$ which is defined as $G_{\pi/4} = (V, E)$, where $V = \{V0, V1, V2, V3, V4, V5, V6, V7\}$ represents sectors (0)–(7) and $E = \{(Vi, Vj)| i \in \{0, \cdots, 7\}, j = i + 3, j = i + 4$ or $j = i + 5 \pmod 8\}$ is a set of edges which represent observation relations. That is, there is an edge (Vi, Vj) on $G_{\pi/4}$, iff there can be a system configuration that if r1 observes r2 in sector (i), r1 is observed in sector (j) by r2 on its own coordinate system . In what follows, we say "r1 corresponds to node Vi" when r1 observed r2 in sector (i).

On this observation-relation graph, if the graph could be 2-colorable, it means that all of the nodes can be colored by two colors such that adjacent nodes must have different colors, a gathering problem can be solved. Let's conceive that 2-coloring with red and blue colors is done for the graph. In such a case, in any configuration, two robots correspond to nodes of two different colors. For instance, if r1 corresponds to node Vi and its color is blue (we say a "blue robot"), then r2 must correspond to the red color node (we say a "red robot"). And we define the following two rules:

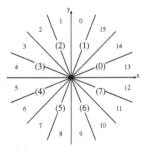

Fig. 4. A 16 divided coordinate system

- A blue robot moves toward the other robot.
- A red robot does not move.

All robots can gather at a single point according to these rules, because of the assumptions 1 and 2 and also the fact that this movement does not change the sector where the other robot is observed.

Unfortunately, $G_{\pi/4}$ cannot be colored with two colors. We need three colors for coloring (Fig.3). Then, we add a following rule for an additional color white.

- A white robot moves toward a point where it becomes a red robot. (Consequently, they will form a pair of red and blue robots.)

Therefore, we can say "the algorithm defines the robots' color and act."

3.2   Correctness of Imazu's Algorithm

We show that Imazu's algorithm can solve the gathering problem with $\pi/4$-absolute error SDC.

For the proof of correctness, we use a local coordinate system divided into 16 sectors indexed from 0 to 15 (Fig.4). In the proof, we use two kinds of indices of Fig.2 (with parentheses) which is also used in the algorithm, and those of Fig.4 (without parentheses). The relation between these two indices is shown in Fig.4. For instance, sector (1) corresponds to the sectors 15 and 0, and so on. Because we are considering $\pi/4$-absolute error SDC, Fig.5(a) and (b) show two extreme cases where the compass (or the local coordinate system) on the robot rotates counter-clockwise and clockwise, respectively. A local coordinate system on any robot can lies between Fig.5(a) and (b). Note that, indicated direction of the compass may vary only at the starting time of any cycle under SDC.

Next, we define an absolute coordinate system and its sectors. The absolute coordinate system is defined on the world such that its positive y-axis points absolute north. This coordinate system is also divided into 16 sectors and indexed same as the local coordinate system. Fig.4 can be considered as the absolute coordinate system when its positive y-axis direction matches with the direction of the absolute north. We call each of those sectors "an absolute sector." Furthermore, we use the term "the absolute sector i of a robot r," which means the

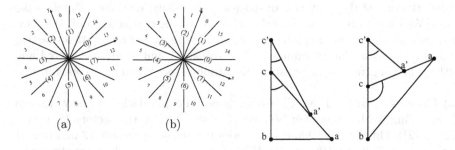

Fig. 5. The most rotated coordinate systems (counter-clockwise(a) and clockwise(b))

Fig. 6. A supplemental figure for lemma 3

absolute sector i of a robot r when assuming that r exists at the origin of the absolute coordinate system.

If a robot r exists in the absolute sector i on the other robot r', r' exists in the absolute sector $j = i + 8$ (mod 16) on r. Let $\mathcal{C}_i$ denote a set of such configurations. Note that if we do not distinguish two robots, $\mathcal{C}_i$ and $\mathcal{C}_{i+8}$ (mod 16) represents the same configuration. So, we get the next lemma.

Lemma 2. *Each configuration which consists of two robots can be represented by one of $\mathcal{C}_0$, $\mathcal{C}_1$, $\mathcal{C}_2$, $\mathcal{C}_3$, $\mathcal{C}_4$, $\mathcal{C}_5$, $\mathcal{C}_6$ and $\mathcal{C}_7$ as long as they do not gather at a single point.*

Table 1. The observation relation for the configurations

Confs.	Local sector		Proof
	r1	r2	case
$\mathcal{C}_0$	15,0,1	7,8,9	(a)
$\mathcal{C}_1$	0,1,2	8,9,10	(a)
$\mathcal{C}_2$	1,2,3	9,10,11	(b)
$\mathcal{C}_3$	2,3,4	10,11,12	(c)
$\mathcal{C}_4$	3,4,5	11,12,13	(d)
$\mathcal{C}_5$	4,5,6	12,13,14	(d)
$\mathcal{C}_6$	5,6,7	13,14,15	(e)
$\mathcal{C}_7$	7,8,9	14,15,0	(e)

Lemma 3. *In Fig.6, for $\Delta(a,b,c)$, let c' be any point which is further from b than c on the half line $[b,c)$ and let a' be any point on the segment $[c,a]$. In such a situation, $\widehat{acb} > \widehat{a'c'b}$ is satisfied and $\widehat{a'c'b}$ takes the minimum at $a' = c$.*

Lemma 3 represents a situation where one robot r1 moves from c to c' and r2 moves from a to a'. That is, r1 moves right above and r2 moves toward r1 following the algorithm. In such a case, the angle between the trajectory of r1's movement and the observed direction of r1 by r2 must decrease.

Corollary 1. *Following Imazu's algorithm, there is a time when r1 observes r2 in (its local) sector (6) if r1 observes r2 in sector (0) or (7) (r1 moves toward a point p (Fig.1 line 9)).*

Theorem 1. *Two robots with $\pi/4$-absolute error SDC can gather at a single point by executing Imazu's algorithm.*

Proof. Because of the limitation of space and simplicity, we show a sketch of the proof. We should discuss the asynchrony more for the complete proof. We show that Imazu's algorithm can gather two robots at a single point in spite of starting from any configuration of lemma 2. Two robots r1 and r2 are considered and without loss of generality, r2 exists at the absolute sector i on r1 at configuration C_i.

(a) **Case of C_0 or C_1:** First, C_0 is considered. In C_0, r1 decides to move toward r2 according to the algorithm because r1 observes r2 in the sectors 15, 0 or 1 ((1) or (2)). On the other hand, r2 decides not to move because r2 observes r1 in the sectors 7, 8 or 9 ((5) or (6)). With SDC, the compass does not change its direction during any cycle, so r1 must move directly toward r2 and the sector does not change in which each robot observes the other one. From lemma 1, r1 and r2 can gather at a point (r2's point) in a finite time.

The case C_1 can be proved similarly.

(b) **Case of C_2:** In this case, r1 always decides to move toward r2, and r2 decides not to move or to move right above. If r2 decides not to move continually, the robots can gather at a single point by the same argument with (a). Now, we need to prove the case that r2 decides to move right above.

If r2 decides to move right above, it reaches its goal (from lemma 1) or stops on the way to its goal. In both cases, the direction of r1 observed by r2 changes (from lemma 3) because the robot moves at least the distance δ.

When r2 decides to move right above, r2 must observe r1 in the sector 11. So, the positive direction of r2's y-axis must be rotated clockwise not less than 0 and at most $\pi/8$ from the absolute north. Hence, the direction of r2's movement is from 0 (includes 0) to $\pi/8$ (includes $\pi/8$) rotated from the absolute north (its positive y-axis direction). If r2 moves along the absolute north (the positive direction of r2's y-axis), the absolute sector on r2 in which r1 exists is shifted to the absolute sector 9 on r2 from lemma 3 (because r1 moves towards r2's (past) position). On the other hand, if r2 moves along the direction which is rotated clockwise $\pi/8$ from the absolute north, the direction of r1 from r2 is shifted to the absolute sector 8 on r2 by the same argument. Hence, the direction of r1 from r2 is shifted from the absolute sector 11 to the absolute sector 9 or 8 by passing through 10. That is, during finite cycles, r2 moves right above and reaches the point where r1 exists in the absolute sector 9 (or 8) on r2. This is a configuration where r2 exists in the absolute sector 1 (or 0) on r1.

So, by executing Imazu's algorithm, starting from C_2, two robots can gather or the configuration is changed from C_2 to C_0 or C_1.

(c) **Case of C_3:** The robot r1 always decides to move toward r2 at any cycle, while r2 decides to move right above or not to move according to its own compass. As same as the case (b), we need to prove the case where r2 has cycles deciding to move right above.

We assume that r2's compass differs from the absolute north by clockwise where the difference is not less than 0 and at most $\pi/8$. In this case, by the same argument with the case (b), the configuration will become C_0 or C_1 in a finite time.

Next, we assume that r2's compass differs from the absolute north by counterclockwise where the difference is not less than 0 and at most $\pi/8$. From lemma 3, if r2 moved to right above, r1 comes to exist in the absolute sector 10 on r2. Then r2 comes to exist in the absolute sector 2 on r1.

Hence, starting from the configuration C_3, Imazu's algorithm makes the configuration C_0, C_1, C_2, or our objective (gathering) configuration.

(d) **Case of C_4 or C_5:** At both configurations, r2 always decides to move right above, while r1 decides to move toward r2 or not to move according to its own compass.

If r1 arrived at r2's position before r2 finished its own observation (Look state), they gathered. If r1 did not arrived at r2's position or decided not to move, r2 must decide to move right above. Hence, the configurations will become C_3 or C_2 in a finite time from lemma 3.

(e) **Case of C_6 or C_7:** At both configurations, r1 always decides not to move while r2 decides to move right above or toward r1.

If r2 continually decides to move toward r1, in other words, r2 always observes r1 in sector 15 or 0, they gathered at a single point (from lemma 1).

On the other hand, if r2 decides to move right above, by the same arguments as the case (c), the configurations will become C_2, C_3, C_4 or C_5 in a finite time (from lemma 3).

From these (a)–(e) arguments, Imazu's algorithm can solve the gathering problem with $\pi/4$-absolute error SDC.   □

4 $\pi/3$-Relative Error FXC Algorithm

In this section, we introduce a $\pi/3$-relative error FXC algorithm for the gathering problem. The algorithms that solve $\pi/4$-relative error FXC are known [7,8]. Our algorithm improves the difference of the angle on relative error FXC.

This algorithm is shown in Fig.7 and uses 6 sectors depicted in Fig.8. Each sector has the same central angle $\pi/3$ and each border is included in the sector

```
1. If (no robot is observed) then
     // gathering has been finished.
2.   no move;
3. else if (other robot is observed in
                 (4) or (5)) then
4.    p := the coordinate where
             the other robot exists at;
5.    Move(p); // travels toward p.
6. else if (other robot is observed in
                 (0) or (1)) then
7.    no move;
8. else if (other robot is shown in
                 (2) or (3)) then
9.    p := the coordinate at which is
             on $l_1$ and I can observe
             the other robot in (1);
10.   Move(p);
```

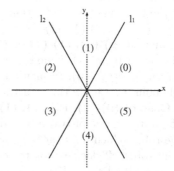

Fig. 7. $\pi/3$-relative error FXC algorithm   **Fig. 8.** A 6 divided coordinate system

adjacent on the counter-clockwise side. We define a line l_1 which is a border dividing sets of sectors $\{(1), (2), (3)\}$ and $\{(4), (5), (0)\}$. A line l_2 is defined such that a border dividing sets of sectors $\{(2), (3), (4)\}$ and $\{(5), (0), (1)\}$.

We call the robot's move of line 9 and 10 in Fig.7 "alignment move." In section 3, we knew Imazu's algorithm has the similar move. In Imazu's algorithm, a robot moves to right above, while a robot moves along the line l_1 in this algorithm.

This $\pi/3$-relative error FXC algorithm is designed by the same idea as Imazu's algorithm. The observation-relation graph $G_{\pi/3}$ is used (Fig.9) which can be defined as same as $G_{\pi/4}$. This graph can be colored by three colors.

4.1   Correctness of the $\pi/3$-Relative Error FXC Algorithm

We prove the correctness of the algorithm that can solve the gathering problem with $\pi/3$-relative error FXC. Lemma 3 also holds for alignment moves of the algorithm. So, we can get lemma 4.

Lemma 4. *When the robots r1 and r2 execute the algorithm (Fig.7), if r1 observes r2 in sector (2) or (3), there exists a time when r1 observes r2 in sector (1).*

Theorem 2. *Two robots equipped with $\pi/3$-relative error FXC can gather at a single point by executing the $\pi/3$-relative error FXC algorithm(Fig.7).*

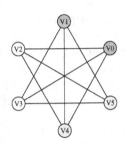

Fig. 9. The observation-relation graph $G_{\pi/3}$

Proof. Because of the limitation of space and simplicity, we show a proof sketch. We should discuss the asynchrony more for the complete proof.

Starting from any configuration, we show that the robots can gather at a single point by using the observation-relation graph $G_{\pi/3}$ (Fig.9).

The notation (i)-(j) means that the configuration that r1(r2) observes r2(r1) in sector (i) and r2(r1) observes r1(r2) in sector (j). Now, it is easy to be verified that the following 9 configurations are enough to be considered: (0)-(2), (0)-(3), (0)-(4), (1)-(3), (1)-(4), (1)-(5), (2)-(4), (2)-(5), (3)-(5).

(a) Case of (0)-(4), (1)-(4) or (1)-(5): If a robot r1 observes the other robot r2 in sector (0) or (1), it does not move. On the other hand, since r2 observes r1 in sector (4) or (5), r2 moves toward r1. Hence, the robots can gather at a single point in a finite time starting from these three configurations.

(b) Case of (0)-(2), (0)-(3), (1)-(3), (2)-(4), (2)-(5) or (3)-(5): At these configurations, one of two robots does the alignment move. We assume that r1 is the robot which does the alignment move and r2 is another one.

The robot r2 decides not to move or to move toward r1 according to the algorithm. If r2 arrived at r1 before r1 finish its observation (Look state), the problem was solved. If not, that is, in the case of which r2 decides not to move or r1 observes r2 before r2 arrives at r1's position, r1 does the alignment move. From lemma 4, r1 moves to the point at which r1 can observe r2 in sector (1) in a finite time. Then, r2 observes r1 in sectors (3), (4), or (5) from the observation-relation graph $G_{\pi/3}$ (Fig.9). It means that this configuration is (1)-(3), (1)-(4) or (1)-(5).

(b-1) **Case of (1)-(4) or (1)-(5):** These cases can be proved by same way as (a).

(b-2) **Case of (1)-(3):** In this case, r2's compass should rotate counter-clockwise by more than 0, at most $\pi/3$. In other words, r1's compass should rotate clockwise by more than 0, at most $\pi/3$.

After such a case, from lemma 4, the configuration is changed to the new one where r2 observes r1 in sector (1). In this configuration, it is easy to say that r1 never observes r2 in sector (0), (1), (2) nor (3). Thus, r1 should observe r2 in sector (4) or (5), that is, the configuration is (1)-(4) or (1)-(5). This is the case of (b-1). These arguments are true under the assumption where r1 does not observe r2 before r2 arrives at its goal. If this assumption does not hold, we can prove the correctness by the same argument as the case (b).   □

5 Conclusion and Remark

In this paper, we have considered the gathering problem when robots are equipped with inaccurate compasses. We classified the models of inaccurate compasses and introduced three results. First, there is no algorithm which can solve the gathering problem with FDC. And we propose two algorithms which can solve the problem with $\pi/4$-absolute error SDC and $\pi/3$-relative error FXC, respectively. Especially, the $\pi/4$-absolute error SDC algorithm is the first algorithm considering a time-varying compass.

We have focused on dynamic (inaccurate) compasses, it is very interesting to declare the difference among our compass models (relative v.s. absolute, EFC-SDC v.s. FXC, and so on). Also, we did not discuss about upper and lower bound of differential of compasses. But now we have got some results about it. The results will be published at another opportunity.

By our personal communications with Ms. Souissi, Prof. Défago and Prof. Yamashita, we have known that there is a π-relative error FXC algorithm for the gathering problem. Their and our results was developed independently, but an approach to the problem has some similarities.

Acknowledgments. We wish to express our gratitude to Prof. Yamashita, Prof. Défago and Ms. Souissi for helpful discussions.

This work was supported in part by the Japan Society for the Promotion of Science, Grant-in-Aid for Scientific Research (C), 1750036, 2006, Scientific Research on Priority Area New Horizons in Computing(C08), and the Telecommunication Advancement Foundation in 2006.

References

1. Prencipe, G.: CORDA: distributed coordination of a set of autonomous mobile robots. ERSADS 2001, pp. 185–190 (2001)
2. Prencipe, G.: Distributed coordination of a set of atuonomous mobile robots PhD thesis Università di Pisa (2002)
3. Prencipe, G.: On the feasibility of gathering by autonomous mobile robots SIROCCO 2005. In: Pelc, A., Raynal, M. (eds.) SIROCCO 2005. LNCS, vol. 3499, pp. 246–261. Springer, Heidelberg (2005)
4. Cieliebak, M., Flocchini, P., Prencipe, G., Santoro, N.: Solving the robots gathering problem ICALP 2003, pp. 1181–1196 (2003)
5. Flocchini, P., Prencipe, G., Santoro, N., Widmayer, P.: Gathering of asynchronous oblivious robots with limited visibility. Theoretical Computer Science 337(1–3), 147–168 (2005)
6. Souissi, S., Défago, X., Yamashita, M.: Using eventually consistent compasses to gather oblivious mobile robots with limited visibility. In: Datta, A.K., Gradinariu, M. (eds.) SSS 2006. LNCS, vol. 4280, pp. 484–500. Springer, Heidelberg (2006)
7. Souissi, S., Défago, X., Yamashita, M.: Gathering asynchronous mobile robots with inaccurate compasses. In: Shvartsman, A.A. (ed.) OPODIS 2006. LNCS, vol. 4305, pp. 333–349. Springer, Heidelberg (2006)
8. Imazu, H., Itoh, N., Katayama, Y., Inuzuka, N., Wada, K.: A gathering problem for autonomous mobile robots with disagreement in compasses 1st Workshop on Theoretical Computer Science in Izumo(Japanese) pp. 43–46 (2005)
9. Suzuki, I., Yamashita, M.: Distributed anonymous mobile robot – formation and agreement problems SIROCCO 96, pp. 313–330 (1996)
10. Suzuki, I., Yamashita, M.: Distributed anonymous mobile robots: formation of geometric patterns. SIAM J. Comput. 28(4), 1347–1363 (1999)
11. Ando, H., Oasa, Y., Suzuki, I., Yamashita, M.: Distributed memoryless point convergence algorithm for mobile robots with limited visibility. IEEE Trans. 15(5), 818–828 (1999)
12. Agmon, N., Peleg, D.: Fault-tolerant gathering algorithms for autonomous mobile robots SODA 2004, pp. 1070–1078 (2004)
13. Défago, X., Gradinariu, M., Messika, S., Raipin-Parvedy, P.: Fault-tolerant and self-stabilizing mobile robots gathering (feasibility study). In: Dolev, S. (ed.) DISC 2006. LNCS, vol. 4167, pp. 46–60. Springer, Heidelberg (2006)
14. Cohen, R., Peleg, D.: Convergence of autonomous mobile robots with inaccurate sensors and movements STACS 2006, pp. 549–560 (2006)

Fault-Tolerant Simulation of Message-Passing Algorithms by Mobile Agents

Shantanu Das[1], Paola Flocchini[1], Nicola Santoro[2], and Masafumi Yamashita[3]

[1] School of Information Technology and Engineering, University of Ottawa, Canada
{shantdas,flocchin}@site.uottawa.ca
[2] School of Computer Science, Carleton University, Canada,
santoro@scs.carleton.ca
[3] Dept. of Computer Science and Communication Engg., Kyushu University, Japan
mak@csce.kyushu-u.ac.jp

Abstract. The recently established computational equivalence between the traditional message-passing model and the mobile-agents model is based on the existence of a mobile-agents algorithm that simulates the execution of message-passing algorithms. Like most existing protocols for mobile agents, this simulation protocol works correctly only if the agents are fault-free.

We consider the problem of performing the simulation of message-passing algorithms when the simulating agents may crash unexpectedly. We show how to simulate any distributed algorithm for the message-passing model in a mobile-agents system with k agents, tolerating up to $f \leq k - 1$ crashes during the simulation. Two fault-tolerant simulation algorithms are presented, one for non-anonymous settings (i.e., where either the networks nodes or the agents or both have distinct identities), and one for anonymous systems (where both the network nodes and the agents are anonymous). In both cases, the simulation overhead is polynomial.

Unlike the existing fault-free simulation algorithm, both our protocols are able to detect termination even if the simulated algorithm has no explicit termination detection.

1 Introduction

1.1 The Framework

A distributed computing environment typically consists of a collection of autonomous computational entities that can communicate among each-other to perform a common task. The most common model of distributed computation is the *message-passing* model, in which the entities are connected through point-to-point links, according to some fixed topology (often represented as a graph); the entities are stationary and communicate by sending and receiving bounded sequences of bits (called messages), through the incident links (called ports).

Another model of distributed computation that has been studied recently is the *mobile agents* model. In this model the entities are mobile (rather than stationary) and their movement is constrained by the topology of the environment

G. Prencipe and S. Zaks (Eds.): SIROCCO 2007, LNCS 4474, pp. 289–303, 2007.

(which is again represented by a graph). The entities, called *agents* or *robots*, have computing and storage capabilities, and can move along the edges of the graph going from one node to an adjacent node. Each node of the network (called a host), provides a storage area (called whiteboard) for incoming agents, whose access is held in fair mutual exclusion. The communication between the agents occurs by writing notes on and reading notes from the whiteboards.

Mobile agents have been extensively studied for several years by researchers in Artificial Intelligence and in Software Engineering. They offer a simple and natural way to describe distributed settings where mobility is inherent, and an explicit and direct way to describe the entities of those settings, such as mobile code, software agents, viruses, robots, web crawlers, etc. Further, they allow to express immediately notions such as selfish behavior, negotiation, cooperation, etc. arising in the new computing environments. As a programming paradigm, the model allows a new philosophy of protocol and communication software design. As a computational universe, the model opens a variety of new challenging problems (e.g., *rendezvous, intruder detection, network decontamination*, etc.), most of them with immediate practical relevance and applicability.

In addition to the study of the new problems opened by the mobile agents model, intriguing research questions have naturally arisen on the differences (or similarities) between computing with mobile agents and computing with stationary agents. The two environments were initially compared from a systems engineering point of view by Fukuda et al.[13]. An insight on the *computational* relationship between the two models was provided by Barrière et al. [4], who noticed that any mobile agent algorithm can be simulated in the message-passing model; this immediately implies that all the impossibility results under the message-passing model hold also for the mobile-agents model. The reverse direction has been an open problem for quite a while. The question has been answered recently by Chalopin et al. [8], who proved that indeed the two models are computationally equivalent; in fact, they showed how to construct a mobile-agents algorithm Y from a message-passing algorithm X in such a way that every possible outcome of Y is a valid outcome for algorithm X. While the consequences of this theoretical result are not yet fully realized, already several results have been transferred between the two models [8]. Like most previous results on mobile agent computing, also these equivalence results and the simulation algorithm assume a *fault-free* environment.

In this paper, we consider instead systems where agents may fail by crashing at any time, and we investigate the problem of developing a *fault-tolerant* simulation. Thus the model considered in this paper is more realistic with respect to the typical networked environments where transmission errors or inconsistencies may cause some agents to be dropped or deleted from the system. We show that even in such an environment, we can simulate any message-passing algorithm, while tolerating an arbitrary number of agent crashes.

Moreover, we show that the simulation can be done with explicit *termination detection*; that is, if the simulated algorithm terminates (explicitly or not), the simulation will be able to detect this within a finite time and all (surviving) agents would reach terminal states.

1.2 Main Results

In this paper, we will focus on the most common type of fault that can occur in mobile agent systems—*agent crashes*. This kind of fault occurs when an agent is destroyed while moving from one node to another, for example due to network congestion, or unavailability of a host. Let us denote by k the number of agents and by f the number of crashes in the system.

We first consider the *non-anonymous* setting; that is when the network nodes have distinct identifiers. Notice that, in this case, each agent can choose a distinct identity for itself (if it does not already have one); thus, also agents are non-anonymous. Further note that the case when the network is anonymous but the agents have distinct identities is equivalent to the above one (since agents can assign a distinct names to each node using the whiteboards).

For this setting, we design a simulation algorithm that is robust against any number of crashes, short of the collapse of the entire system. We show that, if the simulated algorithm terminates, the simulation will explicitly terminate within finite time; this will happen regardless of the number $f \leq k-1$ of crashes that occur during the simulation, and even if the simulated computation does not have termination detection. Our simulation algorithm, *DisSimulate*, has an overhead of at most $O(n)$ moves per agent, for each message transmitted in the original algorithm, where n is the number of nodes. The local memory required by each agent is $O(n \log \Delta)$, where Δ is the maximum degree in the network.

We then consider the *anonymous* setting; that is when neither the network nodes nor the agents have distinct identifiers. Assuming some knowledge about the size of the system, the algorithm *DisSimulate* can be executed on anonymous networks too. However, as expected, the cost is much higher in this case and depends on how efficiently the network can be traversed. In the worst case, an exponential overhead can be incurred for simulating each message transmission. To overcome this difficulty, we propose another algorithm, *AnSimulate*, that simulates any message-passing computation on an anonymous network even when $f \leq k-1$ of agents crash during the simulation. The overhead of *AnSimulate* is $O(m + n \cdot k)$ agent moves in total for each message exchanged in the simulated computation (where m is the number of edges in the graph/network). The local memory requirement is same as before. The algorithm *AnSimulate* requires knowledge of the knowledge of the number of agents in the system[1] and as before the simulation explicitly terminates within finite time, even if the simulated computation does not have an explicit termination detection mechanism.

Due to the space constraint, most of the proofs have been omitted from this paper and can be found in the full version.

---

[1] The knowledge of the network size instead, would also suffice.

1.3   Related Work

Although most of the classical results on distributed computing are based on the message-passing stationary-agent model, some of these algorithms (e.g. [1,16]) use the concept of mobile objects called *tokens* moving among the stationary processes in the network. Algorithms for network exploration (e.g. [1,18]) are often described in terms of a travelling process or entity (sometimes modelled as a finite automata, e.g. [12]). These algorithms can be thought of as "mobile-agent" algorithms, where the mobile agent (or travelling token or automaton) is created by a stationary process to perform a particular task (e.g. exploration) after which it returns to process that created it and is destroyed there.

The model considered in this paper, consists of fully autonomous entities that are not associated with any stationary process and these entities are continuously performing computational tasks while moving among a network of data-repository nodes called whiteboards. Certain problems which are specific to this model, and have been studied recently are *agent-rendezvous*, (e.g., [17,23]), *intruder capture* (e.g., [3,6]), *network decontamination* (e.g., [11,19]), and *black hole search* (e.g., [10,15]). The problem of leader election or spanning tree construction have been studied under both the mobile-agents model (e.g., [4,9]) and the message-passing model (e.g., [14,16]). However, there has not been many studies on the differences (or similarities) between computing with mobile agents and computing with stationary agents, except for [4,8,13] as mentioned before.

There have been numerous studies [2,7,20,22] on computability in anonymous networks under the message-passing model, in terms of which tasks can be computed on a given network and under what conditions. Most of these results would hold under the mobile-agents model too, however the complexity of doing tasks in the mobile-agents model would be different.

The study of mobile-agents systems has mostly been limited to fault-free environments. One exception is the investigation of systems with a *black hole*, i.e. a highly harmful node that destroys any agent arriving at the node. In this case, the research has mostly focussed on the problem of locating the black hole (e.g., see [10,15]). The issue of computing with agents that can disappear (i.e. crash) anywhere in the network has not been considered before, to the best of our knowledge.

2   Terminology and Definitions

The S.A. model: In the message-passing model, the computational entities are stationary and are connected by point-to-point communication links. We call this the *stationary-agents* (SA) model, which can be described as follows. The computing environment is modelled by a connected undirected graph $G(V, E)$. Each vertex[2] of the graph G is associated with a fixed computational entity, called a stationary agent. Each node also contains a local memory which is accessible only to the agent associated with this vertex. Each agent can perform

---

[2] We use the terms 'vertex' and 'node' interchangeably.

any number of computational steps, it can read from and write to the local memory of the node and it can send messages and receive messages on each edge connected to the node. The agents are reactive entities, i.e. they react in response to external stimuli, e.g. the receipt of a message. The state of an agent is defined by the contents of the local memory. Initially some of the agents would be in the special state "Initiator"; such agents would start the computation process spontaneously. The initial value stored in the local memory at a node defines its identity. Thus, the nodes have distinct identities only if the initial contents of memory at each node is different.

The edges incident to a node are labelled by local orientation $\lambda = \{\lambda_v : v \in V\}$, where for each vertex u, $\lambda_u : \{(u, v) \in E : v \in V\} \rightarrow \{1, 2, \ldots, degree(u)\}$ defines the labelling on its incident edges. We shall use $\lambda(e)$ to denote the labels on the edge $e = (u, v)$ i.e. the pair $(\lambda_u(u, v), \lambda_v(u, v))$. The agent at node u can send a message m on any incident edge $e = (u, v)$ using the primitive SEND$(m, \lambda_u(e))$. In this case, the agent at node v receives the information $(m, \lambda_v(e))$, within a finite amount of time.

The M.A. model: The *mobile-agents* (MA) model is similar to the above model, with the following differences. As before the environment is modelled by the labelled graph (G, λ); however the nodes of the graph are not associated with any fixed computational entities. Instead, there are k computational entities (called mobile agents) each of which may be located at any of the nodes of the graph at any time during the computation. The agents can perform computations and they can move along the edges of the graph. Each agent also has its individual memory that is accessible only by that agent and is called its *notebook*; This local memory moves with the agent when it travels from one node to another. The contents of the *notebook* defines the state of the agent (in particular, it contains a special variable called *Next-Node*). Each agent starts in the same initial state with the notebook containing only the algorithm to be executed. The node where an agent starts from is called its *homebase*.

In this model, the nodes of the graph are just repositories of data, with no "intelligence" (i.e. computational ability) associated with them. The local memory at each node, called the *whiteboard* of that node, is accessible to any agent that is physically present at this node. The contents of the *whiteboard* defines the state of the node. Access to a whiteboard occurs in fair mutual-exclusion.

An agent at any node v can read and modify the contents of the whiteboard of node v as well as its own notebook memory, and it can leave node v through any incident edge $e(v, w)$. When an agent leaves a node v through an edge $e(v, w)$, the agent either reaches node w within a finite amount of time or permanently disappears (i.e. crashes). Initially the agents are all identical and indistinguishable from each other (no unique identities), and they execute the same protocol. In general, a mobile agent at any node v would repeatedly execute the following cycle of steps:

1. [READ] Read the contents of the whiteboard to the agent's notebook.
2. [COMPUTE] Perform a sequence of computations modifying the contents of the agent's notebook.

3. [WRITE] Write to the whiteboard part of the results of the computation.
4. [MOVE] If *Next-Node*= $x > 0$, then leave the node v through the edge labelled x. Otherwise if *Next-Node*=0, remain at node v.

During the time an agent A executes steps 1 to 3 at node v, the whiteboard of node v would be exclusively accessible to agent A. (Prior to executing step-1, the agent obtains a lock on the whiteboard and it releases the lock at end of step-3.)

Assumptions: We make the following assumptions.

[A1] Each communication link—represented by an edge of the graph G—satisfies the first-in-first-out property (i.e. agents traversing the same edge may not overtake one-another).
[A2] An agent may crash while traversing an edge (i.e. during the MOVE step) but it cannot die while performing some computation at a node (i.e. during the READ-COMPUTE-WRITE steps) .
[A3] At most $k - 1$ agents may crash (i.e. $f \leq k - 1$). Agent crashes are permanent; Once an agent crashes, it may not become alive again.

3   Simulation in Non-anonymous Systems

In this section we consider the case when the network nodes are non-anonymous; that is each node of the graph G is provided with a distinct identity that is initially stored in the local memory(whiteboard) of the node.

Consider an arbitrary (message-passing) algorithm—we shall call it algorithm Z—being executed on a network (G, λ), in the conventional stationary-agent (S.A.) model. Such an algorithm can be described as follows:

Algorithm Z
- *An initiator, upon starting the algorithm, executes the following steps:*
Step-1 Initiate the algorithm and perform local computation, possibly generating messages to be sent.
Step-2 Send zero or more messages through some of the ports.
Step-3 Wait for messages to arrive from one or more ports.

- *Any entity, on receiving a message m, executes the following steps:*
Step-1 Read message m and perform local computation, possibly generating some messages to be sent.
Step-2 Send zero or more messages through some of the ports.
Step-3 Wait for messages to arrive from one or more ports.

The algorithm is said to have terminated when every node is in Step-3 (passive mode) and there are no messages in transit.

To simulate such an algorithm in a mobile agent system, we need to execute the active steps (Step-1 and Step-2) at each node. This involves performing local computation at the nodes and delivering messages between nodes. The idea of the simulation is simple— the mobile agents can move from node to node delivering the messages; An agent that delivers a message m to a node x, can also perform the local computation at node x that results from the receipt of the message m. For effectively simulating the message-passing computation using mobile agents that can fail at any time, we need to address the following issues:

[FR] Fairness: Every message that is generated at a node should be delivered within a finite time to its destination.

[TD] Termination Detection: Once all messages have been delivered and the execution of the original algorithm Z terminates, then each agent should be able to detect this and stop the simulation.

Since an agent may crash while trying to deliver a message, we need to keep track of which messages have been delivered successfully. For this purpose, we maintain two message queues at each node– the *To-Be-Delivered*(TBD) message queue and the *Messages-Received*(MR) queue. If message m generated at a node v is to be sent through port p, the pair (m,p) is added to the TBD queue at node v. Once the message is received at the destination node, say w through port q, the pair (m,q) is added to the MR queue at node w and the entry in the TBD queue of node v is removed. An entry in the MR queue can be removed when the message is read and the actions corresponding to the receipt of this message are executed. We keep such messages in a third queue called *Messages-Executed*(ME) queue.

The simulation of the message-passing algorithm Z would be started at those nodes which are in the special state "*Initiator*". In order to ensure that the agents can detect when the computation has terminated, we would maintain a dynamic forest-like structure among the nodes of the graph, using an idea of Shavit and Francez (see [21]). The initiator nodes would be the root nodes and every other (active) node x would contain a link to its parent node i.e. the node y which send the first message to node x to activate it. We would maintain at each node x a child-list containing the links to all its (active) children, i.e. all those nodes which were activated by messages from node x and are still active. Once a leaf node completes its local computation and it does not have any messages to send, then it is removed from the child-list of its parent node and becomes inactive. When a root node has no children and no further messages to send, it becomes inactive too. If all the root nodes become inactive then the simulation is terminated.

The simulation algorithm executed by each agent A is given below:

Algorithm *DisSimulate*

I Explore the network G and construct a traversal path P_A that starts and ends at the homebase of A, and visits every other node at least once.

II Walk through path P_A and at each node v which is in the state *Initiator* do the following -

    1. Initiate the algorithm Z at node v and perform all local computation steps until the first time a message needs to be sent or received.

    2. Update the state of node v by writing to the whiteboard the current state of execution. In particular, set *node-state* to "Processing" if a message needs to be sent, and to "Waiting" if a message needs to be received.

    3. If a message m has to be sent through port p, then add (MID(m),p) to the TBD queue.

III Walk through path P_A and at each node x do the following -

    1. If the TBD queue is not empty, then execute procedure *Deliver-Message*.

    2. If node-state = "Processing", then continue with the local computation at current node until a message needs to be sent or received. Update the state of node x by writing to the whiteboard.

    3. While the MR queue is not empty, execute procedure *Receive-Message*.

    4. If there are no messages neither in the TBD queue nor in the MR queue, the Child-List is empty, and the node-state is "waiting", then write 'TERM' (for "terminated") on the whiteboard of x. Then go to the parent node y(if any) and delete x from the Child-List at node y. Now return back to node x.

IV If all the nodes visited in the previous step had a 'TERM' symbol written on their whiteboard then terminate the algorithm. Else repeat previous step.

where the procedures *Deliver-Message* and *Receive-Message* are as follows:

Procedure *Deliver-Message*

    1. Let (m_i, p) be the first entry in the TBD queue. Leave the current node x through port p to reach node y (say, through port q),

    2. If the message (m_i, q) is not present in the MR queue (or, the ME queue) at node y
       Add the message (m_i, q) to the MR queue,
       Delete any 'TERM' symbol (if present) from the whiteboard of node y,
       If the parent-link of node y is not set, set it to q.
       Set result to *Success*.
       Else set result to *Fail*.

    3. Return back to node x and delete message (m_i, p) from the TBD queue (if present). If the parent-link of node y was set to q, then add the link $p = \lambda_x(e)$ to the Child-list of node x.

    4. If the result is *Fail* and the TBD queue is not empty, then go back to the first step. Otherwise, return the result.

Procedure *Receive-Message*

    1. Let (m_i, q) be the first entry in the MR queue. Remove the message (m_i, q) from the MR queue and perform the local computation at the current node that results from receiving this message from port q.

    2. If a message m has to be sent through port p, then add $< MID(m), p >$ to the TBD queue. Update the state of the current node by writing to the whiteboard.

    3. Add (m_i, q) to the ME queue

We denote by L_A the length of the traversal path constructed by an agent A and we define L to be the maximum length of such a path. Notice that we can always ensure that L is $O(n)$. The following lemmas show that the algorithm satisfies the *Fairness* and *Termination-Detection* properties.

Lemma 1. *During the algorithm* DisSimulate *the following holds: (i) each message generated at some node x and added to the TBD queue at position r is delivered to its destination after at most $3 \cdot L \cdot r$ moves by any single agent and (ii) every message is delivered and executed exactly once (i.e. no message is repeated).*

Lemma 2. *During any execution of the algorithm DisSimulate, the size of the TBD queue at any node v would be at most 1 and the size of the MR queue would be at most $(k \cdot L)$. The size of ME queue at any node v would be at most the degree of that node.*

Lemma 3. *When any agent A terminates the simulation, then the following conditions hold: (i) No node has any more messages to send and (ii) there are no messages in transit.*

Also notice that once the execution of the original algorithm Z terminates (i.e. when all messages have been delivered and executed), then every agent that is alive would terminate after at most $L \cdot n$ moves.

Theorem 1. *The result obtained (i.e. the final state of the nodes) in any possible execution of algorithm* DisSimulate, *would be exactly identical to the result of some possible execution of the original algorithm Z in the S.A. model.*

Notice that the above theorem holds, irrespective of the number of agents failing or crashing, if at least one agent is alive. In other words, we can say the following:

Remark 1. A mobile agent system with at least one agent in a network of n nodes, is computationally as powerful as a stationary agent system with n agents.

Let us now measure the cost of the simulation.

Theorem 2. *During the simulation, the total number of moves made by the agents is at most $O(k \cdot L)$ per message exchanged in an execution of the original algorithm Z.*

The amount of local memory required by each agent (for storing a copy of the traversal path P_A) is $O(L \log \Delta)$, where Δ is the maximum degree of the graph. The amount of additional memory required at the nodes needs to be enough for storing the message queues.

4   Simulation in Anonymous Systems

In the previous section, we investigated systems having distinct identifiers for either the nodes of the network or the agents, or both. In this section, we consider the simulation of message-passing algorithms in anonymous networks by anonymous agents.

4.1   Employing the Existing Solution

The same simulation algorithm of the previous section can possibly be executed on anonymous networks too. Notice that even if the graph is anonymous, it is still possible for an agent to find a traversal path in the graph that visits every node, provided that the agent knows the size n of the graph or at least an upper bound on n.

Lemma 4. *An agent A starting at a node v of an anonymous graph G, can construct a path P (represented by a sequence of edge labels) of finite length that starts and ends at v and is guaranteed to visit each node at least once.*

However, the path constructed by an agent may be very large when the graph is anonymous and the agent has no knowledge other than size of the graph. In the worst case, this path may of length exponential in n. In fact, given any arbitrary anonymous graph G with k agents placed among the nodes of G, there is no known (deterministic) algorithm that will enable the agents to construct a path of length polynomial in n, visiting all the vertices of G. Thus, we have the following corollary of Theorem 2.

Corollary 1. *The algorithm* DisSimulate *when executed on anonymous networks, has an exponential overhead, in terms of agent moves, for delivering each message.*

4.2   In Absence of Good Traversal Paths

When it is not possible to construct efficient traversal paths, we can use a different approach for simulating a message-passing computation on a graph G. We partition the graph G among the k agents and each agent is responsible for the subgraph of G that it owns (we call this the territory of the agent). Each agent simulates the computation within its territory, while periodically checking if any of its neighboring agents have crashed; in that case it annexes the territory of the dead agent. The initial territories are obtained using the procedure EXPLORE (as in [9]). The algorithm given below, simulates a given message-passing algorithm Z, on the graph G, using k mobile agents. We use the notation $\mathbf{id}_v(e)$ to denote an identifier[3] for an edge e at node v.

ALGORITHM *AnSimulate*:

Phase 0: Each agent A executes procedure EXPLORE to obtain its territory T_A. The territory T_A is a tree rooted at the homebase, and all the other nodes contain a link marked as home-link which connects this node to its parent in the tree T_A. Let n_A be the size of T_A. Agent A then traverses its territory T_A and at each node v that it visits — if v in the state "initiator", then agent A initiates the algorithm Z at node v performing all local computation steps until

---

[3] Notice that such an identifier can be obtained by simply using the sequence of edge-labels for the path from v to e.

the first time a message needs to be sent or received; Agent A then updates the state of node v and if any message m is to be sent through the port p, then the pair (m, p) is added to the *To-Be-Delivered*(TBD) queue at node v.

Phase $i \geq 1$: Agent A, (if alive) executes the following steps:

STEP 1: Agent A does a depth-first traversal of its territory T_A. During the traversal, for each node u that it visits, agent A does the following:

- If the TBD queue of node u is not empty, then execute procedure *Deliver-Message*;
- If node-state = "Processing", then continue with the local computation at current node until a message needs to be sent or received. Update the state of node x by writing to the whiteboard.
- While the MR queue is not empty, execute procedure *Receive-Message*.
- If there are no messages in both the TBD queue and MR queue, the Child-List is empty, and the node-state is "waiting", then write 'TERM' on the whiteboard of u. Then go to the parent node v(if any) and delete u from the Child-List at node v. Now return back to node u.
- Write DONE(i, n_A) on the whiteboard of u.

If during Step-1, agent A finds an ANNEXED$(j, n_B, \textbf{id}_h(e'))$ mark in its home-base h, then agent A goes to the edge e' and marks this edge as Tree-edge (For the next phase, $n_A \leftarrow n_A + n_B$ and $T_A \leftarrow T_A + \{e'\} + T_B$.) In this case, agent A skips STEP-2 and jumps to STEP-3 to update its territory.

STEP 2: Agent A starts a depth-first traversal of its territory. During the traversal, for each external edge $e = (u, v)$ incident to some node u in its territory, it traverses the edge e to reach the other end v, reads DONE(j, n_B) from whiteboard[4] at v and takes the following actions:

- If $(j < i)$ or, $(j = i$ AND $n_B < n_A)$, then go to the root-node x of the tree T_B containing v and write ANNEXED$(i, n_A, \textbf{id}_x(e))$ (only if there is no other ANNEXED mark at node x).
- If successful in writing the ANNEXED mark, then return to e and mark this edge as a Tree-edge.(For the next phase $n_A \leftarrow n_A + n_B$ and $T_A \leftarrow T_A + \{e\} + T_B$.)

STEP 3: Agent A updates its territory T_A to include all territories that it annexed and those annexed by the agents that it defeated. If agent A itself was defeated, then it adds the territory of the agent C that defeated it and all territories annexed by C. The home-links of the nodes in the territory are updated and the value of n_A is modified accordingly. If all nodes in its territory had 'TERM' written on the whiteboards then agent A executes procedure *Termination-Detection* to check if the termination condition has been reached and if so, stops. Otherwise agent A goes to phase $i + 1$.

---

[4] If no such mark is found, it reads DONE$(j = 0, n_B = 0)$.

Procedure *Termination-Detection*

For $r = 1$ to k, do

    If there is some node in T_A which does not have a 'TERM' mark on its whiteboard,
        Return false;

    Else write 'TERM(r)' on the whiteboard of every node in T_A.

    For each non-tree edge $e = (u, v)$ incident to a node $u \in T_A$,

        Traverse edge e to reach node $v \in T_B$ (say),

        If node v has no 'TERM' mark, then return false;

        Else if found a 'TERM(j)' mark and $j < r$ then,

            Go to the root of tree T_B and check if it is marked 'TERM(r)'

            If not, mark 'TERM(r)' and merge T_A with T_B using the edge e;

If $r = k$, then return true;

Procedure *EXPLORE*

1. Set *Path* to empty; Mark the homebase as explored and include it in territory T;
2. While there are unexplored edges at the current node u,

        select an unexplored edge e,

        mark link $l_u(e)$ as explored and then traverse e to reach node v;

        If v is already marked (or v contains another agent),

            mark e as a non-tree edge;

            return back to u;

        Otherwise

                mark node v as explored and mark link $l_v(e)$ as home-link;

                Add link $l_v(e)$ to *Path*;

                Add the edge e and node v to the territory T;

3. When there are no more unexplored edges at the current node,

    If Path is not empty then,

        remove the last link from Path, traverse that link and repeat Step 2;

    Otherwise, Stop and return the territory T;

The procedures *Deliver-Message* and *Receive-Message* are same as before.

In the above algorithm, an agent A annexes the territory of another agent B, if either (i) B has died (or B is slower than A) or, (ii) if during some phase i, B has a smaller territory than A. After agent A annexes the territory of agent B, these two territories are merged and both the agents(if alive) continue the simulation in the bigger territory. The algorithm ensures that the territory of an agent is always a tree, so that a single traversal by any agent can be completed in $O(n)$ moves.

Lemma 5. *During algorithm AnSimulate, the following always holds:*

1. *The edges marked as tree-edges form a spanning forest of G, containing at most k trees (each rooted at some homebase).*
2. *Every node is visited by an alive agent at least once in every k phases.*
3. *An alive agent completes each phase within a finite amount of time.*
4. *Whenever an alive agent visits a node v, the top-most message in the TBD queue of v is delivered to its destination (unless the queue is empty).*

5. *Every message generated at a node v is delivered within a finite amount of time.*

Lemma 6. *When an agent A completes procedure* Termination-Detection *with a return value of true, then every message corresponding to the execution of algorithm Z, has been delivered and there are no messages in transit.*

Lemma 7. *When every message corresponding to the execution of algorithm Z, has been delivered, every alive agent terminates within at most k phases.*

Based on the above lemmas, we can say that the algorithm satisfies the both fairness and termination detection conditions. Finally we have the following theorem:

Theorem 3. *Algorithm* AnSimulate *correctly simulates any given message-passing algorithm Z, even if up to $f \leq k - 1$ agents crash.*

Let us now analyze the cost of the simulation. Notice that during every phase in which no agents crash, at least one message is delivered unless there are no more pending messages. The total number of moves per phase is $O(m \cdot k)$. Using a slight modification (to ensure that in every phase, any non-tree edge is traversed by at most one agent from each side) we can reduce the moves per phase to $O(m + n \cdot k')$ for k' alive agents.

Theorem 4. *For the (modified) algorithm AnSimulate, the overhead for delivering each message is $O(m + n \cdot k')$, where k' is the number of surviving agents.*

We would like to remark that the large overhead for message delivery is due to the fact that we have to deal with failures of any number of agents at any time during the simulation. In environments without agent failures, it is always possible to simulate a message-passing computation much more efficiently. For example, the algorithm given by Chalopin et al.[8] for the fault-free environment requires $O(n)$ agent moves per message delivery in the worst case.

5   Conclusions and Open Problems

We proposed and studied methods for simulating any message-passing computation in a mobile agent system with faulty agents. In particular, we have shown how to simulate a given message-passing algorithm, in a mobile agent system, while tolerating the crash-fault of any number of agents, provided that at least one agent is alive. Thus, agent crashes do not restrict the computational power of a mobile agent system. Another interesting observation is that a mobile agent system of n nodes with at least one agent is computationally as powerful as a stationary agent system with n agents.

We presented an algorithm *DisSimulate*, for simulating a message-passing computation in a labelled network (or, a network which can be explored efficiently). In this algorithm, the agents work independently of one another, with no communication among the agents and this makes this algorithm very robust.

For anonymous networks where distinct node identities are not available, we gave another algorithm *AnSimulate*, which partitions the network into disjoint parts that are serviced by different groups of agents. Even though the agents may crash at any time, the algorithm ensures that the simulation proceeds flawlessly irrespective of the agent crashes and the system always stabilizes to a state where the workload is equally distributed among the remaining agents.

In the present paper, we have only considered agent crash faults, but not node crashes. In the mobile agent model, a node crash (i.e. a whiteboard crash) implies that all data stored in the corresponding whiteboard would be deleted. Such faults can be dealt with using the known fault tolerance techniques for the message-passing (S.A.) model. In particular, given any algorithm for the S.A. model that is t-crash tolerant, the same algorithm can be simulated in a mobile agent system (using our proposed method), to tolerate up to t crash faults of the nodes, irrespective of the number of agents failing.

One of the limitations of our results is the assumption that agents can only fail while traversing an edge. We have not considered the possibility of an agent failing while performing computation at a node, because in such a case, the whiteboard of the node may remain locked and thus inaccessible to all other agents, which can create a deadlock. Future studies on this problem should be directed towards finding a way to deal with crashes of agents inside a node, while avoiding the deadlock situation.

References

1. Afek, Y., Gafni, E.: Distributed algorithms for unidirectional networks. SIAM Journal on Computing 23(6), 1152–1178 (1994)
2. Angluin, D.: Local and global properties in networks of processors. In: Proc. 12th ACM Symp. on Theory of Computing (STOC '80), pp. 82–93 (1980)
3. Barrière, L., Flocchini, P., Fraigniaud, P., Santoro, N.: Capture of an intruder by mobile agents. In: Proc. 14th ACM Symp. on Parallel Algorithms and Architectures (SPAA'02), pp. 324–332 (2002)
4. Barrière, L., Flocchini, P., Fraigniaud, P., Santoro, N.: Can we elect if we cannot compare. In: Proc. 15th ACM Symp. on Parallel Algorithms and Architectures (SPAA'03), pp. 200–209 (2003)
5. Bender, M., Fernandez, A., Ron, D., Sahai, A., Vadhan, S.: The power of a pebble: Exploring and mapping directed graphs? In: Proc. 30th ACM Symp. on Theory of Computing (STOC'98), pp. 269–287 (1998)
6. Blin, L., Fraigniaud, P., Nisse, N., Vial, S.: Distributed chasing of network intruders. In: Proc. 13th Colloquium on Structural Information and Communication Complexity (SIROCCO'06), pp. 70–84 (2006)
7. Boldi, P., Vigna, S.: An effective characterization of computability in anonymous networks. In: Proc. of 15th Int. Conference on Distributed Computing (DISC'01), pp. 33–47 (2001)
8. Chalopin, J., Godard, E., Métivier, Y., Ossamy, R.: Mobile agents algorithms versus message passing algorithms. In: Proc. 10th Int. Conf. on Principles of Distributed Systems (OPODIS'06), pp. 187–201 (2006)

9. Das, S., Flocchini, P., Nayak, A., Santoro, N.: Distributed exploration of an unknown graph. In: Proc. 12th Coll. on Structural Information and Communication Complexity (SIROCCO'05), pp. 99–114 (2005)
10. Dobrev, S., Flocchini, P., Prencipe, G., Santoro, N.: Finding a black hole in an arbitrary network: optimal mobile agents protocols. Distributed Computing, to appear. Preliminary version. In: Proc. 21st ACM Symposium on Principles of Distributed Computing (PODC'02), pp. 153–162 (2002)
11. Flocchini, P., Luccio, F.L., Huang, M.: Decontamination of chordal rings and tori using mobile agents International Journal of Foundation of Computer Science (To appear)(2007)
12. Fraigniaud, P., Ilcinkas, D.: Digraph exploration with little memory. In: 21st Symp. on Theoretical Aspects of Computer Science (STACS'04), pp. 246–257 (2004)
13. Fukuda, M., Bic, L.F., Dillencourt, M.B., Merchant, F.: Messages versus messengers in distributed programming. In: Proc. 17th international Conference on Distributed Computing Systems (ICDCS '97), pp. 347–354 (1997)
14. Gallager, R.G., Humblet, P.A., Spira, P.M.: A distributed algorithm for minimum-weight spanning trees. ACM Trans. Program. Lang. Syst. 5(1), 66–77 (1983)
15. Klasing, R., Markou, E., Radzik, T., Sarracco, F.: Hardness and approximation results for black hole search in arbitrary networks. Theoretical Computer Science (To appear) (2007)
16. Korach, E., Kutten, S., Moran, S.: A modular technique for the design of efficient distributed leader finding algorithms. ACM Trans. Program. Lang. Syst. 12(1), 84–101 (1990)
17. Kranakis, E., Krizanc, D., Rajsbaum, S.: Mobile agent rendezvous: A Survey. In: Proc. 17th international Conference on Distributed Computing Systems (SIROCCO '06), pp. 1–9 (2006)
18. Kutten, S.: Stepwise construction of an efficient distributed traversing algorithm for general strongly connected directed networks or: Traversing one way streets with no map. In: Proc. of 9th Int. Conference on Computer Communication (ICCC'88), pp. 446–452 (1988)
19. Luccio, F., Pagli, L., Santoro, N.: Network decontamination in presence of local immunity. International Journal of Foundation of Computer Science (To appear) (2007)
20. Sakamoto, N.: Comparison of initial conditions for distributed algorithms on anonymous networks. In: Proc. of the 18th annual ACM Symp. on Principles of Distributed Computing (PODC '99), pp. 173–179 (1999)
21. Shavit, N., Francez, N.: A new approach to detection of locally indicative stability. In: Proceedings of 13th International Colloquium on Automata, Languages and Programming, (ICALP'86), pp. 344–358 (1986)
22. Yamashita, M., Kameda, T.: Computing on anonymous networks: Part I–Characterizing the solvable cases. IEEE Transactions on Parallel and Distributed Systems 7(1), 69–89 (1996)
23. Yu, X., Yung, M.: Agent rendezvous: A dynamic symmetry-breaking problem. In: Int. Coll. on Automata Languages and Programming (ICALP'96), pp. 610–621 (1996)

Optimal Conclusive Sets for Comparator Networks

Guy Even[1], Tamir Levi[2], and Ami Litman[2]

[1] School of Electrical Engineering, Tel-Aviv University, Tel-Aviv 69978, Israel
guy@eng.tau.ac.il
[2] Faculty of Computer Science, Technion, Haifa 32000, Israel
{levyt,litman}@cs.technion.ac.il

Abstract. A set of input vectors S is conclusive if correct functionality for all input vectors is implied by correct functionality over vectors in S. We consider four functionalities of comparator networks: sorting, merging of two equal length sorted vectors, sorting of bitonic vectors, and halving (i.e., separating values above and below the median). For each of these functionalities, we present tight lower and upper bounds on the size of conclusive sets. Bounds are given both for conclusive sets composed of binary vectors and of general vectors. The bounds for general vectors are smaller than the bounds for binary vectors implied by the 0-1 principle. Our results hold also for comparator networks with unbounded fanout.

Assume the network at hand has n inputs and outputs, where n is even. We present a conclusive set for sorting that contains $\binom{n}{n/2}$ nonbinary vectors. For merging, we present a conclusive set with $\frac{n}{2} + 1$ nonbinary vectors. For bitonic sorting, we present a conclusive set with n nonbinary vectors. For halving, we present $\binom{n}{n/2}$ binary vectors that constitute a conclusive set. We prove that all these conclusive sets are optimal.

Keywords: Zero-One Principle, Comparator Networks, Sorting Networks, Bitonic Sorting, Merging Networks.

1 Introduction

Comparator networks are combinational circuits with fanout one built only from comparators, where a comparator is a gate that sorts a pair of numbers. When the fanout is not restricted, we call such a network a *min-max network*. The 0-1 principle introduced by Knuth [5] states that a comparator network is a sorting network if and only if it sorts all binary inputs. Sorting is not the only functionality that comparator networks and min-max networks are useful for. Additional functionalities include merging two sorted vectors, halving vectors (i.e., separating values above and below the median), and sorting restricted sets of vectors (e.g., bitonic sorting). Since its introduction in 1973, the 0-1 principle was extensively used for proving the correctness of various types of networks.

In this paper we address the following questions: How many vectors are needed to verify the functionality of a given a min-max network? The 0-1 principle states that a comparator network with n inputs is a sorting network if it correctly sorts every binary

G. Prencipe and S. Zaks (Eds.): SIROCCO 2007, LNCS 4474, pp. 304–317, 2007.

vector. Can sorting be verified using fewer vectors? Assume n is even. A similar question can be asked for merging networks where the input consists of two sorted vectors of length $n/2$. The number of binary inputs for a merging network is $(n/2 + 1)^2$. Can merging be verified using fewer vectors? We also ask: does the verification of comparator networks require fewer vectors than min-max networks?

We refer to a set of vectors that verifies a specific functionality as a *conclusive set*. Our goal is to find small conclusive sets for various functionalities. The main motivation for smaller conclusive sets is for testing the functionality of a given min-max network; the smaller the conclusive set, the faster the test runs.

So far, only binary vectors were considered for conclusive sets. We introduce the usage of nonbinary vectors (i.e., vectors in $\{0, \ldots, n - 1\}^n$) for conclusive sets. Interestingly, our main result is that smaller conclusive sets are possible if nonbinary vectors are allowed. In addition, we prove lower bounds on the size of conclusive sets that imply the optimality of our constructions. We also prove lower bounds on the sizes of conclusive sets consisting solely of binary vectors.

Previous work. Previous work falls into two main categories: extensions of the 0-1 principle to functionalities other than sorting (mainly merging) and an attempt to prove lower bounds on the size of binary conclusive sets for sorting.

The main application of the 0-1 principle is to facilitate the design and verification of sorting and merging networks. We review some of the applications of the 0-1 principle from the literature. Miltarsen et. al. [10] used a variant of the 0-1 principle to prove the correctness of a merging network. Liszka and Batcher [8] used it to prove the the correctness a merging network called the modulo merger. Bender and Williamson [3] used it to prove structure theorems for recursively constructed merging networks. Batcher and Lee [6] used it to prove the correctness of a k-merger network whose input consists of k sorted vectors of equal length. Nakatani et. al. [11] used it to prove the correctness of a bitonic sorter. Rajasekaran and Sen [12] generalized the 0-1 principle to networks that sort almost all 0-1 inputs. They proved bounds on the fraction of correctly sorted general vectors based on the fraction of correctly sorted binary vectors.

Rice [13] investigated a computational model called *continuous in-place functions* (CIP-functions). The set of functions computable by comparator networks (with fanout one) is strictly contained in the set of CIP-functions. Rice [13] proved that a CIP-function sorts all vectors if and only if it sorts all the binary vectors. In addition, Rice proved the following lower bound. If $S \subseteq \{0, 1\}^n$ is a conclusive set for sorting with respect to CIP-functions, then $\{0, 1\}^n \setminus \{0^n, 1^n\} \subseteq S$. Rice proves this by presenting, for every binary vector $v \notin \{0^n, 1^n\}$, a witness CIP-function f_v that sorts all binary vectors except for v. Rice's result does not imply lower bounds on the size of binary conclusive sets with respect to comparator networks (with fanout one). We strengthen Rice's result by presenting, for every 0-1 vector v, a witness comparator network N_v whose fanout equals one that sorts all 0-1 vectors except for v. Hence, we obtain lower bounds on the size of binary conclusive sets even when the fanout is one. Our construction is also quite simple and relies only on sorting networks. A proof that the set of CIP-function equals the set of functions computable by min-max networks appears in [2].

Our results. Table 1 summarizes our results. The first column in the table lists the four functionalities that we deal with (see Sec. 3 for formal definitions). We consider both conclusive sets that consist only of binary vectors and general conclusive sets (i.e., conclusive sets that contain vectors in $\{0, \ldots, n-1\}^n$). The second column lists the sizes of binary conclusive sets that follow from the 0-1 principle. The third columns lists optimal sizes of binary conclusive sets that are proved in Sec. 6 and 5.3. The fourth column lists optimal sizes of general conclusive sets and the fifth column lists the type of conclusive set that achieves each bound. These general conclusive sets are presented in Sec. 5. Their optimality is proved in Sec. 6.

Table 1. Summary of results: sizes of conclusive sets for various functionalities

	size of conclusive set implied by the 0-1 principle	optimal size of binary conclusive set	optimal size of general conclusive set	description
Sorting	$2^n - 2$ [5]	$2^n - 2$	$\binom{n}{\lceil n/2 \rceil}$	covering permutations
Merging	$(\frac{n}{2} + 1)^2$ [10]	$(\frac{n}{2} + 1)^2$	$\frac{n}{2} + 1$	sandwiches
Bitonic Sorting	$(n-1) \cdot n$ [11]	$(n-1) \cdot n$	n	unitonic
Halving	$2^n - 2$	$\binom{n}{n/2}$	$\binom{n}{n/2}$	balanced binary vectors

Techniques. The 0-1 principle is originally stated for sorting networks, and it has been common to informally extend it to other functionalities such as merging [10] and bitonic sorting [11]. With each vector v, we attach a set of binary vectors that are called the 0-1 images of v. A binary vector b is a 0-1 image of v if there exists a monotonic function f such that $b = \langle f(v_0), \ldots, f(v_{n-1}) \rangle$. In Lemma 1, we present a variant of the 0-1 principle that deals with a single input vector and all its 0-1 images. This variant forms the basis in Theorem 3 for proving a formal extension of the 0-1 principle for each of the functionalities considered in Table 1.

Upper bounds on the size of conclusive sets are obtained by presenting sets of vectors whose 0-1 images constitute a binary conclusive set. Since a nonbinary vector of length n may have up to $n + 1$ different 0-1 images, a reduction in the size of conclusive sets is achieved for certain functionalities.

Lower bounds are based on Lemma 6 that proves, for every binary vector v, the existence of a comparator network (with fanout one) N_v that sorts all binary vectors except for v. This lemma obviously proves lower bounds on the size of binary conclusive sets. In the case of nonbinary conclusive sets, lower bounds are obtained by focusing on balanced binary vectors (i.e., vectors that contain the same number of zeros and ones). Since every vector has at most one balanced 0-1 image, the number of nonbinary vectors in a conclusive set cannot be smaller than the number of balanced binary vectors in a binary conclusive set.

Organization. This paper is organized as follows. In Section 2, comparator networks and min-max networks are formally defined. Various functionalities of min-max networks are presented in Section 3. In Section 4 the well known 0-1 principle for sorting networks is presented along with some variants. These variants enable extending the 0-1 principle to the functionalities presented in Section 3. In Section 5 we present smaller conclusive sets for each of these functionalities. In Section 6 we prove lower bounds on the sizes of binary and general conclusive sets. These general lower bounds match the upper bounds presented in Section 5. We conclude with a discussion and two open problems.

2 Comparator Networks and Min-Max Networks

A *comparator* is a combinational gate with 2 input ports a_1, a_2 and 2 output ports b_{min}, b_{max}. Each port may carry a single number (e.g., a k-bit string that is the binary representation of a number in the range $[0, 2^k - 1]$). We denote by $v(x)$ the value carried by port x. A comparator sorts the pairs of numbers in the following sense. Suppose the input values are $v(a_1)$ and $v(a_2)$, where the number $v(a_i)$ is input to port a_i, for $i = 1, 2$. The output values satisfy

$$v(b_{min}) = \min\{v(a_1), v(a_2)\}$$
$$v(b_{max}) = \max\{v(a_1), v(a_2)\}.$$

Note that when restricted to Boolean inputs, a comparator simply consists of one AND-gate and one OR-gate.

A *min-max network* is a combinational circuit, all the gates of which are comparators. This means that the topology of a min-max network is a directed acyclic graph with 3 types of vertices: (i) A set of input vertices X that serve as external input ports. (ii) A set of output vertices Y that serve as external output ports. (iii) A set of comparators C. The in-degree of input vertices and the out-degree of output vertices are zero. Comparators have two incoming edges, one edge per input port. Every edge emanates from an input vertex or an output port of a comparator. Every edge enters an input port of comparator or an output vertex. Exactly one edge incomes every input port of a comparator and every output vertex. Note that when restricted to binary vectors, every output of a min-max network N is computable by a monotonic boolean circuit (i.e., a circuit that contains only AND-gates and OR-gates, and lacks inverters).

A *comparator network* is a min-max network in which the fanout of every input vertex and every output port of a comparator is one. All our results (i.e., upper and lower bounds) hold both for min-max networks and comparator networks.

We focus on min-max networks in which the number of input vertices equals the number of output vertices, namely $|X| = |Y|$. We denote by n the number of input vertices and assume that n is even. We also assume that the range of valid input/output values contains the set $\{0, \ldots, n - 1\}$.

Often, min-max networks are used for sorting. To be able to define such functionality, one must label the output vertices (e.g., which output vertex outputs the maximum value?). The output vertices are labeled $y_0, \ldots, y_{n-1}$. Similarly, the input vertices are labeled $x_0, \ldots, x_{n-1}$.

3 Functionality

Since every min-max network is a combinational circuit, the functionality is well defined. Let N denote a min-max network with input vertices $X = \{x_0, \ldots, x_{n-1}\}$ and output vertices $Y = \{y_0, \ldots, y_{n-1}\}$. We now introduce notation for the relation between the input and output values of min-max networks.

An *input vector* is a function $v : X \to \mathbb{N}$, where $v(x_i)$ denotes the value fed by the input vertex x_i. An *output vector* is a function $w : Y \to \mathbb{N}$, where $w(y_i)$ denotes the value that is received by the output vertex y_i. Given an input vector v, we denote by $N(v)$ the output vector obtained when the min-max network N is input the vector v. We often refer to input and output vectors as sequences of length n rather than functions, namely, $v = \langle v(x_0), \ldots, v(x_{n-1}) \rangle$.

We say that a vector $w = \langle w_0 \ldots, w_{n-1} \rangle$ is *sorted* if $w_i \leq w_j$ whenever $i \leq j$. We now define four functionalities of a min-max networks.

sorting: A min-max network is a *sorting network* if $N(v)$ is sorted for every input vector v.

bitonic sorting: We first define ascending-descending vectors and bitonic vectors. A vector v is *ascending-descending* if it is a concatenation of a non-decreasing vector and a non-increasing vector (the two vectors need not be of equal length, in fact, one of these vectors may even be empty). A vector v is *bitonic* if it is a cyclic rotation of an ascending-descending vector.

A min-max network is a *bitonic sorter* if $N(v)$ is sorted for every input vector v that is bitonic [5, p. 232].

merging: We first define bi-sorted vectors. A vector v is *bi-sorted* if v is the concatenation of two sorted vectors of equal length, namely, $v_{i+1} \geq v_i$ for every $i \in \{0, \ldots, n - 2\} \setminus \{\frac{n}{2} - 1\}$.

A min-max network is a *merging network* if $N(v)$ is sorted for every input vector v that is bi-sorted.

halving: We first define halved vectors. A vector v is *halved* if $v_j \geq v_i$ for every $0 \leq i < \frac{n}{2} \leq j < n$.

A min-max network is a *halver* if $N(v)$ is halved for every input vector v.

4 The 0-1 Principle

Let $f : \mathbb{N} \to \mathbb{N}$ denote a (non-decreasing) monotonic function (i.e., $a \leq b$ implies $f(a) \leq f(b)$). Given a vector v and a function f, $f(v)$ denotes the vector $\langle f(v_0), \ldots, f(v_{n-1}) \rangle$.

We now cite two important theorems on comparator networks.

Theorem 1 ([5],p. 224). *For every comparator network N, every monotonic function f, and every input vector v,*

$$f(N(v)) = N(f(v)).$$

Theorem 1 can be proved for the value transmitted along every edge in N by induction on its "depth" (i.e., maximum distance from an input node). Theorem 1 has several applications. One application shows that a comparator network is a sorting network if and only if it sorts every input vector that is one-to-one (i.e, vector with distinct values). The most useful application of Theorem 1 is the 0-1 principle.

Theorem 2 (The 0-1 principle). *Let N denote a comparator network with n inputs and n outputs. The network N is a sorting network if and only if it sorts every input vector in $\{0,1\}^n$.*

We present a variant of Theorem 2 that deals with a single vector instead of the set of all vectors. A vector in $\{0,1\}^n$ is called a 0-1 *vector*. A *threshold* function is a function $\tau_k : \mathbb{N} \to \{0,1\}$ defined by

$$\tau_k(i) \triangleq \begin{cases} 0 & i < k \\ 1 & i \geq k. \end{cases}$$

When f is a threshold function, we refer to the vector $f(v)$ as a 0-1 *image* of v. Clearly, every v with k distinct values has exactly $k+1$ different 0-1 images. Two trivial 0-1 images are the vectors 0^n and 1^n.

The following lemma implies the 0-1 principle.

Lemma 1. *Let N denote a min-max network with n inputs and n outputs. Let v denote an input vector. The output vector $N(v)$ is sorted (halved) if and only, for every threshold function f, the output vector $N(f(v))$ is sorted (halved).*

Proof. We prove only the sorting version; the halving version is proved analogously. The easy direction is to show that if $N(v)$ is sorted, then $N(f(v))$ is sorted for every threshold function f. Indeed, by Theorem 1, $N(f(v)) = f(N(v))$. Since $N(v)$ is sorted, so is $f(N(v))$, and therefore $N(f(v))$ is sorted.

The other direction is proved by contradiction. Assume $N(v)$ is not sorted and $N(f(v))$ is sorted for every threshold function. Let i denote an index such $N(v)_i > N(v)_{i+1}$. Consider the threshold function τ_k for $k = N(v)_i$. Let $w = \tau_k(N(v))$. By Theorem 1, $w = N(\tau_k(v))$. Note that $w_i = 1$ while $w_{i+1} = 0$. Hence w is not sorted, contradicting the assumption.

We now state a 0-1 principle for merging networks, bitonic sorters, and halvers (this principle appears in [10, pp 152] for merging networks).

Theorem 3. *Let N denote a min-max network with n inputs and n outputs.*

- *The network N is a merging network iff $N(v)$ is sorted for every 0-1 bisorted vector v.*
- *The network N is a bitonic sorter iff $N(v)$ is sorted for every 0-1 bitonic vector v.*
- *The network N is a halver iff $N(v)$ is halved for every 0-1 vector v.*

Theorem 3 follows directly from Lemma 1 and from the following lemma.

Lemma 2. *A vector v is bisorted/bitonic/halved iff all its 0-1 images are bisorted/bitonic/halved.*

Proof. We prove the lemma only for bitonic case; the other cases are proved similarly. Every threshold function is monotonic, and therefore, if v is bitonic, then all its 0-1 images are also bitonic.

The converse direction is proved by contradiction. Assume that v is not bitonic. That means that the cyclic rotations of v are also not bitonic. By applying rotations, we may assume that the value of the first component v_0 of v is minimum, namely $v_0 = \min_i v_i$. Since v is not bitonic, it is also not ascending-descending.

We claim that there exist $0 < i < j < n - 1$ such that $v_0 < v_i$, $v_j < v_i$, and $v_j < v_{j+1}$ (see Fig. 1). Indeed, the index i is chosen to be the maximal index such that the subsequence $\langle v_0, \ldots, v_i \rangle$ is ascending. Since v is not ascending, it follows that $i < n - 1$. By the choice of i, it follows that $v_{i+1} < v_i$. By the minimality of v_0, it follows that $v_i > v_0$, since otherwise $v_{i+1} < v_0$. The index j is chosen to be the maximal index such that the subsequence $\langle v_i, \ldots, v_j \rangle$ is descending. Since $v_j \leq v_{i+1}$, it follows that $v_j < v_i$. Since v is not ascending-descending, $j < n - 1$. Finally, by the choice of j, $v_{j+1} > v_j$.

Let τ_k be the threshold function with $k = \min\{v_i, v_{j+1}\}$. Note that $\tau_k(v)$ contains $\langle 0, 1, 0, 1 \rangle$ as a subsequence, which implies that $\tau_k(v)$ is not bitonic, contradicting the assumption that all the 0-1 images of v are bitonic.

Fig. 1. A vector which is not ascending-descending on the left and the corresponding 0-1 vector on the right

5 Smaller Conclusive Sets

Definition 1. *A set of vectors C is conclusive for sorting if every min-max network that sorts all vectors in C is a sorting network.*

One can also define conclusive sets for other functionalities, such as merging networks, bitonic sorters, or halvers. For each functionality, a conclusive set serves as a proof of the correct functionality of the min-max network. Obviously, the set of all valid inputs is a conclusive set (e.g., the set of all bisorted vectors is conclusive for merging). The 0-1 principle implies that the set of all valid 0-1 vectors is a conclusive sets for the sorting, merging, bitonic sorting, and halving. Our goal is to present even smaller conclusive sets for these functionalities.

5.1 Sandwiches for Merging

We refer to a vector over $\{0, \ldots, n - 1\}$ of length n with distinct values as a *permutation*. We now define a special type of bisorted vectors called sandwiches (Figure 2 depicts a sandwich).

Definition 2. *A sandwich v is obtained from the sorted vector $\langle 0, \ldots, n-1 \rangle$ by choosing a block of length $n/2$ that serves as the second half of v. The components of the vector outside the block constitute the first half of v.*

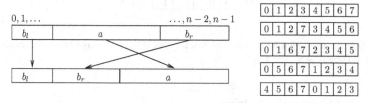

Fig. 2. On the left, a construction of a sandwich by choosing an interval a of length $n/2$ and then swapping the interval a and the interval b_r. On the right are all 5 sandwiches of length 8.

Note that a vector $v = \langle v_0, \ldots, v_{n-1} \rangle$ is a *sandwich* if and only if v satisfies 3 conditions: (i) v is a permutation, (ii) v is bisorted, and (iii) the second half of v is an interval (i.e., there exists an $0 \le i \le \frac{n}{2}$ such that $\langle v_{n/2} \ldots, v_{n-1} \rangle = [i, i + \frac{n}{2} - 1]$).

Since there are $n/2 + 1$ blocks of length $n/2$ in the sorted vector, we conclude with the following observation.

Observation 4. *There are exactly $n/2 + 1$ different sandwiches of length n.*

Lemma 3. *Every bisorted 0-1 vector is a 0-1 image of a sandwich.*

Proof. Let v be a bisorted 0-1 vector. Let p and q denote the number of zeros in the first and second halves of v, respectively. (See Figure 3). Let s denote the sandwich obtained when the block that defines the second half of s starts in position p. It follows that $\tau_{p+q}(s) = v$.

Lemma 4. *A min-max network is a merging network iff it sorts all sandwiches.*

Proof. Since sandwiches are bisorted, they are sorted by a merging network. We now prove that the set of sandwiches is conclusive for merging. Let N denote a min-max network with n inputs and outputs that sorts all sandwiches of length n.

Since N sorts all sandwiches, by Lemma 1, N sorts all 0-1 images of sandwiches. By Lemma 3, this means that N sorts all bisorted 0-1 vectors. By Theorem 3, N is a merging network, and the lemma follows.

Lemma 4 implies the following corollary.

Corollary 1. *The set of $n/2 + 1$ sandwiches is a conclusive set for merging networks of width n.*

5.2 Unitonic Vectors for Bitonic Sorting

A vector is *unitonic* if it is a cyclic rotation of the vector $\langle 0, 1, \ldots, n-1 \rangle$. In the full version we prove the following theorem.

Theorem 5. *The set of n unitonic vectors is a conclusive set for bitonic sorters of width n.*

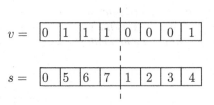

Fig. 3. A 0-1 bisorted vector v and the sandwich s that is the pre-image of v with respect to the threshold 4. Here $n = 8, p = 1$ and $q = 3$.

5.3 Balanced Vectors for Halving

A binary vector is *balanced* if it contains the same number of zeros and ones. In the full version we prove the following theorem.

Theorem 6. *The set of $\binom{n}{n/2}$ balanced vectors is a conclusive set for halvers.*

5.4 Conclusive Sets for Sorting

Let $\mathbb{P}$ denote the partially ordered set consisting of all the subsets of $\{0, \ldots, n-1\}$ ordered by inclusion. A sequence $A_1, A_2, \ldots, A_k$ of subsets in $\mathbb{P}$ is a *chain* if it satisfies

$$\emptyset \subseteq A_1 \subsetneq A_2 \subsetneq \cdots \subsetneq A_k \subseteq \{0, \ldots, n-1\}.$$

An *antichain* is a family of subsets $\{B_i\}_{i \in I}$ no two of which are related (i.e., $B_i \not\subseteq B_j$ for every $i \neq j \in I$). The *indicator vector* $\chi(A)$ of a subset A is the vector $\langle \chi(A)_0, \ldots, \chi(A)_{n-1} \rangle \in \{0,1\}^n$ defined by $\chi(A)_i = 1$ iff $i \in A$.

Lemma 5. *For every chain $\{A_c\}_{c \in C}$, there exists a permutation vector π such that, for every $c \in C$, the indicator vector $\chi(A_c)$ is a 0-1 image of π.*

Proof. Without loss of generality, C is an interval $\{0, \ldots, |C|-1\}$. Let $A_0 = \{i_{0,1}, \ldots, i_{0,k_0}\}$. Similarly, for $0 \leq j < |C|-1$, let $A_{j+1} \setminus A_j = \{i_{j+1,1}, \ldots, i_{j+1,k_{j+1}}\}$. Define $\pi(i_{0,\ell}) = n - \ell$ for $1 \leq \ell \leq k_0$. For $j \geq 1$, define $\pi(i_{j,\ell}) = n - \sum_{m < j} k_m - \ell$. If the maximal set in the chain does not contain all the elements, then we augment π to be a permutation by adding the missing values to the unassigned components.

Every indicator vector $\chi(A_c)$ is a 0-1 image of π since $\chi(A_c) = \tau_t(\pi)$, where $t = n - \sum_{i \leq c} k_i$.

Theorem 7. *There exists a conclusive set of size $\binom{n}{n/2}$ for sorting networks of width n.*

Proof. By Sperner's theorem [9], the size of every antichain is at most $\binom{n}{n/2}$. The collection of subsets of size $n/2$ is an antichain of size $\binom{n}{n/2}$, and hence, it is an antichain of maximum cardinality. By Dilworth's theorem [4], there exist $\binom{n}{n/2}$ chains that cover all the subsets in $\mathbb{P}$. By Lemma 5, associate a permutation to each chain in the cover. Let Ψ be the set of $\binom{n}{n/2}$ permutations associated with the chains in the cover. Lemma 5 implies that every indicator vector of every subset in $\mathbb{P}$ is a 0-1 image of a permutation in Ψ. Note that the set of indicator vectors of subsets in $\mathbb{P}$ is simply $\{0,1\}^n$. By Lemma 1, if a network N sorts all members of Ψ, then N sorts all the 0-1 images of Ψ, and hence all 0-1 vectors. By Lemma 2, if N sorts all members of Ψ, then N is a sorting network; hence, Ψ is a conclusive set for sorting networks.

6 Lower Bounds for Conclusive Sets

In this section we prove lower bounds on the size of conclusive sets for the tasks considered in Section 5. These lower bounds prove the optimality of all conclusive sets presented in Section 5 even if the fanout is one. A similar result in a stronger model, called continuous in-place mappings, is proven in [13]. (Note that a "counter-example" in a weak model implies the existence of a counter-example in a stronger model.) The lower bounds rely on the following lemma.

Lemma 6 (The witness network). *For every 0-1 vector $v \notin \{0^n, 1^n\}$, there exists a comparator network N_v in which all fanouts equal 1 such that N_v sorts all 0-1 vectors except v.*

Proof. The network N_v is depicted in Figure 4. Given v, let $L \triangleq \{x_i \mid v_i = 0\}$ and $H \triangleq \{x_i \mid v_i = 1\}$. Let ℓ' denote the maximal value in L, and let h' denote the minimal value in H. The inputs in L are fed into a sorting network $S_{|L|}$ of width $|L|$. The outputs of the sorting network $S_{|L|}$ are separated into the output that carries the maximal value ℓ' and the remaining $|L| - 1$ outputs denoted by L'. Similarly, the inputs in H are fed to a sorting network $S_{|H|}$ of width $|H|$. The outputs of $S_{|H|}$ are separated into the output that carries the minimal value h' and the remaining $|H| - 1$ outputs denoted by H'. The $n - 2$ outputs in $L' \cup H'$ are input to a sorting network S_{n-2}. The outputs of S_{n-2} are split into the lower $|L| - 1$ outputs denoted by L'' and the upper $|H| - 1$ outputs denoted by H''. Finally, L'' together with h' is input to a sorting network $S_{|L|}$ to output the outputs $y_0, \ldots, y_{|L|-1}$ of N_v. Similarly, H'' together with ℓ' is input to a sorting network $S_{|H|}$ to output the outputs $y_{|L|}, \ldots, y_{n-1}$ of N_v. Since there exists sorting networks with fanout one (e.g., Batcher's sorting [1]), the fanout of all ports in N_v equals one.

It remains to show that the network N_v fails in sorting a 0-1 vector u if only if $u = v$. Note that $u \neq v$ if and only if $\ell' = 1$ or $h' = 0$. For two 0-1 vectors a and b (not necessarily of the same length), we say that b *dominates* a if $\max_i a_i \leq \min_i b_i$. We denote the relation "b dominates a" by $a \preceq b$. Note that $L'' \preceq H''$ since L'' is the lower part and H'' is the upper part of the outputs of S_{n-2}.

If $u \neq v$, there are two cases. We prove the case $\ell' = 1$ (the case in which $h' = 0$ is similar). We claim that

$$L'' \cdot h' \preceq H'' \cdot \ell'. \tag{1}$$

Equation 1 obviously holds if $h' = 0$. If $h' = 1$, then H is all ones, and therefore, so are H' and H''. It follows that $\min_i (H'' \cdot \ell')_i = 1$, and Eq. 1 holds. Since Eq. 1 holds, it follows that $N(u)$ is sorted, as required.

The comparator network N fails in sorting v since $\ell' = 0$ while H' is all ones, therefore $y_{|L|} = 0$. On the other hand, $h' = 1$ while L'' is all zeros, hence $y_{|L|-1} = 1$, and the lemma follows.

We note that an attempt to design a witness network simply by flipping two (adjacent) output vertices of a sorting network fails. The reason is that a sorting network in which the outputs y_i and y_{i+1} are flipped fails in sorting all binary vectors whose weight is $i+1$. If non-adjacent outputs y_i and y_j are flipped, where $i < j$, then all binary vectors whose weight is greater than i and at most j are not sorted.

Furthermore, there do not exist witness networks for permutations as stated in the following claim.

Claim. For every permutation vector v, there does not exist a min-max network N_v such that N_v sorts all permutation vectors except v.

Proof. By Lemma 1, if N_v does not sort v, then there exists a 0-1 image b of v such that N_v does not sort b. The number of permutations w such that b is a 0-1 image of w is $k!(n-k)!$, where k is the weight of b. Hence, N_v fails in sorting many permutations if it fails in sorting one.

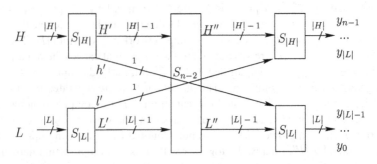

Fig. 4. A comparators network N_v that sorts all 0-1 vectors except for v. The building blocks of N_v are sorting networks $S_{|L|}, S_{|H|}$ and S_{n-2}.

If v is a balanced 0-1 vector, we obtain the following corollary of Lemma 6.

Corollary 2. *For every balanced 0-1 vector v, there is a network N_v that halves every 0-1 vector except v.*

The following lemma states necessary and sufficient conditions for a set to be conclusive with respect to min-max networks (and therefore, also with respect to comparator networks).

Lemma 7. *Let C be a set of vectors.*

- *C is conclusive for sorting iff every 0-1 vector is a 0-1 image of a vector of C.*
- *C is conclusive for merging iff every bisorted 0-1 vector is a 0-1 image of a vector of C.*

- C is conclusive for bitonic sorting iff every bitonic 0-1 vector is a 0-1 image of a vector of C.
- C is conclusive for halving iff every balanced 0-1 vector is a 0-1 image of a vector of C.

Proof. We consider the task of sorting.

($\Rightarrow$) Assume that there exists a 0-1 vector z that is not a 0-1 image of any vector in C. Consider the network N_z (guaranteed by Lemma 6) that sorts all 0-1 vectors except for z. We claim that N_z sorts all vectors in C. Indeed, N_z sorts all 0-1 images of vectors in C, and hence, by Lemma 1 sorts all vectors in C. However, N_z is not a sorting network, implying that C is not a conclusive set for sorting.

($\Leftarrow$) Suppose every 0-1 vector is a 0-1 image of a vector of C. Let N denote a min-max network. If N sorts every $v \in C$, then N sorts every 0-1 image of $v \in C$, and hence, N sorts every 0-1 vector. By Theorem 2, N is a sorting network. Hence, sorting all vectors in C implies that N is a sorting network, and thus, C is a conclusive set for sorting.

The proof for merging, bitonic sorting, and halving is similar.

The following lemma states exact sizes of 0-1 balanced vectors of three different kinds.

Lemma 8. *The following statements hold for every $n > 0$:*

- *There are exactly $\binom{n}{n/2}$ balanced 0-1 vectors of length n.*
- *There are exactly $n/2 + 1$ balanced bisorted 0-1 vectors of length n.*
- *There are exactly n balanced bitonic 0-1 vectors of length n.*

Proof. The number of balanced 0-1 vectors follows from the fact that there are $\binom{n}{n/2}$ possible choices for the indexes of the ones. Every 0-1 balanced bisorted vector is of the form $0^i \cdot 1^{n/2-i} \cdot 0^{n/2-i} \cdot 1^i$ There are $n/2 + 1$ possible values for i. Finally, every 0-1 balanced bitonic vector is a rotation of $0^{n/2} \cdot 1^{n/2}$. There are n possible rotations, and the lemma follows.

The following lemma states lower bounds on the size of conclusive sets for sorting, merging, bitonic sorting, and halving.

Lemma 9. *Let C be a set of vectors of length n.*

- *If C is conclusive for sorting then $|C| \geq \binom{n}{n/2}$.*
- *If C is conclusive for merging then $|C| \geq n/2 + 1$.*
- *If C is conclusive for bitonic sorting then $|C| \geq n$.*
- *If C is conclusive for halving then $|C| \geq \binom{n}{n/2}$.*

Proof sketch: The proof relies on the observation that every vector has at most one 0-1 image that is balanced. If C is conclusive for sorting, then by Lemma 7 every 0-1 vector is a 0-1 image of a vector in C. In particular, every balanced 0-1 vector is a 0-1 image of a vector in C. By the above observation, the number of vectors in C is not less than the number of 0-1 balanced vectors. Hence, by Lemma 8 $|C| \geq \binom{n}{n/2}$, as required. The proof of the other three lower bounds is similar.   □

7   Discussion and Open Problems

We presented upper bounds and lower bounds on the size of conclusive sets for sorting, merging, halving, and bitonic sorting (see Table 1). Separate bounds are presented for binary vectors and vectors over $\{0, \ldots, n-1\}$. We show that the use of nonbinary vectors reduces the size of conclusive sets in all cases, except for halving.

Knuth [5, ex. 2, p. 218] proved the following property about selection. If the output y_t of a comparator network outputs the tth smallest input, then it has $t-1$ outputs that output the $t-1$ smallest inputs and $n-t$ outputs that output the $n-t$ largest inputs. Let $0 < i_1, i_2$ and $i_1 + i_2 < n$. A vector v is (i_1, i_2)-separated if $v_j \leq v_k \leq v_\ell$ for every $j < i_1 \leq k < i_1 + i_2 \leq \ell$. A min-max network N is called an (i_1, i_2)-separator if $N(v)$ is (i_1, i_2)-separated for every vector v. Knuth's statement about selection implies that if the output y_t of a comparator network N outputs the tth smallest input, then N is a $(t-1, 1)$-separator. Note that every halver is an $(n/2, n/2)$-separator. Our techniques for conclusive sets for halvers can be extended to (i_1, i_2)-separators. The set of all binary vectors of weights i_1 or $i_1 + i_2$ is an optimal conclusive set. This result can be extended to networks that separate the input into any number of "blocks".

It seems reasonable that a min-max network with n inputs and outputs should accept values in the set $\{0, \ldots, n-1\}$. The question of finding upper bounds and lower bounds for the case in which only $2 < k < n$ values are accepted remains open. Obviously, as k decreases from n to 2, the size of conclusive sets increases in all cases except for halving.

Another open problem is to formalize a neater characterization of $\binom{n}{n/2}$ permutations that constitute a conclusive set for sorting. The characterization in the proof of Theorem 7 is based on chains that cover all the subsets in the poset $\mathbb{P}$.

Acknowledgments

We thank Tuvi Etzion for the suggestion to use Sperner's Theorem in the proof of Theorem 7.

References

1. Batcher, K.E.: Sorting Networks and their Applications. In: Proc. AFIPS Spring Joint Computer Conference, vol. 32, pp. 307–314 (1968)
2. Even, G., Levi, T., Litman, A.: A complete characterization of functions that commute with monotonic functions (in preparation)
3. Bender, E.A., Williamson, S.G.: Periodic Sorting Using Minimum Delay Recursively Constructed Merging Networks. Electronic Journal Of Combinatorics, vol. 5 (1998)
4. Dilworth, R.P.: A Decomposition Theorem for Partially Ordered Sets. Annals of Mathematics 51, 161–166 (1950)
5. Knuth, D.E.: The art of computer programming Sorting and searching, vol. 3. Addison-Wesley, London (1973)
6. Lee, D.L., Batcher, K.E: A Multiway merge sorting network. IEEE Transactions on Parallel and Distributed Systems 6, 211–215 (1995)
7. Levi, T.: Minimal depth merging networks, M.Sc. Thesis, Technion (March 2006)

8. Liszka, K.J., Batcher, K.E.: A Modulo merge sorting network, Symposium on the Frontiers of Massively Parallel Computation (1992)
9. Lubell, D.: A short proof of Sperner's theorem. J. Combin. Theory 1, 299 (1966)
10. Miltersen, P.B., Paterson, M., Tarui, J.: The asymptotic complexity of merging networks. J. ACM 43(1), 147–165 (1996)
11. Nakatani, T., Huang, S.T., Arden, B.W., Tripathi, S.K.: K-way bitonic Sort. IEEE Trans on Computers 38, 283–288 (1989)
12. Rajasekaran, S., Sen, S.: A generalization of the 0-1 principle for Sorting. IPL 94, 43–47 (2005)
13. Rice, W.D.: Continuous Algorithms. Topology Appl. 85, 299–318 (1998)

Selfish Routing with Oblivious Users

George Karakostas[1,*], Taeyon Kim[1,**], Anastasios Viglas[2],
and Hao Xia[1,***]

[1] McMaster University, Dept. of Computing and Software, 1280 Main St. West,
Hamilton, Ontario L8S 4K1, Canada
{karakos,kimt22,xiah}@mcmaster.ca
[2] University of Sydney, School of Information Technologies, Madsen Building F09,
The University of Sydney, NSW 2006, Australia
tasos@it.usyd.edu.au

Abstract. We consider the problem of characterizing user equilibria
and optimal solutions for selfish routing in a given network. We extend
the known models by considering users *oblivious to congestion*. While
in the typical selfish routing setting the users follow a strategy that
minimizes their individual cost by taking into account the (dynamic)
congestion due to the current routing pattern, an *oblivious* user ignores
congestion altogether. Instead, he decides his routing on the basis of
cheapest routes on a network without any flow whatsoever. These
cheapest routes can be, for example, the shortest paths in the network
without any flow. This model tries to capture the fact that routing
tables for at least a fraction of the flow in large scale networks such as
the Internet may be based on the physical distances or hops between
routers alone. The phenomenon is similar to the case of traffic networks
where a certain percentage of travelers base their route simply on the
distances they observe on a map, without thinking (or knowing, or
caring) about the delays experienced on this route due to their fellow
travelers. In this work we study the price of anarchy of such networks,
i.e., the ratio of the total latency experienced by the users in this setting over the optimal total latency if all users were centrally coordinated.

Keywords: Selfish routing, price of anarchy, oblivious users.

1 Introduction

The general framework of a system of non-cooperative users can be used to
model many different optimization problems such as network routing, traffic or
transportation problems, load balancing and distributed computing, auctions
and many more. Game-theoretical techniques can be used to model and analyze
such systems in a natural way. The performance of a system of non-cooperative

---

* Research supported by an NSERC Discovery grant and MITACS.
** Research supported by an NSERC Discovery grant and MITACS.
*** Research supported by an NSERC Discovery grant and MITACS.

G. Prencipe and S. Zaks (Eds.): SIROCCO 2007, LNCS 4474, pp. 318–327, 2007.
© Springer-Verlag Berlin Heidelberg 2007

users is measured by an appropriate cost function which depends on the behaviour, or strategies of the users. For example in the case of network routing, the total, system-wide cost can be defined as the total routing cost, or the total latency experienced by all the users in the network. On the other hand, there is also a cost associated with each individual user (for example the latency experienced by the user). It is a well known fact that if each user optimizes her own cost, then they might choose a strategy that does not give the optimal total cost for the entire system, also known as *social cost*. In other words, the *selfish* behaviour of the users leads to a sub-optimal performance.

Koutsoupias and Papadimitriou [3] initiated the study of the *coordination ratio* (also referred to as the price of anarchy): How much worse is the performance of a system of selfish users where each user optimizes her own cost, compared to the best possible performance that can be achieved on the same system? In particular, this question was first studied in the setting of selfish network routing by Roughgarden and Tardos [5]. In this model, the network users experience edge latencies that depend on the congestion on each edge according to some latency function. Given a particular flow pattern, the users decide to route their flow through paths of minimum latency. A *traffic equilibrium* is an assignment of traffic to paths so that no user can unilaterally switch her flow to a path of smaller cost. Wardrop's principle [6] for selfish routing postulates that

> at equilibrium, for each origin-destination pair the travel costs on all the routes actually used are equal, or less than the travel costs on all unused routes.

In the past, several variations of this basic model have been considered. For example, Roughgarden [4] studied the case of combining selfish and centrally coordinated users on the same network, proposing *Stackelberg strategies* for the latter that would improve the price of anarchy. Karakostas and Viglas [2] studied the combination of selfish and malicious users: A malicious user will choose a strategy that will cause the worst possible performance for the entire network. These models try to capture a richer set of paradigms in networks such as the Internet, where traffic does not consist by users of the same profile or behavior. In this work we introduce a new paradigm that is based on the following observation: A fundamental assumption in the basic selfish routing model is that each user is able to measure the latencies of *all* paths available to him at any moment, in order to pick the best possible path currently available for his flow. It is clear that in very large networks this assumption is probably quite unrealistic, since it may not be possible to measure these latencies or measure them as often as needed. Hence it may be easier for a fraction α of the network users to consult predefined routing tables based on *non-dynamic* parameters of the network, such as the physical distances between nodes. The price these users pay for the convenience is a degree of naivety in their decisions, since they are completely *oblivious* to congestion phenomena. We call such users *oblivious*.

More specifically, we consider oblivious users that route their flow through the shortest path connecting its origin to its destination, as measured in the network without flow. We study the price of anarchy in case of linear edge latency functions, first for a (single commodity) single pair of nodes connected by a set of parallel edges, and then for general topologies with an arbitrary number of origin-destination pairs. Unlike the case of selfish routing without oblivious users where the price of anarchy is bounded by 4/3 [5], our bounds are not independent of the network parameters. For both the cases of parallel links and general topologies, the bounds depend on the coefficients a_e of the linear latency functions $l_e(f_e) = a_e f_e + b_e$ for edges e, where f_e is the total flow through edge e. In addition, the general topology bound depends on the minimum fraction of total demand that the optimal routing sends through any edge. Although these bounds can be very large, if, for example, there are network edges with vastly different behavior under congestion (as is the case in a traffic network with both highways and side-streets), this seems to be unavoidable in view of the fact that the myopic behavior of oblivious users may lead to great congestion of 'wide' edges by them, and in that way directing the selfish users to non-congested but 'narrow' paths. Indeed, we provide an example exhibiting such behavior for the simple case of parallel links in Section 3.1. In addition, the dependence of the general topology bound on the 'spread' of the optimal flow seems to be necessary, given that the oblivious flow concentrates the oblivious users on specific (initially fastest, but possibly very slow after the selfish users have been added) paths, which may be orthogonal to what the optimal flow does.

Organization: In Section 2 we define the model, in Section 3 we study linear latency functions in simple networks of two nodes connected by parallel links, and in Section 4 we study linear functions in general, multicommodity networks. We conclude with a discussion in Section 5.

2   Preliminaries

We are given a directed network $G = (V, E)$ with a latency function $l_P(\cdot)$ associated to each path P. For a flow f on G, $l_P(f)$ is the latency (cost) of path P for this particular flow. Notice that in general this latency depends on the whole flow f, and not only on the flow f_e through each edge $e \in P$. In this paper we adopt the *additive model* for the path latencies, i.e., $l_P(f) = \sum_{e \in P} l_e(f_e)$, where l_e is the latency function for edge e and f_e is the amount of flow that goes through e. We also let $\mathcal{P}$ be the set of all available paths in the network and assume that for every source-sink pair there is at least one path joining the source to the sink. In this work we assume that the latency functions are linear functions of the edge flow f_e, i.e., $l_e(f_e) = a_e f_e + b_e$, $\forall e \in E$. The total cost of a flow f is defined as $C(f) = \sum_{e \in E} f_e l_e(f_e)$.

We consider the case where for every origin-destination commodity of demand d, a fraction α of it consists of an infinite number of *oblivious* users, each carrying an infinitesimal amount of flow through the shortest path connecting the source to the destination when there is no flow routed on G. If there are more than one shortest paths, we will assume that all these users pick the smallest in a lexicographic ordering. The rest $(1 - \alpha)d$ of the demand consists of an infinite number of *selfish* users, each carrying an infinitesimal amount of flow.

3   Parallel Links

Let G be a network consisting of parallel links connecting two nodes s, t. We will assume that the edge latency functions are strictly increasing, i.e., for every edge e with $l_e(f_e) = a_e f_e + b_e$ we have $a_e > 0$. Note that in this setting, both the traffic equilibrium flow and the optimal flow are unique. In what follows, we use $0 \leq \alpha \leq 1$ to denote the fraction of total flow from s to t that is oblivious. We will use the term 'traffic equilibrium' for flows with $\alpha = 0$ that are at traffic equilibrium, while we reserve the term 'oblivious equilibrium' for flows with $\alpha > 0$ and with their selfish users at traffic equilibrium in the network that results *after* routing the oblivious users. The following observation is true due to Wardrop's principle and since the latency functions are increasing:

Proposition 1. *Let f^d and $f^{d+\delta}$ be flows at traffic equilibrium, of demand d and $d + \delta$ respectively, with $\delta \geq 0$. Then we have $f_e^d \leq f_e^{d+\delta}$, $\forall e \in E$.*

In what follows, we denote the total demand from s to t with d, and the optimal flow of demand d with f^{opt}. We denote the flow of demand d at traffic equilibrium with f^*, and the flow of demand d at oblivious equilibrium with $\tilde{f}$, where the flow of oblivious users with total demand αd is denoted with $\tilde{f}^o$ and the flow of selfish users (with total demand $(1 - \alpha)d$) with $\tilde{f}^*$, i.e., $\tilde{f} = \tilde{f}^* + \tilde{f}^o$. Obviously, the oblivious flow will be routed through the edge e with the smallest $l_e(0) = b_e$ (or the first such edge in a lexicographic ordering, if there are more than one). Let e_s be this edge.

Proposition 2. $f_e^* \geq \tilde{f}_e^*$, $\forall e \in E$.

Proof: If $f_{e_s}^* \geq \alpha d$ then $f^* = \tilde{f}$. Otherwise we have $\alpha d > f_{e_s}^*$. In this case, no selfish users will flow through e_s because of Wardrop's principle, i.e., $\tilde{f}_{e_s}^* = 0$. By removing e_s from G together with the portion of flow on it, we get two new Nash flows $f^{*'}$ and $\tilde{f}^{*'}$, of demand $d - f_{e_s}^*$ and $d - \alpha d$. As $\alpha d > f_{e_s}^*$, from Proposition 1, we have $f^{*'} \geq \tilde{f}^{*'}$. Then since $\tilde{f}_{e_s}^* = 0$, we get $f^* \geq \tilde{f}^*$. □

Lemma 1. $C(\tilde{f}^*) \leq \frac{4}{3}(1 - \alpha)C(f^{opt})$.

Proof: Since latency functions are increasing and $f^* \geq \tilde{f}^*$ from Proposition 2, we know that $\forall e$, $l_e(f_e^*) \geq l_e(\tilde{f}_e^*)$. Also Wardrop's principle for f^* implies that $l_e(f_e^*) = L(f^*)$, $\forall e : f_e^* > 0$ and $l_e(f_e^*) \geq L(f^*)$, $\forall e$, where $L(f^*)$ is the common latency of the paths used by the traffic equilibrium flow f^*. Then

$$C(f^*) - C(\tilde{f}^*) = \sum_e \left(l_e(f_e^*)f_e^* - l_e(\tilde{f}_e^*)\tilde{f}_e^* \right)$$

$$\geq \sum_e (f_e^* - \tilde{f}_e^*)l_e(f_e^*)$$

$$\geq \sum_e (f_e^* - \tilde{f}_e^*)L(f^*) = \alpha dL(f^*) = \alpha C(f^*).$$

Then Theorem 4.5 of [5] implies the lemma.   □

The Karush-Kuhn-Tucker conditions imply that f^{opt} is a traffic equilibrium for latency functions $l_e^*(x) = \frac{\partial l_e}{\partial f_e}(x)$, i.e., for $l_e^*(f_e) = 2a_e f_e + b_e$. Then Wardrop's principle implies that

$$l_{e_1}^*(f_{e_1}^{opt}) = l_{e_2}^*(f_{e_2}^{opt}), \; \forall e_1, e_2 : f_{e_1}^{opt}, f_{e_2}^{opt} > 0$$

and

$$l_e^*(f_e^{opt}) \geq l_{e_1}^*(f_{e_1}^{opt}), \; \forall e_1 : f_{e_1}^{opt} > 0.$$

We will use this fact in what follows.

Proposition 3. $f_{e_s}^{opt} > 0$.

Proof: Let e be an edge with $f_e^{opt} > 0$. Then we have

$$l_{e_s}^*(f_{e_s}^{opt}) \geq l_e^*(f_e^{opt}) = 2a_e f_e^{opt} + b_e$$
$$> b_e$$
$$\geq b_{e_s} = l_{e_s}^*(0).$$

Therefore $f_{e_s}^{opt} > 0$, since functions $l_e^*(x)$ are increasing.   □

In what follows, let $\mathcal{E}^{opt} = \{e : f_e^{opt} > 0\}$.

Proposition 4. For any edge e, we have $l_{e_s}(f_{e_s}^{opt}) \leq l_e(f_e^{opt})$. For any edge $e \in \mathcal{E}^{opt}$, we have $a_{e_s} f_{e_s}^{opt} \geq a_e f_e^{opt}$.

Proof: For any edge e, we have

$$l_e(f_e^{opt}) = \frac{1}{2} \left(l_e^*(f_e^{opt}) + b_e \right)$$
$$\geq \frac{1}{2} \left(l_{e_s}^*(f_{e_s}^{opt}) + b_e \right)$$
$$\geq \frac{1}{2} \left(l_{e_s}^*(f_{e_s}^{opt}) + b_{e_s} \right)$$
$$= l_{e_s}(f_{e_s}^{opt}).$$

where the first inequality is due to Proposition 3, and the second is due to the definition of e_s.

Similarly, we get the second part of the proposition for any edge $e \in \mathcal{E}^{opt}$.   □

Lemma 2. $C(\tilde{f}^o) \leq \max\{\alpha, \alpha^2 r\} C(f^{opt})$, where $r = \sum_{e \in \mathcal{E}^{opt}} (a_{e_s}/a_e)$.

Proof: We have

$$C(f^{opt}) = \sum_e l_e(f_e^{opt}) f_e^{opt}$$

$$\geq l_{e_s}(f_{e_s}^{opt}) \sum_e f_e^{opt} \qquad \text{(Proposition 4)}$$

$$= (a_{e_s} f_{e_s}^{opt} + b_{e_s}) d.$$

From the second part of Proposition 4 we have that $(a_{e_s}/a_e) f_{e_s}^{opt} \geq f_e^{opt}$, $\forall e \in \mathcal{E}^{opt}$. By summing over all $e \in \mathcal{E}^{opt}$, we get $a_{e_s} f_{e_s}^{opt} \geq a_{e_s} d/r$. Thus

$$C(f^{opt}) \geq \left(\frac{a_{e_s}}{r} d + b_{e_s}\right) d.$$

Therefore

$$\frac{C(\tilde{f}^o)}{C(f^{opt})} \leq \frac{(a_{e_s} \alpha^2 d + \alpha b_{e_s}) d}{\left(\frac{a_{e_s}}{r} d + b_{e_s}\right) d} \leq \max\{\alpha, \alpha^2 r\}.$$

$\square$

Theorem 1. If $\tilde{f}_{e_s}^o = \alpha d \geq f_{e_s}^*$, then $\frac{C(\tilde{f})}{C(f^{opt})} \leq \frac{4}{3}(1 - \alpha) + \max\{\alpha, \alpha^2 r\}$, otherwise $\frac{C(\tilde{f})}{C(f^{opt})} \leq \frac{4}{3}$.

Proof: If $\tilde{f}_{e_s}^o = \alpha d < f_{e_s}^*$ then $\tilde{f} = f^*$ and the second part of the theorem follows. In the case $\alpha d \geq f_{e_s}^*$, edge e_s which is used by the oblivious users is no longer attractive to selfish users, i.e., $\tilde{f}_{e_s}^* = 0$. Thus $\tilde{f}^*$ and $\tilde{f}^o$ are actually orthogonal, i.e., $\tilde{f}^{*T} \tilde{f}^o = 0$. Then, if $A > 0$ is the $|E| \times |E|$ diagonal matrix whose diagonal elements are the a_e's, we have

$$C(\tilde{f}) = (A(\tilde{f}^* + \tilde{f}^o) + b)^T (\tilde{f}^* + \tilde{f}^o)$$

$$= \tilde{f}^{*T} A \tilde{f}^* + \tilde{f}^{o^T} A \tilde{f}^o + 2\tilde{f}^{*T} A \tilde{f}^o + b^T(\tilde{f}^* + \tilde{f}^o)$$

$$= (A\tilde{f}^* + b)^T \tilde{f}^* + (A\tilde{f}^o + b)^T \tilde{f}^o$$

and the first part of the theorem now comes from Lemmata 1 and 2. $\square$

3.1 A Bad Example for Parallel Links

We provide an example to show that in networks with parallel links, the price of anarchy can be as bad as our bound in Theorem 1 in case $\alpha = 1$, i.e., all users are oblivious. The network has only two links, namely e_1 and e_2, with latency functions $l_1(x) = 10x$ and $l_2(x) = x + \epsilon$, where $0 < \epsilon < 1$. The total demand is $d = 1$.

The optimal cost in this setting is $C_{opt} = 10/11 + (40\epsilon - \epsilon^2)/44$. When $\alpha \geq (1+\epsilon)/11$, the cost of the oblivious equilibrium is $C_{eq} = 11\alpha^2 - (2+\epsilon)\alpha + (1+\epsilon)$.

One can see that when α is one (all users are oblivious), and ϵ tends to zero, the price of anarchy is

$$\lim_{\epsilon \to 0} \frac{C_{eq}}{C_{opt}} = 11,$$

which is exactly the bound we get in Theorem 1.

However, this example is not tight when $\alpha < 1$. The loss of tightness comes from the 4/3 which is the upper bound for the selfish routing price of anarchy [5]. We used this result directly in the last step of Lemma 1 and in the first case of Theorem 1. While this example is a tight example for Lemma 2, it is not tight for the 4/3. The price of anarchy here is very close to 1. Thus a real tight example for our bound would be one that is tight for both 4/3 and Lemma 2. Unfortunately such an ideal example does not exist since the tightness of Lemma 2 requires very small b_e/a_e for all links, but in order to make 4/3 tight we need a relatively large b_e/a_e to make a distinction between the selfish flow and the optimal flow. This implies that the bound in Theorem 1 is not tight, and a tighter bound remains as an open problem.

4   General Topologies

In this section we study the price of anarchy of oblivious equilibria for general topologies, arbitrary number of origin-destination pairs (commodities) and linear latency functions. We will use the concept of β-function defined in [1]. Let $\mathcal{L}$ be a family of continuous and non-decreasing latency functions. For every function $l \in \mathcal{L}$ and every value $v \geq 0$, let us define:

$$\beta(v, l) := \frac{1}{vl(v)} \max_{x \geq 0} \{x(l(v) - l(x))\}.$$

In addition, let us define

$$\beta(l) := \sup_{v \geq 0} \beta(v, l),$$

and

$$\beta(\mathcal{L}) := \sup_{l \in \mathcal{L}} \beta(l).$$

We will denote the inner product of two vectors x, y by $\langle x, y \rangle$.

We will also use an alternative characterization of a traffic equilibrium f^* of demand d, as a solution to the following variational inequality:

$$\langle l(f^*), f - f^* \rangle \geq 0, \ \forall f \in \{f : f \text{ is a flow satisfying demand } d\}.$$

By applying this formulation to the selfish part of an oblivious equilibrium $\tilde{f} = \tilde{f}^* + \tilde{f}^o$ in the network obtained after the oblivious users have been routed[1], we get

$$\langle l(\tilde{f}), f - \tilde{f}^* \rangle \geq 0, \ \forall f \in \{f : f \text{ is a flow satisfying demand } (1 - \alpha)d\}.$$

---

[1] Note that the new latency functions in this network are $l'_e(f_e) = a_e(\tilde{f}_e^o + f_e) + b_e = l_e(f_e + \tilde{f}_e^o)$.

By setting $f := (1 - \alpha)f^{opt}$ we have

$$\langle l(\tilde{f}), \tilde{f}^* \rangle \leq (1 - \alpha)\langle l(\tilde{f}), f^{opt} \rangle. \tag{1}$$

Lemma 3. $\langle l(\tilde{f}), \tilde{f}^* \rangle \leq (1 - \alpha)\beta(\mathcal{L})C(\tilde{f}) + (1 - \alpha)C(f^{opt})$.

Proof:

$$\langle l(\tilde{f}), \tilde{f}^* \rangle \overset{(1)}{\leq} (1 - \alpha)\langle l(\tilde{f}), f^{opt} \rangle$$

$$\leq (1 - \alpha)\left(\sum_e \beta(\tilde{f}_e, l_e)l_e(\tilde{f}_e)\tilde{f}_e + \sum_e l_e(f_e^{opt})f_e^{opt} \right)$$

$$\leq (1 - \alpha)\left(\beta(\mathcal{L})C(\tilde{f}) + C(f^{opt}) \right).$$

$\square$

Lemma 4. $\langle l(\tilde{f}), \tilde{f}^o \rangle \leq \frac{n\alpha d\gamma_a}{f_{min}^{opt}}C(f^{opt})$, where $n = |V|$, $\gamma_a = \frac{\max_e a_e}{\min_e a_e}$, and $f_{min}^{opt} = \min_e f_e^{opt}$.

Proof: Let P_{s_i} be the path used by the oblivious users corresponding to the i-th origin-destination pair (commodity). Note that this is the shortest path amongst all possible paths $\mathcal{P}_i$ connecting this pair when we define the edge distances as $l_e(0) = b_e$. Also let d_i be the demand for this pair, therefore αd_i is the amount of oblivious flow routed through P_{s_i}. Let also $a_{min} = \min_e a_e$, $a_{max} = \max_e a_e$.

The key observation is that the oblivious flow $\tilde{f}^o$ is a traffic equilibrium for the original network, if we define its latency functions as $l_e^o(f) = b_e$. From the discussion above, this implies that

$$\langle l(0), f - \tilde{f}^o \rangle \geq 0, \ \forall f \in \{f : f \text{ is a flow satisfying demand } \alpha d\},$$

or, if we set $f := \alpha f^{opt}$, we get

$$\sum_{e \in E} b_e \tilde{f}_e^o \leq \alpha \sum_{e \in E} b_e f_e^{opt}. \tag{2}$$

Then

$$\langle l(\tilde{f}), \tilde{f}^o \rangle = \sum_{e \in E} \left(a_e(\tilde{f}_e^* + \tilde{f}_e^o)\tilde{f}_e^o + b_e\tilde{f}_e^o \right)$$

$$\overset{(2)}{\leq} \alpha \sum_i d_i \sum_{e \in P_{s_i}} a_e(\tilde{f}_e^* + \tilde{f}_e^o) + \alpha \sum_{e \in E} b_e f_e^{opt}. \tag{3}$$

To get a upper bound of the first term:

$$\alpha \sum_i d_i \sum_{e \in P_{s_i}} a_e(\tilde{f}_e^* + \tilde{f}_e^o) = \alpha \sum_i d_i \sum_{e \in P_{s_i}} a_e(\tilde{f}_e^* + \alpha d_i)$$

$$\leq n\alpha^2 a_{max} \sum_i d_i^2 + \alpha \sum_i d_i \sum_{e \in P_{s_i}} a_e \tilde{f}_e^*$$

$$\leq n\alpha^2 a_{max} \sum_i d_i^2 + n\alpha a_{max}(1-\alpha) \sum_i d_i^2$$

$$\leq n\alpha d\gamma_a(a_{min}d)$$

$$\leq n\alpha d\gamma_a \sum_{e \in E} a_e f_e^{opt}$$

$$\leq \frac{n\alpha d\gamma_a}{f_{min}^{opt}} \sum_{e \in E} a_e f_e^{opt2}. \qquad (4)$$

Since $\frac{n\alpha d\gamma_a}{f_{min}^{opt}} \geq \alpha$ the combination of (3),(4) proves the lemma.

□

Theorem 2.
$$\frac{C(\tilde{f})}{C(f^{opt})} \leq \frac{4\left(1 - \alpha + n\alpha d\gamma_a/f_{min}^{opt}\right)}{3 + \alpha}$$

Proof: By combining Lemma 3 with Lemma 4, we have

$$C(\tilde{f}) = \langle l(\tilde{f}), \tilde{f}^o \rangle + \langle l(\tilde{f}), \tilde{f}^* \rangle$$

$$\leq (1-\alpha)C(f^{opt}) + (1-\alpha)\beta(\mathcal{L})C(\tilde{f}) + \frac{n\alpha d\gamma_a}{f_{min}^{opt}}C(f^{opt}),$$

hence

$$\frac{C(\tilde{f})}{C(f^{opt})} \leq \frac{1 - \alpha + n\alpha d\gamma_a/f_{min}^{opt}}{1 - (1-\alpha)\beta(\mathcal{L})}.$$

For $\mathcal{L}$ being the set of non-decreasing linear functions $\beta(\mathcal{L}) = \frac{1}{4}$ [1], and the theorem follows.

□

5 Discussion and Open Problems

The obvious open problem is the tightening of the bounds of Theorems 1,2. One method of doing so seems to be the avoidance of relating the cost of oblivious equilibria to the optimal cost via traffic equilibria. It is precisely this intermediate step that doesn't allow us yet to have a tight analysis for Theorem 1.

Especially Theorem 2 for general topologies may be possible to be improved by removing its dependence on the minimum optimum path flow f_{min}^{opt}. Although the optimum flow is a parameter of the network, it may be very difficult to be determined by the network designer, while the other parameters of the network (G, n, d, a_e, b_e) can be set directly. Finally, it would be interesting to get non-trivial bounds (if they exist) for general latency functions.

References

1. Correa, J.R., Schulz, A.S., Stier, N.E.: Selfish routing in capacitated networks. Mathematics of Operations Research 29, 961–976 (2004)
2. Karakostas, G., Viglas, A.: Equilibria for networks with malicious users. Mathematical Programming Series A, published online (July 29, 2006) DOI: 10.1007/s10107-006-0015-2
3. Koutsoupias, E., Papadimitriou, C.: Worst-case equilibria. In: Proceedings of the 16th Annual Symposium on Theoretical Aspects of Computer Science, pp. 404–413 (1999)
4. Roughgarden, T.: Stackelberg scheduling strategies. SIAM Journal on Computing 33, 332–350 (2004)
5. Roughgarden, T., Tardos, É.: How bad is selfish routing? Journal of the ACM 49, 236–259 (2002)
6. Wardrop, J.G.: Some theoretical aspects of road traffic research. In: Proc. Inst. Civil Engineers, Part II, vol. 1, pp. 325–378 (1952)

Upper Bounds and Algorithms for Parallel Knock-Out Numbers

Hajo Broersma, Matthew Johnson, and Daniël Paulusma

Department of Computer Science, Durham University,
Science Laboratories, South Road, Durham DH1 3LE, U.K.
{hajo.broersma,matthew.johnson2,daniel.paulusma}@durham.ac.uk

Abstract. We study parallel knock-out schemes for graphs. These schemes proceed in rounds in each of which each surviving vertex simultaneously eliminates one of its surviving neighbours; a graph is reducible if such a scheme can eliminate every vertex in the graph. We show that, for a reducible graph G, the minimum number of required rounds is $O(\sqrt{\alpha})$, where α is the independence number of G. This upper bound is tight and the result implies the square-root conjecture which was first posed in MFCS 2004. We also show that for reducible $K_{1,p}$-free graphs at most $p-1$ rounds are required. It is already known that the problem of whether a given graph is reducible is NP-complete. For claw-free graphs, however, we show that this problem can be solved in polynomial time.

Keywords: parallel knock-out schemes, claw-free graphs, computational complexity.

1 Introduction

In this paper, we continue the study on *parallel knock-out schemes* for finite undirected simple graphs introduced in [7] and studied further in [2,3,4]. Such a scheme proceeds in rounds: in the first round each vertex in the graph selects exactly one of its neighbours, and then all the selected vertices are eliminated simultaneously. In subsequent rounds this procedure is repeated in the subgraph induced by those vertices not yet eliminated. The scheme continues until there are no vertices left, or until an isolated vertex is obtained (since an isolated vertex will never be eliminated).

A graph is *KO-reducible* if there exists a parallel knock-out scheme that eliminates the whole graph. The *parallel knock-out number* of a graph G, denoted by $\text{pko}(G)$, is the minimum number of rounds in a parallel knock-out scheme that eliminates every vertex of G. If G is not reducible, then $\text{pko}(G) = \infty$.

Knock-out schemes have an obvious relationship with games on graphs, a topic which has received considerable attention in the last decades ([6]). But unlike many games on graphs, knock-out schemes can be motivated by practical settings, e.g., in which objects exchange entities that inactivate the receiving objects, like viruses that paralyse or block computers, or computational tasks that disable processors or sensors from other tasks. Especially in the relatively new

G. Prencipe and S. Zaks (Eds.): SIROCCO 2007, LNCS 4474, pp. 328–340, 2007.
© Springer-Verlag Berlin Heidelberg 2007

area of sensor networks, knock-out schemes for the underlying graph structures can model practical situations in which sensors exchange data with neighbouring sensors that temporarily disables the receiving sensors from their main monitoring tasks. This happens, e.g., in situations where sensors have a low battery and limited computational power. They share measured and processed data with other sensors in their close vicinity as well as with more powerful PCs, laptops or mainframes at larger distances. Consider a setting with a number of sensors that perform simple measurements, for instance on temperature, humidity, smoke levels, movements, or the like. Data sharing is important for two reasons: in order to rule out erroneous data (by comparisons with data gathered at a neighbouring sensor) and in order to preprocess the data before sending it to a more powerful computer. During the preprocessing stage in a sensor no new data can be collected by that sensor, so the chosen neighbouring sensors are out of order for the time being, while the other sensors continue collecting data, sharing it with other active neighbouring sensors, and so on, until all sensors are out of order or run out of available neighbouring sensors. Then a new round of data collection and sharing starts. In the ideal case all sensors have shared their data with at least one neighbouring sensor and have performed some preprocessing of their data. In order to keep the time intervals between successive rounds of data collection as short as possible, the number of stages within one round should be kept to a minimum. This problem setting can be modelled by parallel knock-out schemes and the parallel knock-out number comes up naturally.

Our main motivation for studying knock-out schemes, though, is the intimate relationship between this concept and well-studied structural graph theoretical concepts like perfect matchings, hamiltonian cycles and 2-factors (they all yield knock-out schemes of one round). Apart from these structural aspects, we are interested in complexity aspects. Whereas the classical complexity problems related to matchings and hamiltonian cycles have been settled many years ago, the analogous problems related to knock-out schemes have been resolved recently, and only for general graphs and graphs of bounded tree-width. For many interesting classes, however, these problems on knock-out schemes are still open [3].

1.1 Our Results

In [3], a number of results, conjectures and questions on upper bounds for knock-out numbers were presented. For trees, the problem was resolved by showing that the knock-out number of a tree on n vertices was $O(\log n)$ and by exhibiting a family of trees that met this bound. They also presented a family of bipartite graphs whose knock-out numbers grow proportionally to the square root of the number of vertices, and conjectured that for any KO-reducible graph on n vertices the knock-out number is at most $2\sqrt{n}$. In this paper, in Section 3, we prove this conjecture.

In [3], a polynomial algorithm was also given that would determine the parallel knock-out number of any tree. In [4] it was shown that the problem of finding parallel knock-out numbers is, for general graphs, NP-complete. In this paper, in Section 4, we present a polynomial algorithm that finds the knock-out number of claw-free graphs, that is, graphs that do not contain an induced $K_{1,3}$; these form a well-studied class of graphs, see [5] for a survey. We also give a tight bound on the knock-out number of reducible $K_{1,p}$-free graphs, generalizing a result of [3] on claw-free graphs.

2   Preliminaries

Graphs in this paper are denoted by $G = (V, E)$. An edge joining vertices u and v is denoted by uv. If not stated otherwise a graph is assumed to be undirected and simple. If a graph G is directed then an arc from a vertex u to a vertex v is denoted by (u, v). In the *null graph*, $V = E = \emptyset$. For graph terminology not defined below, we refer to [1].

For a vertex $u \in V$ we denote its *neighbourhood*, that is, the set of adjacent vertices, by $N(u) = \{v \mid uv \in E\}$. The *degree* of a vertex is the number of edges incident with it, or, equivalently, the cardinality of its neighbourhood. A subset $U \subseteq V$ is called an *independent set* of G if no two vertices in U are adjacent to each other. The *independence number* α of a graph G is the number of vertices in a maximum independent set of G.

A *complete bipartite* graph $K_{|X|,|Y|}$ is a bipartite graph with the maximum number of edges between its bipartite classes X and Y. If $|X| = 1$, then it is a *star* and the vertex in X is the *centre vertex* and the vertices in Y are *leaves*. If $|X| = 1$ and $|Y| = 1$ we arbitrarily choose one of the star's two vertices to be the centre vertex. A graph G that does not contain a $K_{1,p}$ as an induced subgraph for some $p \geq 1$ is said to be $K_{1,p}$-*free*. A $K_{1,3}$-free graph is also called *claw-free*.

For a graph G, a *KO-selection* is a function $f : V \to V$ with $f(v) \in N(v)$ for all $v \in V$. If $f(v) = u$, we say that vertex v *fires at* vertex u, or that vertex u *is knocked out* by vertex v. We also say that u is a *victim* of v. For each $u \in f(V)$, we denote the set of vertices that fire at u by $K(u)$, i.e., $v \in K(u)$ if and only if $f(v) = u$. If $K(u) = \{v\}$, that is, vertex v is the only vertex that fires at u, then we call u the *unique* victim of v. For a subset $U \subseteq f(V)$ we use the shorthand notation $K(U) = \bigcup_{u \in U} K(u)$, and we say that such a subset U is *knocked out* by a subset $W \subseteq V$ if $K(U) \subseteq W$, that is, if every vertex in U is knocked out by a vertex in W.

For a KO-selection f, we define the corresponding *KO-successor* of G as the subgraph of G that is induced by the vertices in $V \setminus f(V)$; if H is the KO-successor of G we write $G \rightsquigarrow H$. Note that every graph without isolated vertices has at least one KO-successor. A graph G is called *KO-reducible*, if there exists a finite sequence

$$G \rightsquigarrow G^1 \rightsquigarrow G^2 \rightsquigarrow \cdots \rightsquigarrow G^r,$$

where G^r is the null graph. If no such sequence exists, then $\mathrm{pko}(G) = \infty$. Otherwise, the *parallel knock-out number* of G, $\mathrm{pko}(G)$, is the smallest number r for which such a sequence exists. A sequence S of KO-selections that transform G into the null graph is called a *KO-reduction scheme*. A single step in this sequence is called a *round* of the KO-reduction scheme. We denote the number of rounds in S by $r(S) = r$.

For a KO-reduction scheme S we denote the set of vertices that are victims of a vertex v by $L(v)$. For a subset $W \subseteq V$, we use the shorthand notation $L(W) = \bigcup_{v \in W} L(v)$.

An *in-tree* is a directed tree that contains a *root* u that can be reached from any other vertex by a directed path. Note that a graph containing only one vertex is an in-tree. For $i = 1, \ldots, r$, we denote the subset of vertices knocked out in round i by R_i. Let G_i be the directed graph with vertex set R_i and an arc from a vertex u to a vertex v if and only if u fires at v in round i. We may also use G_i to denote the underlying undirected graph; it will always be clear which from the context). Also, observe that G_i and G^i denote two different graphs. As each vertex in a round has exactly one edge oriented away from it, we can make the following observation (which is illustrated in Fig. 1).

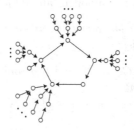

Fig. 1. A component of a graph G_i

Observation 1. *Let S be a KO-reduction scheme for a graph G. For $i = 1, \ldots, r$, each component of G_i is formed by a directed cycle D on at least two vertices, such that each vertex on D is the root of some pendant in-tree.*

Another observation we will use is the following.

Observation 2. *If a graph G contains two distinct vertices of degree 1 that share the same neighbour, then G is not KO-reducible.*

Note that when referring to, for example, G_i, it is implicit that we know with respect to which KO-reduction scheme this graph is defined (we wish to avoid the cumbersome notation necessary to make it explicit). Sometimes we will be considering pairs of schemes and will write, for instance, that G_2 has fewer vertices under S' than under S. The meaning of this should be clear.

3   Resolving the Square-Root Conjecture

Let S be a KO-reduction scheme for a KO-reducible graph G. It turns out that the square-root conjecture can be solved by considering schemes that knock out vertices "as early as possible". Hence, we define

$$w(S) = \sum_{i=1}^{r(S)} i|R_i|,$$

and we say that S is a *minimal* KO-reduction scheme for G if

$$w(S) = \min\{w(S) \mid S \text{ is a KO-reduction scheme for } G\}.$$

For a minimal KO-reduction scheme S of a graph G, we can make a number of further assumptions. We use the following terminology. If G_i has a component C that consists of two vertices u and v we call C a *two-component* of G_i. Note the existence of arcs (u, v) and (v, u) between the vertices u and v of a two-component C. If G_i has a component C that consists of vertices $u, v_1, \ldots, v_p$ for some $p \geq 2$ with arcs $(u, v_1), (v_1, u), (v_2, u), \ldots, (v_p, u)$ then we call C a *star-component* of G_i with *centre vertex* u. The vertices $v_1, \ldots, v_p$ are called the *leaves* of C, and v_1 is called the *centre-victim*, and the other leaves are called *centre-free*. Finally, if G_i has a component that is a directed cycle with an odd number of vertices then we call such a component an *odd cycle-component* of G_i.

Lemma 1. *If G is KO-reducible, then G admits a minimal KO-reduction scheme S with the following properties:*

(i) *Each component C of G_1 is either a two-component, a star-component or an odd cycle-component.*

(ii) *For $2 \leq i \leq r - 1$, every component of G_i is either a two-component or a star-component.*

(iii) *Every component of G_r is a two-component.*

(iv) *If C is an odd cycle-component (in G_1) then no vertices of $R_2, \ldots, R_r$ fire at vertices of C in round 1.*

(v) *For $1 \leq i \leq r - 1$, there is no edge in G between any two leaves of the same star-component or of two different star-components in G_i.*

Proof. Let G be a KO-reducible graph. Then G admits a KO-reduction scheme S. Let C be a component in G_i for some $1 \leq i \leq r$. We start the proof by showing that if S is minimal, then we can assume that C is either a two-component, a star-component or an odd cycle-component. By Observation 1, C is formed by a directed cycle D on vertices $u_1, \ldots, u_p$ for some $p \geq 2$, such that each u_i is the root of some pendant in-tree T_i.

Suppose p is even and $p \geq 4$. We adjust the firing by letting the vertices of V_D fire at each other according to a perfect matching of D. Hence, we may assume that this case does not occur.

Suppose $p \geq 3$ is odd. If D contained a vertex that is knocked out by some vertex v in its corresponding pendant in-tree, then we can adjust the firing by letting the vertices of $V_D \cup \{v\}$ fire at each other according to a perfect matching of this subgraph. Hence, we may assume that $C = D$ is an odd cycle-component.

Suppose that $p = 2$. Then the underlying undirected graph of C is a tree, and it is obvious that it can be decomposed into two-components and star-components (and that we can let these components define the firing).

By Observation 2, we have that G_r cannot contain any star-components.

To complete the proof of (i)–(iii), we must show that odd cycle-components only occur in G_1. To do this we shall first prove a claim which also immediately implies (iv): for any odd cycle-component D we may assume that $K(D) = D$; that is, vertices in D are only knocked out by each other. Suppose D is an odd cycle-component on vertices $u_1, \ldots, u_p$ in some G_i for $i \geq 1$, such that there exists a vertex $v \in K(D) \backslash D$ and v fires at u_1. We adjust the firing by replacing the arc (u_p, u_1) by (u_p, u_{p-1}) and return to a previous case. Hence, we may assume that this case does not occur.

Now suppose that a graph G_i, $i \geq 2$, contains an odd cycle-component D. First suppose that in round $i - 1$ all vertices in D fire at vertices in R_{i-1} that either are centre vertices of star-components, or else belong to two-components or odd cycle-components. Since we just saw that no vertices in $R_{i+1} \cup \ldots \cup R_r$ fire at D, we can move D to G_{i-1} (since all victims of D in R_{i-1} are not unique, it does not matter if the vertices of D fire at each other instead). This way we obtain a KO-reduction scheme S' with $w(S') < w(S)$. This contradicts the minimality of S. In the remaining case, there exists a vertex u in D that fires at a leaf w in a star-component in R_{i-1}. We let u and w fire at each other in round $i - 1$, so we are able to move u to R_{i-1} as $K(D) = D$. We let the other vertices in D fire at each other in round i according to a perfect matching of $D - u$. This way we again obtain a KO-reduction scheme S' with $w(S') < w(S)$, contradicting the minimality of S.

To finish the claim we prove (v). Suppose u and v are leaves in G_i for some $1 \leq i \leq r - 1$, such that u and v are adjacent in G. In case u and v are leaves of different star-components, we adjust the firing by letting u and v fire at each other, and, if necessary, changing the centre-victims to be vertices other than u and v. Suppose u and v are leaves of the same star-component C. Let z be the centre vertex of C. If C has a third leaf, then we again let u and v fire at each other and let another leaf be the centre-victim. Otherwise we can form an odd cycle-component and return to a previous case. $\qquad\square$

We call a minimal KO-reduction scheme S of a graph G that satisfies the properties (i)-(v) of Lemma 1 a *simple* KO-reduction scheme of G. We will continue to find further properties of simple KO-reduction schemes.

Observation 3. *Let S be a simple KO-reduction scheme for a graph G. Let u, v be, respectively, vertices of R_i and R_j, $i < j$, such that u is the unique victim of v. Then u is a centre-free leaf of a star-component in G_i.*

Proof. By Lemma 1, u cannot be a vertex of an odd cycle-component. If u is in a two-component, or u is the centre vertex or centre-victim of a star-component, then there are at least two vertices firing at u. Hence u must be a centre-free leaf of a star-component. $\qquad\square$

Lemma 2. *Let S be a simple KO-reduction scheme for a graph G with $r \geq 2$. Let C be a two-component in G_r. Then in rounds $1, \ldots r - 1$ all victims of one of the two vertices of G_r are not unique, and all victims of the other one are unique.*

Proof. For $i = 1, \ldots, r - 1$, let x_i be the victim of u in round i, and let y_i be the victim of v in round i.

Suppose both x_{r-1} and y_{r-1} are not unique victims. We show that this means that it is possible to move u and v to R_{r-1}. If $x_{r-1} \neq y_{r-1}$ or $x_{r-1} = y_{r-1}$ is the victim of vertices other than u and v, then let u and v fire at each other in round $r - 1$. If $x_{r-1} = y_{r-1}$ is fired at by only u and v, then it is a centre-free vertex of a star-component and we can adjust the firing to let u, v and x_{r-1} form an odd cycle-component in G_{i-1}. Either way we obtain a new KO-reduction scheme S' with $w(S') < w(S)$, contradicting the minimality of S. Hence we can assume that y_{r-1} is a unique victim.

We show that all victims of u are not unique by contradiction. Let h be the largest index such that x_h is unique. By Observation 3, vertices x_h and y_{r-1} are centre-free leaf vertices of star-components. Since centre vertices are not unique victims, we can let u and x_h fire at each other in round h, and we can let v and y_{r-1} fire at each other in round $r - 1$. This way we obtain a new KO-reduction scheme S' with $w(S') < w(S)$. This contradicts the minimality of S.

Now we again find a contradiction to show that all victims of v are unique. Let h be the largest index such that y_h is not a unique victim. Then we let v fire at y_j in round $j-1$ for $j = h+1, \ldots, r-1$ (so we move those vertices from R_j to R_{j-1}), and v does not fire at y_h anymore. Since x_{r-1} is not a unique victim, we can then let u and v fire at each other in round $r - 1$. This way we obtain a new KO-reduction scheme S' with $w(S') < w(S)$. This contradicts the minimality of S and completes the proof of the lemma. $\square$

Lemma 3. *Let S be a simple KO-reduction scheme for a graph G with $r \geq 2$. For each $i \geq 2$, R_i contains a vertex v_i whose victims in round $1, \ldots, i - 1$ are all unique. Let u_r be the (unique) neighbor of v_r in G_r. Then $\bigcup_{i=2}^{r} L(v_i) \cup \{u_r\}$ is an independent set of cardinality $\frac{r^2-r+2}{2}$ in G.*

Proof. Since R_r is non-empty, there exists a two-component C in G_r. Let u_r and v_r be the two vertices of C. By Lemma 2, we may assume that all victims of u_r in rounds $i = 1, \ldots, r-1$ are not unique, and all victims of v_r are unique. Denote the victims of v_r in rounds $i = 1, \ldots, r - 1$ by $y_1^r, \ldots, y_{r-1}^r$, respectively. By Observation 3, every y_i^r is a centre-free leaf vertex of a star-component C_i^r. For $i = 2, \ldots, r - 1$, let v_i be the centre vertex of C_i^r and for $h = 1, \ldots i - 1$, let y_h^i be the victim of v_i in round h. We claim that these victims y_h^i are all unique. For $i = r$, this is already shown. We prove the rest of the statement by contradiction. Let $2 \leq i \leq r - 1$. Let h be the largest index such that y_h^i is not a unique victim of v_i. We adjust the firing as follows. Since y_h^i is not a unique victim of v_i, we do not have to let v_i fire at it. Then we let v_i fire at y_j^i in round $j - 1$ for $j = h+1, \ldots, i-1$, so we move y_j^i to R_{j-1} for $j = h+1, \ldots, i-1$. In round $i-1$ we let v_i fire at y_i^r, so we move y_i^r to R_{i-1}. Then we do not have to let v_r fire at y_i^r. Hence, we can let v_r fire at y_j^r in round $j - 1$ for $j = i+1, \ldots, r-1$, so we move y_j^r to round $j - 1$ for $j = i + 1, \ldots, r - 1$. Finally, we let u_r and v_r fire at each other in round $r - 1$. This is possible, because the victim of u_r in round $r - 1$ is not unique, due to Lemma 2. This way we have obtained a new KO-reduction scheme S' with $w(S') < w(S)$, contradicting the minimality of S.

We will now prove that

$$L = \bigcup_{i=2}^{r} L(v_i) = \bigcup_{i=2}^{r} \bigcup_{h=1}^{i-1} y_h^i$$

is an independent set. We first note that

$$|L| = \left| \bigcup_{i=2}^{r} \bigcup_{h=1}^{i-1} y_h^i \right| = \sum_{i=2}^{r} \sum_{h=1}^{i-1} 1 = \frac{r^2 - r}{2},$$

since all vertices in L are unique victims.

Because S is simple, by Lemma 1, there is no edge between any two vertices y_h^i and y_h^j. Suppose there were an edge $y_h^i y_j^r$, where $h \neq j$. If $h < j$, then we move y_j^r to R_h, each y_k^r for $k = j + 1, \ldots, r - 1$ to R_{k-1}, and finally u_r and v_r to R_{r-1}. We can adjust the firing and obtain a new KO-reduction scheme S' with $w(S') < w(S)$. This contradicts the minimality of S. If $h > j$, then we move y_h^i to R_j, each y_k^r for $k = i, \ldots, r - 1$ to R_{k-1}, and finally u_r and v_r to R_{r-1}. We adjust the firing and obtain the same contradiction as before. Suppose there exists an edge between two vertices y_h^i and y_j^k with $h < j$ and $r \notin \{i, j\}$. We move y_j^k to R_h, each y_l^r for $\ell = j, \ldots, r - 1$ to $R_{\ell-1}$, and finally u_r and v_r to R_{r-1}. We adjust the firing and obtain the same contradiction as before.

Now suppose u_r is adjacent to a vertex y_h^i of L. By Lemma 2, all victims of u_r are not unique. Then we can let u_r fire at y_h^i in round i. Then y_h^i is no longer a unique victim and we find a KO-reduction scheme S' with $w(S') < w(S)$ as before. This final contradiction completes the proof. □

We are now ready to state our main theorem, which proves (and strengthens) the square-root conjecture posed in [3].

Theorem 1. *Let G be a KO-reducible graph. Then*

$$pko(G) \leq \min \left\{ -\frac{1}{2} + \sqrt{2n - \frac{7}{4}}, \; \frac{1}{2} + \sqrt{2\alpha - \frac{7}{4}} \right\}.$$

Proof. It is straightforward to check that the statement holds for a graph G with $pko(G) = 1$. Let S be a simple KO-reduction scheme for a graph G with $r \geq pko(G) \geq 2$. By Lemma 3, we find an independent set L' of G that has cardinality $|L'| = \frac{1}{2}(r^2 - r + 2) \leq \alpha$. Note that R_1 contains a centre vertex of a star-component. This, together with Lemmas 2 and 3, implies that $n \geq |L'| + r - 1 + 1 = \frac{1}{2}(r^2 - r + 2) + r$. Solving both inequalities gives us the required upper bound. □

We note that the bound mentioned in Theorem 1 is asymptotically tight. In [3], it has been proven that for all $p \geq 1$, $pko(K_{p,q}) = p = \Theta(\sqrt{n}) = \Theta(\sqrt{\alpha})$ for all complete bipartite graphs on $n = p + q$ vertices with $q = \frac{1}{2}p(p + 1)$.

4   Claw-Free Graphs

It is known that claw-free graphs can be knocked out in at most two rounds [3] if they are KO-reducible (not all claw-free graphs are, take for example an isolated vertex or a path on three vertices). We generalize this result for $K_{1,p}$-free graphs for any $p \geq 2$. This solves a question in [3].

Theorem 2. *Let $p \geq 1$. If a $K_{1,p}$-free graph G is KO-reducible then $pko(G) \leq p - 1$.*

Proof. The case $p = 1$ is trivial. For $p \geq 2$, the statement follows directly from Lemma 3.   □

This result is the best possible. In [3, Section 4], a tree Y_ℓ is defined for each integer $\ell \geq 1$, and it is shown that $pko(Y_\ell) = \ell$. It is also easy to check that Y_ℓ is $K_{1,\ell+1}$-free. We omitted the details.

In the rest of this section, we suppose that $G = (V, E)$ is a claw-free graph and show that $pko(G)$ can be determined in polynomial time. We need the following lemma.

Lemma 4. *Let G be a connected claw-free graph with $pko(G) = 2$. Then there is a simple KO-reduction scheme in which only two vertices u and v survive to the second round.*

Proof. By Lemma 1 and claw-freeness, we know there is a simple two-round KO-reduction scheme for G such that

  (i) each component of G_1 is a two-component, star-component or odd cycle,
  (ii) each component of G_2 is a two-component,
  (iii) in the first round the vertices of G_2 do not fire at vertices that belong to odd cycles in G_1, and
  (iv) the leaves of the star-components in G_1 are not adjacent.

As the leaves of the star-components are not adjacent, we can, by claw-freeness and Lemma 1, further suppose that each star-component is a path on three vertices which we shall call a *three-component*.

Note that among all schemes that satisfy these properties, S is the one with the fewest number of components in G_2 (as it is minimal). To prove the lemma, we show that if, for S, G_2 contains more than one component, then we can find a scheme S' that admits fewer components to G_2.

For S, let the vertex sets of the two-components of G_2 be $\{\{u_i, v_i\} \mid i = 1, \ldots, q\}$. By Lemma 2, we can assume that the victim of u_i in G_1 is not unique, but that of v_i is unique. By Observation 3, v_i fires at the centre-free leaf of a three-component, say y_i. Let x_i be the victim of u_i. Suppose that x_i is the centre vertex of a three-component. Then there is also an edge from u_i to one of the leaves, say w, of the three-component (else, by (iv), x_i, u_i and the leaves of the three-component induce a claw). Let z be the other leaf of the three-component.

Suppose that $y_i = w$. Then let S' be a scheme identical to S except that in the first round

- v_i fires at y_i,
- y_i fires at u_i,
- u_i fires at v_i,
- x_i and z fire at each other.

Thus S' has one fewer two-component in G_2 than S.

Suppose that $y_i = z$. Then let S' be a scheme identical to S except that in the first round

- v_i and y_i fire at each other,
- u_i fires at x_i,
- x_i fires at w,
- w fires at u_i.

Thus S' has one fewer two-component in G_2 than S.

Suppose $y_i \notin \{w, z\}$. Then let S' be a scheme identical to S except that in the first round

- v_i and y_i fire at each other,
- u_i and w fire at each other, and
- x_i and z fire at each other.

Thus S' has one fewer two-component in G_2 than S. Hence, we have proven that x_i is not the centre-vertex of a three-component.

Suppose that x_i is the leaf of a three-component. If y_i also belongs to this three-component, then, since $x_i \neq y_i$, we have that u_i, v_i and the three-component of their victims lie on a 5-cycle in G. Then let S' be a scheme identical to S except that in the first round these five vertices fire according to an orientation of this 5-cycle. Thus S' has one fewer two-component in G_2 than S.

If x_i is the leaf of a three-component that does not contain y_i, then u_i, v_i and the components containing their first round victims lie on a path of length 8 in G so can be matched. So let S' be a scheme identical to S except that in the first round these eight vertices fire according to this matching. Thus S' has one fewer two-component in G_2 than S.

Thus x_i is not the leaf of a three-component, and, by (iii), x_i belongs to a two-component.

Thus u_i and v_i combined with the components of G_1 containing their victims lie on a path of length 7 in G. We call such a path a *seven-component*. Let us motivate this choice of name by showing that the seven-components are vertex-disjoint.

The vertices v_i, $1 \leq i \leq r$, fire at distinct three-components in the first round (as their victims are unique and one of the leaves of each three-component is the centre-victim). We must also show that the victims x_i of the vertices u_i, $1 \leq i \leq r$, belong to distinct two-components. Suppose that x_i and x_j, $i \neq j$, are distinct but belong to the same two-component in G_1. Then let S' be a scheme identical to S except that in the first round

- v_i and y_i fire at each other,
- v_j and y_j fire at each other,
- u_i and x_i fire at each other, and
- u_j and x_j fire at each other.

Again S' has fewer two-components in G_2 than S. Now suppose that $x_i = x_j$. If either u_i or u_j is adjacent to the other vertex in x_i's two-component, then we have the previous case. Otherwise, there is an edge $u_i u_j$ (else there is a claw). So let S' be a scheme identical to S except that in the first round

- v_i and y_i fire at each other,
- v_j and y_j fire at each other, and
- u_i and u_j fire at each other.

Again S' has fewer two-components in G_2 than S.

We have shown that the seven-components are vertex-disjoint. Note that all the three-components in G_1 contain a victim of a vertex in G_2 and so must be a subgraph of a seven-component. Thus we can represent S as a collection of vertex-disjoint seven-components, two-components and odd cycles that span G. We denote such a representation G^*. Note that the number of two-components in G_2 is equal to the number of seven-components in G^*. Thus to prove the lemma we show that if for S there is more than one seven-component in G^*, then we can find another scheme with fewer seven-components.

Let $A = a_1 \cdots a_7$ and $B = b_1 \cdots b_7$ be a pair of seven-components in G^*. First we consider the case where, in G, A and B are joined by an edge $a_i b_j$ for some i, j. We shall show that this implies that the vertices of A and B admit a perfect matching; thus we can replace two seven-components by seven two-components. If i and j are both odd, then we match a_i with b_j and the remaining vertices and edges of A and B form paths of even length, so can clearly be matched. If i is even and j is odd, then, if either a_{i-1} or a_{i+1} is adjacent to b_j, we have the previous case. Otherwise, by claw-freeness, there is an edge $a_{i-1} a_{i+1}$ and we include both this and $a_i b_j$ in the matching, and, again, what remains of A and B are paths of even length. Finally suppose that i and j are both even. If there are any other edges from a vertex in $\{a_{i-1}, a_i, a_{i+1}\}$ to a vertex in $\{b_{j-1}, b_j, b_{j+1}\}$, then we have an earlier case. Otherwise, claw-freeness implies edges $a_{i-1} a_{i+1}$ and $b_{j-1} b_{j+1}$, and we include these and $a_i b_j$ in the matching to again leave only even length paths.

So we can assume that no pair of seven-components in S are joined by an edge in G. Now let us assume that S is such that we can find seven-components A and B such that the length of the shortest path in G between them is minimum (that is, there is no pair of seven-components in any other simple scheme separated by a shorter path).

Suppose a shortest path from A to B meets A at a_i and the next vertex along is w. In G^*, w must belong to either a two-component or an odd cycle.

First suppose w is in a two-component C whose other vertex is z. We describe how to use the vertices of A and C to find a seven-component A' and two-component C' such that w is in A'; thus A' is closer to B than A contradicting our choice of A and B. By symmetry, there are four cases according to which vertex of A neighbours w. Suppose a_1 is adjacent to w. Then replace A and C with $A' = zwa_1 \cdots a_5$ and $C' = a_6a_7$. If a_2 is adjacent to w, then claw-freeness implies one of the edges a_1a_3, a_1w or a_3w is present. Let C' be, respectively, a_6a_7, a_6a_7 or a_1a_2, and in each case we find a path of length 7 on the remaining vertices to be A'. If a_3 is adjacent to w, then let $A' = zwa_3 \cdots a_7$ and $C' = a_1a_2$. If a_4 is adjacent to w, then one of a_3a_5, a_3w or a_5w is present. Let C' be, respectively, a_1a_2, a_1a_2 or a_6a_7, and in each case we find a path of length 7 on the remaining vertices to be A'.

Finally suppose that w belongs to an odd cycle. If a_i, i odd, is joined to w, then there is a perfect matching on the vertices of A and the cycle and we have a scheme with fewer seven-components. Suppose a_i, i even, is adjacent to w. If either a_{i-1} or a_{i+1} is joined to w, then we have the previous case. Otherwise, there must be an edge $a_{i-1}a_{i-1}$, and if we match both this pair of vertices and a_i and w, then the remaining vertices of A and the cycle induce even-length paths and a perfect matching can again be found. $\square$

Theorem 3. *Computing the parallel knock-out number of a claw-free graph can be done in polynomial time.*

Proof. By Theorem 2, it is sufficient to present methods for checking whether or not $\text{pko}(G)$ is equal to 1 or 2, since if it is neither it must be ∞. Deciding whether a graph can be knocked-out in a single round can be solved in polynomial time ([3]). So we need only show how to check whether G can be knocked out in two rounds.

Suppose that $\text{pko}(G) = 2$. By Lemma 4, we can assume that there is a two-round simple KO-reduction scheme for G in which only two vertices, say u and v, survive to the second round, and, by the proof of the lemma, there is exactly one three-component in G_1.

Let w be the first round victim of v. Then $G - \{u, v, w\}$ has a spanning subgraph comprising two-components and odd cycles (that is, $G_1 - w$) and can thus be knocked out in one round. Therefore the following is a necessary condition for $\text{pko}(G) = 2$: there are three vertices u, v and w in V such that

- there are edges uv and vw,
- u and w have neighbours other than v and each other, and
- $\text{pko}(G - \{u, v, w\}) = 1$

It is easy to see that this condition is also sufficient. Therefore to decide whether or not $\text{pko}(G) = 2$, we look for a set of three vertices that satisfies this condition. This can be done in polynomial time. $\square$

As noted before any graph with $\text{pko}(G) = 1$ has a spanning subgraph consisting of a number of mutually disjoint matchings edges and disjoint cycles. For *claw-free* graphs we have found the following characterization, which directly follows from the proof of Lemma 4.

Corollary 1. *Let G be a connected claw-free graph with $\text{pko}(G) = 2$. Then G has a spanning subgraph consisting of a number of vertex-disjoint matching edges, odd cycles and one path on seven vertices.*

5   Conclusions

We solved the square-root conjecture of [3] by giving a tight upper bound on the parallel knock-out number of a KO-reducible graph G. We also showed that the parallel knock-out number of a KO-reducible $K_{1,p}$-free graph is at most $p - 1$, and that this bound is tight. For claw-free graphs we showed that their parallel knock-out number can be computed in polynomial time. The question of whether the parallel knock-out number for $K_{1,p}$-free graphs with $p \geq 4$ can also be computed in polynomial time remains open.

References

1. Bondy, J.A., Murty, U.S.R.: Graph Theory with Applications. Macmillan, London and Elsevier, New York (1976)
2. Broersma, H., Fomin, F.V., Woeginger, G.J.: Parallel knock-out schemes in networks, In: Proceedings of the 29th International Symposium on Mathematical Foundations in Computer Science (MFCS 2004), LNCS 3153, 204–214 (2004)
3. Broersma, H., Fomin, F.V., Královič, R., Woeginger, G.J.: Eliminating graphs by means of parallel knock-out schemes. Discrete Applied Mathematics 155, 92–102 (2007)
4. Broersma, H., Johnson, M., Paulusma, D., Stewart, I.A.: The computational complexity of the parallel knock-out problem. In: Proceedings of the 7th Latin American Theoretical Informatics Symposium (LATIN 2006), LNCS 3887, 250–261 (2006)
5. Faudree, R., Flandrin, E., Ryjáček, Z.: Claw-free graphs—a survey. Discrete Mathematics 164, 87–147 (1997)
6. Fraenkel, A.S.: Combinatorial games: selected bibliography with a succinct gourmet introduction, Electronic Journal of Combinatorics (2007) http://www.emis.ams.org/journals/EJC/Surveys/ds2.pdf
7. Lampert, D.E., Slater, P.J.: Parallel knockouts in the complete graph. American Mathematical Monthly 105, 556–558 (1998)

Author Index

Lecture Notes in Computer Science

For information about Vols. 1–4421

please contact your bookseller or Springer